RISK ANALYSIS IN
NUCLEAR WASTE MANAGEMENT

ON RELIABILITY AND RISK ANALYSIS

A series devoted to the publication of courses and educational seminars given at the Joint Research Centre, Ispra Establishment, as part of its education and training program.
Published for the Commission of the European Communities,
Directorate-General Telecommunications, Information Industries and Innovation,
Scientific and Technical Communications Service.

The publisher will accept continuation orders for this series which may be cancelled at any time and which provide for automatic billing and shipping of each title in the series upon publication.
Please write for details.

RISK ANALYSIS IN NUCLEAR WASTE MANAGEMENT

Proceedings of the ISPRA-Course held at the Joint Research Centre,
Ispra, Italy, 30 May – 3 June 1988

Edited by

A. SALTELLI,
D. A. STANNERS
and
M. D'ALESSANDRO

Commission of the European Communities, Institute of the Environment,
Joint Research Centre, Ispra, Italy

KLUWER ACADEMIC PUBLISHERS
DORDRECHT / BOSTON / LONDON

Library of Congress Cataloging in Publication Data

Risk analysis in nuclear waste management : proceedings of the ISPRA
-Course held at the Joint Research Centre, Ispra, Italy, 30 May-3
June 1988 / edited by A. Saltelli, D.A. Stanners and M.
D'Alessandro.
       p.   cm. -- (Ispra courses on reliability and risk analysis)
    ISBN-13:978-94-010-6963-2   e-ISBN-13:978-94-009-1049-2
    DOI: 10.1007/978-94-009-1049-2
    1. Radioactive waste disposal--Risk assessment--Congresses.
I. Saltelli, A. (Andrea), 1953-   . II. Stanners, D. A., 1955-
III. D'Alessandro, Marco.   IV. Series.
TD898.14.R57R57   1989
363.72'89--dc20   89-20110

ISBN-13:978-94-010-6963-2   *Printed on acid-free paper*

Commission of the European Communities Joint Research Centre Ispra (Varese), Italy

Publication arrangements by
Commission of the European Communities,
Directorate-General Telecommunications, Information Industries and Innovation, Scientific and
Technical Communications Service, Luxembourg
EUR 11969 EN

© 1989 ECSC, EEC, EAEC, Brussels and Luxembourg
Softcover reprint of the hardcover 1st edition 1989

LEGAL NOTICE
Neither the Commission of the European Communities nor any person acting on behalf of the
Commission is responsible for the use which might be made of the following information.

Published by Kluwer Academic Publishers,
P.O. Box 17, 3300 AA Dordrecht, The Netherlands.

Kluwer Academic Publishers incorporates the publishing programmes of
D. Reidel, Martinus Nijhoff, Dr W. Junk and MTP Press.

Sold and distributed in the U.S.A. and Canada
by Kluwer Academic Publishers,
101 Philip Drive, Norwell, MA 02061, U.S.A.

In all other countries, sold and distributed
by Kluwer Academic Publishers Group,
P.O. Box 322, 3300 AH Dordrecht, The Netherlands.

Preface

One of the arguments most frequently raised by objectors to nuclear power production is the spectrum of the hazard from the nuclear waste. In the public's perception one of the most dangerous features of this waste is the threat to future generations due to the long life of the radio-emitting nuclides. People's concern for this hazard seems to be greater than that for other types of toxic wastes, widely produced and released to the environment (heavy metals, PCBs, arsenic etc.) though these might last even longer.

In fact public attitude toward risk depends very strongly on the nature of the risk, as clearly illustrated in the initial papers of these proceedings. On this basis it appears that the safety of the nuclear practice, including waste disposal, cannot be assessed only by comparison with other practices or risks.

For this reason the analysts have accepted the challenge of investigating the long term safety of this type of waste, taking it much further than for many other known pollution problems affecting mankind. Moreover, for one of the first times this is being done before disposal.

The contributions to these proceedings attempt to describe the specific nature of this type of assessment, presenting new developments which have been made in the field . These include the recourse to system variability analysis, the use of natural and archaeological analogues and predictive geology.

The risk issue is also analysed in its social, political and ethical contexts, in order to provide a proper background to the problem. For this reason the book opens with a discussion on risk policy and risk perception, followed by a lucid and critical analysis of the reasons for the public's lack of confidence in the stated safety of the nuclear industry at large.

The bulk of the mathematical and statistical tools described in the central sections of the proceedings are an attempt to answer some of the doubts raised by this discussion, by explaining the scientific basis of the techniques employed.

A few worked examples should aid the readers to evaluate the success of the risk analysts to date, and to judge the degree of confidence given to the assessment by the use of these new methodologies.

These examples are also aimed at providing some reference cases to scientists becoming involved in risk analysis, even outside the nuclear sector, by illustrating the practice of designing and constructing a safety assessment.

The reader will notice the absence of complex model descriptions, the emphasis being on how the models are used rather than on how they are built.

In the end it is hoped that these proceedings might help the reader in understanding an area of study which is becoming increasingly important in decision making related to nuclear practices at large.

Andrea Saltelli

AKNOWLEDGMENTS

The kind assistance of Mrs. M. G. Giaretta and Mr. J. N. Wells of the JRC Publication Office in the preparation of this manuscript is gratefully aknowledged.

Contributors

Section 1 - Risk perception and risk policy

B. Wynne, Centre for Science Studies and Science Policy, University of Lancaster, Bailrigg, Lancaster (UK)

S. Islam, Research Unit Gottstein, Max Planck Society, Munich (D)

Section 2 - From reactor safety to disposal safety

B. J. Garrick, Pickard, Lowe and Garrick, Inc., Newport Beach California (USA)

Section 3 - The analyst's answer: methodologies and methods

B. W. Goodwin, Atomic Energy of Canada Limited, Whiteshell Nuclear Research Establishment, Pinawa Manitoba (CAN)

A. Saltelli, Commission of the European Communities, Institute of the Environment, Joint Research Centre, Ispra (I)

S. Mobbs, National Radiological Protection Board, Chilton, Didcot, Oxon (UK)

Section 4 - The word to the geologists: what if...

J. Fourniguet, Bureau de Recherches Geologiques et Minieres, Departement Geologie, Orleans Cedex 2 (F)

M. D'Alessandro, Commission of the European Communities, Institute of the Environment, Joint Research Centre, Ispra (I)

Section 5 - Some worked examples

B. G. J. Thompson, Her Majesty's Inspectorate of Pollution, Romney House, London (UK)

J. Lewi, CEA, Institut de Protection et de Surete Nucleaire, Departement d' Analyse de Surete, Fontenay-aux-Roses (F)

D. A. Stanners, Commission of the European Communities, Institute of the Environment, Joint Research Centre, Ispra (I)

F. Girardi, Commission of the European Communities,
Institute of the Environment, Joint Research Centre, Ispra (I)

N. Cadelli, Commission of the European Communities,
DG XII, Bruxelles (B)

Section 6 - A smart approach: using natural
                    and archaeological analogues

I. G. McKinley, Nationale Genossenshaft fur die Lagerung
radioaktiver Abfalle, Baden (CH)

Contents

Section 5 - Some worked examples

Section 6 - A smart approach: using natural
and archaeological analogues

Section 1 - Risk perception and risk policy

The two papers presented in this section address the issue of risk from two different perspectives. In Dr. Wynne paper (Understanding public risk perception) the emphasis is on the divergence between technical and public perception of risk and on the social and psychological reasons behind this difference. In Dr. Islam presentation the reasons of the opponents to nuclear power are reviewed. It is shown that, given the uncertainty affecting the safety analyses, very different conclusions can be drawn by different groups as to the safety of the practice. The ethical implications of the issue are discussed, and the concept of subjective immunity is illustrated. The issue of scientist's objectivity is also touched upon

UNDERSTANDING PUBLIC RISK PERCEPTIONS

Dr. Brian WYNNE
Centre for Science Studies and Science Policy
University of Lancaster
Bailrigg
Lancaster LA1 4YN
United Kingdom

ABSTRACT. The divergence between technical and public perceptions of risks from radioactivity is recognised to be the most important and difficult aspect of radioactive waste management. Understanding the basis of those perceptions becomes an important prerequisite for viable policies and implementation programmes. After providing some historical context to the policy and research interest in public risk perceptions, this chapter reviews several strands of research which have generated different interpretations of their underlying structure and meanings. The main approaches are grouped under three broad headings: psychometric, social psychological, and sociological. The methods and findings of each of these approaches are described, along with some indication of the variants within them. Some comparative observations are then offered, and the chapter finishes with an outline of how decision makers have tried to integrate the findings of risk perception research into technical frameworks of risk management and policity.

1.   INTRODUCTION

"Since 1945 the fear of nuclear radiation has developed into an anxiety as deep-rooted in twentieth century Western culture as was the case with electricity in the nineteenth. Radiation is for most people inexplicable, unseeable, untouchable and almost mystically evil in its association with the appalling destructive power of atomic weapons. It is also associated with that other great twentieth century fear - cancer. Most frightening of all are the unknown effects, the genetic changes which might pass on to future generations. It is hardly surprising if the public's view of things nuclear is at times emotional rather than rational".

This quote from the UK House of Commons Environment Select Committee Report on Radioactive Waste (1986) typifies the predominant attitude to the paralysing hostility which publics all over the world have shown towards plans to site nuclear waste disposal facilities, or even to conduct research for eventual site selection. The public is taken to be

3

A. Saltelli et al. (eds.), Risk Analysis in Nuclear Waste Management, 3–24.
© 1989 ECSC, EEC, EAEC, Brussels and Luxembourg.

"over-reacting" to the "real" risks as defined rationally by science. Several other, extraneous factors are thought to determine the public's view of radioactive risks. The key question is avoided - what is a rational definition of the risks, and are public groups responding on the basis of (mis)perceived size of risks as defined by the experts, or on a different, perhaps more substantial basis altogether? Is the choice, as one scientist put it, between making decisions on the basis of how many will be killed, or how many will be frightened? Or is there more to it than this simple dichotomy?

1.1 Risk Comparisons

It is often forgotten that the units of risk used by scientists are chosen, not given by nature. Risk is usually defined as the product of the frequency or probability of a harmful event and the consequences of each such event: $R = PxC$. However, there are two points about this: first there are different consequences, such as death, injury of various kinds, environmental damage, economic costs, social dislocation in some cases, and even deaths which may vary in their significance, for example if it is a child, or a known person as against an unknown future "statistical" victim; second the risk has to be normalised against some corresponding unit such as population exposed, duration of exposure, unit of output (such as energy), etc.

In each respect the scientific choices made reflect value-commitments, say to ignore injury or delayed deaths, or to evaluate risks per 1000 employees rather than per unit of output (this would alter the relaive risk ranking between a labour-intensive and a capital intensive energy production system for example). It is therefore worth emphasising that although in science, policy making and everyday life we talk of "risks" as if they exist out there in reality in comparable packages, risk, as we know it, is actually a product of our evaluative thinking.

It is common to show tables of risks comparing existing man-made hazards with natural hazards, or different industrial hazards. A typical example is Table 1, taken from the London Royal Society's 1983 Report on Risk Assessment (p.89). Such tables are often used to show the apparent inconsistency of social decisions and attitudes, where for example the level of risk apparently tolerated in mining is much higher than that in the chemical industry. Imposed technological risks are often also similarly compared with commonly "accepted" risks in daily life, such as driving or travelling by air, in order to demonstrate that opposition to such technological risks is unjustified. Whilst such comparisons are sometimes useful first-order illustrations, their normative use in this way has to assume that the only dimension of such activities that people should be interested in is their numerical risk of death. By the end of this chapter the reader should be able to judge the validity of this assumption against a large amount of diverse empirical research which has examined the structures beneath the surface of social reactions and practices.

TABLE 1. Average annual accidental death rates at work in UK
per million at risk (1974-78 except as stated)

Manufacture of clothing and footwear	5
Manufacture of vehicles	15
Manufacture of timber, furniture, etc.	40
Manufacture of bricks, pottery, glass, cement, etc.	65
Chemical and allied industries	85
Shipbuilding and marine engineering	105
Agriculture (employees)	110
Construction industries	150
Railway staff	180
Coal miners	210
Quarries	295
Non-coal miners	750
Offshore oil and gas (1967-76)	1650
Deep sea fishing (accidents at sea only) (1959-68)	2800

1.2 Acceptable Risk

Well before the siting of radioactive waste facilities became a practical issue, the same concern about public perception, or as it was assumed to be, misperception of risks associated with nuclear power had led to the first research programmes on public risk perceptions in the early 1970s. These programmes set out to understand the underlying factors involved in public attitudes, but with a strong surrounding flavour that those attitudes of anxiety and hostility were unfounded. The unstated assumption was that the "causes" of the "error" needed to be identified and "corrected".

This early policy concern to assuage public opposition especially to nuclear power set the tone for most of the ensuing work on risk perceptions, and the field is only slowly overcoming the debilitating effects of this orientation on the creation of a more sensitive and objective understanding of the issues and problems. The early work therefore focused on the patterns of "cognitive bias" thought to lead lay people astray from proper rational understandings and perspectives. The risk perception research was closely tied to the policy belief, amounting to an article of faith, that in decisions about hazardous technologies a level of acceptable risk could be objectively identified and quantitatively defined for standards setting or technology choice. A good historical overview is given in Otway (1985). The "acceptable risk" framework is generally regarded as having begun with Starr's 1969 paper which attempted to offer a scientific basis for observing systematic thresholds of acceptable risk that could be used normatively in decision making. To do this Starr used what he called the "revealed preferences" of the public in defining from existing social behaviour what were to be counted as acceptable risks. Thus, for example, because society does not ban, nor insist upon higher design or operating standards for motor cars, the average level of risk of death from driving (about 1 in 10,000 per year in the UK) can be taken to be accepted by society. According to this approach, this level of risk can thus be used as a norm for the level of acceptable risk for all voluntary risky activities.

Starr's main argument was that society sets a level of acceptable risk by a trade-off between the risks and the benefits - the higher the benefits of a given activity the more risk will be acceptable from it. This does not on its own imply any comparability of acceptability scales from one risk field to another because the benefits are not commensurable. Nevertheless, Starr's work was taken as a normative approach to defining rational perceptions of risks, and this stimulated the huge volume of work devoted to comparing risk magnitudes, for example between different energy supply systems. This comparative approach is examined more critically later.

Starr also claimed to show from empirical data that risks (such as car driving or swimming) deemed to be voluntary, showed a systematically higher level of acceptability than risks imposed upon people (such as air quality standards, or radioactive discharges). He claimed to show that on this voluntary-involuntary dimension alone, regardless of any other aspects of the risks, a constant thousandfold difference in acceptability was shown across a wide range of societal activities. This implied some kind of public recognition of the risk magnitudes as defined by scientists, combined with a universal cognitive "law" which regulates involuntary risks one thousand times more strictly than voluntary ones.

Whilst a difference in perception of voluntary and involuntary risks seems common sense, in all other respects Starr suggested that risks were entirely captured by their numerical magnitude of frequency of death. Indeed, the assumption informing the whole approach was that social behaviour and decisions were motivated only by risk and benefit considerations. Otway and Cohen (1975) failed to replicate Starr's results, and also exposed weaknesses in his data. They also pointed out some of the questionable assumptions underlying the "revealed preferences" approach. In particular, they pointed to the fallacy of assuming that because people are not actively protesting against existing levels of risk in any particular field, they must be happy to accept them. They also noted that Starr's clean distinction between voluntary and involuntary risks is misleading, for example car driving may be virtually unavoidable for some people, but a matter of preference for others. Even in the latter case, however, so many other situation-specific variables enter into decisions about whether to drive that it is impossible to separate the risk (chance of death) as the dominant factor in decisions which Starr's supposedly purely empirical approach assumes to have been made on a bare risk-benefit calculation.

Nevertheless, despite these fundamental criticisms the acceptable risk paradigm gained momentum to dominate expert methods of defining rational levels of risk, and of defining the extent of "deviation" in public risk perceptions. A risk was taken to be an objectively existing and definitively measurable entity rather than an abstraction created and given coherence and shape by a series of expert assumptions and commitments that are not under the control of scientific method. As more modern understanding now sees it, revealing these tacit commitments in the risk-definitions of experts is just as important for constructive decision making as identifying the framework of experiences and assumptions - not merely "error-causing" factors - which shape public definitions of risks (Wynne, 1989). It is now becoming more widely appreciated that the context in which a particular risk arises, such as whether the risk-controlling agency has a reliable track-record, or whether there is perceived to be pressure to cut corners or to enlarge the risk-generating activity, is central to risk perceptions and definitions. This makes general laws of risk-acceptability a false ambition, and renders comparison of risk magnitudes across different areas of activity meaningless, and often provocative when used normatively. These points will be examined again later. First we review the main research traditions.

2. PSYCHOMETRIC RESEARCH

Whereas the revealed preference methods mentioned above have been associated with behavioural psychology and the analysis of actual historical data on societal practice, perhaps the most prolific risk perception research tradition in the late 1970s and 1980s has been cognitive psychology. To avoid the inherent limitations of revealed preference methods, this psychometric approach has attempted to elicit expressed preferences, in controlled laboratory studies using small populations of various social background. The main thrust of such research, best represented in the work of Slovic, Fischoff and co-workers at Decision Research Oregon, has been to identify the extra attributes of perceived risks which have caused people to weight them differently when their numerical magnitudes are the same.

This work therefore began with fairly elementary observations of the divergences and correspondences between expert and lay public estimations of the frequency and significance of risks from a variety of sources, then attempted to find explanations for these patterns. When experts and lay people were asked to rank the estimated frequency of fatalities from a range of risky activities they were found to have reasonable correspondence (apart from fairly minor underestimation by lay people relative to the expert estimates at the higher end, and overestimation at the lower end of the risk spectrum). However, when the same comparative ranking was done according to significance of the risk rather than estimated frequency of fatalities, some major divergences were shown. See also Renn (1983) for European empirical corroboration on this point (though not on others; see later). In other words, people were tacitly introducing other elements besides the favoured scientific dimensions into their assessment of risk; it was not merely ignorance of risk magnitudes which created different perceptions.

2.1 Risk Attributes

In order to identify what these other elements of risk perception might be, Fischoff and colleagues performed various studies which asked people to rate up to ninety hazards on different properties such as:

– 7 –

- voluntariness/involuntariness;
- catastrophic/chronic harm potential (sometimes called "unit kill factor");
- familiarity/unfamiliarity to exposed person;
- familiarity/unfamiliarity to science;
- controllability/incontrollability;
- immediate/delayed effects; and
- certainty/uncertainty of death.

These qualitative risk-properties were referred to as the attributes of a risk. They were taken to be universal risk perception characteristics, though people would be expected to vary in the weighting (utility/disutility) given to each attribute; and a composite risk perception was taken to be made up by integrating the attribute-utilities. Altogether, eighteen such attributes were used, but some were found to overlap with others. For example, a dread/not-dread association was found to discriminate perceptions, but reduced in factor analysis to the catastrophe/chronic harm attribute. Eventually, further factor analysis of the eighteen attributes of risk perceptions managed to reduce them to just two independent dimensions, represented by the orthogonal axes of Dread and Unfamiliarity. Each risk could thus be plotted for its rating in these basic dimensions, which were suggested to represent the emotional and the cognitive elements of the risk perception, respectively. This scheme is shown in Figure 1, for eighty risky activities (Slovic and Fischoff, 1980; Slovic, 1987a; Slovic, Fischoff and Lichtenstein, 1985). Interestingly, quite different social groups showed fairly high convergence. Nuclear power came highly positive in terms of both unfamiliarity and dread association: there was no separate rating for radioactive waste disposal but it would be unlikely to be much different from nuclear power (one might be tempted to argue that the low power intensity in waste disposal compared to reactors would dissolve the catastrophe-dread association, but the well-known radioactive waste accident in the Urals in the Soviet Union in 1958 would perhaps counter such thoughts).

2.2 Interpreting Attributes

These psychometric studies have been important in clarifying the sorts of properties which people include in their perceptions and reactions to risky activities, beyond the reductionist concepts of more quantified methods. Even so, whilst they avoid some of the inherent limitations of the revealed preferences approach, and notwithstanding their immense influence upon thinking about risk perceptions, psychometric methods bring limitations of their own, as outlined later.

Behind this kind of work, influencing how we interpret its practical meaning for policy making, lies an unstated value-question in itself: are the extra dimensions, which this work has identified, to be regarded as extraneous to a properly rational definition of the risks? or are they perfectly rational considerations which people legitimately include, but which scientists wrongly ignore because they cannot be quantified? An uneasy compromise has existed between these opposite positions: experts have come reluctantly to tolerate the stricter

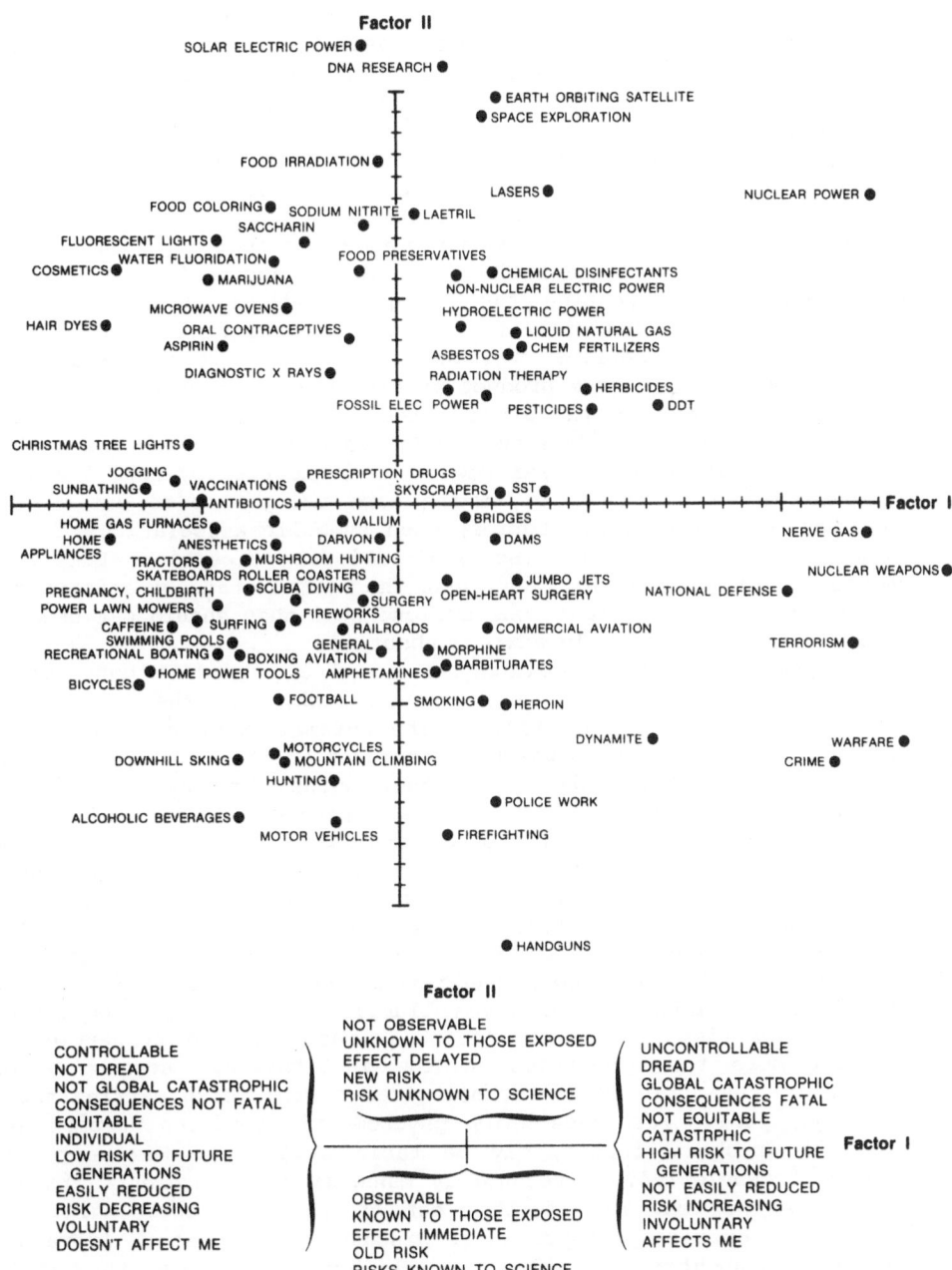

Figure 1. Hazard locations on Factors 1 and 2 of the three-dimensional structure from the interrelationships among 18 risks characteristics. Factor 3 (not shown) reflects the number of people exposed to the hazard and the degree of one's personal exposure. The diagram beneath the figure illustrates the characteristics that comprise the two factors.

Source: Slovic, 1987.

public attitude to nuclear risks, but basically they still treat the perception attributes as extraneous rather than intrinsic.

The different attributes themselves may be treated differently in this respect. For example, scientists and risk managers largely accept the objectivity of the voluntary/involuntary distinction, even though it cannot be quantified. Equally, the reality is accepted of the extra dimension of dread associated with risks that have a high unit kill factor (catastrophic potential) compared to risks that may kill as many overall, but which have small UKF (such as an accident with a power mower). Thus, it is routine for design standards of chemical and nuclear power plants, and aircraft, for example, to reflect this extra severity of risk perceptions over and above the purely quantitative standard computation, of the product of probability of an event and the scale of consequences of the event. Other attributes may be regarded less charitably however. For example, Slovic and colleagues found that the dread association of nuclear power was influenced by recall availability of images in the memory which were thought to be due to the media's way of representing stories; such symbolic associations are by implication ephemeral, distorting more rational perception. Likewise, unfamiliarity might be regarded as just the normal state of affairs for lay people in a technological society, one that they should get used to instead of being neurotically anxious about it.

Later work in this tradition has found an apparent correlation between the dread factor in a perceived risk, and the desire for stricter regulation (Slovic, 1987b). The unfamiliarity of a risk may correlate with the extent to which an accident in that field is seen as a warning signal to society - in other words, it may symbolise a broader public concern about perceived recklessness of technical change based upon inadequate scientific understanding of possible consequences. These more tacit dimensions are inherently difficult, indeed impossible definitively to identify, but they may, nevertheless, be a very substantial - and perhaps highly rational - factor in public perceptions and reactions. It is worth noting at this stage that several of the attributes which the psychometric approach has defined as properties of risks, such as controllability, familiarity, equity of distribution of risks, etc., may not be so much inherent properties of risks, but more the properties of the institutional structures of management of the activity or technological system which carries those risks. Thus, public responses which psychometric research assumes are responses to risk-attributes, may be tacit signals of the unacceptability of technologies, or technology management arrangements, rather than of risks per se. This further set of questions surrounding the interpretation of risk perceptions is discussed again later. It raises a key question about the psychometric approach, which is whether there is any warrant for its founding assumption that risks do have an independent objective existence separable from the more complex contexts in which people experience them, and from which numerical estimates of probabilities of harm are abstracted by technical experts.

2.3 From Attributes to Coping Responses

Research similar to that of the Oregon psychometric approach has been performed by Vlek, Stallen and co-workers (Stallen and Tomas, 1985; Vlek and Stallen, 1981). This work focused rather more upon the differences between individually managed and societally managed risks, and also on the individual as the source of behaviour rather than structural properties of the environment (such as "risk-attributes"). This "expressive orientation" as its authors called it, allows more freedom from the precommitment to the assumption that perceptions are driven by the features of "the risk" as constructed in the psychometric framework. More general and diffuse concerns are recognisable in this approach, and also perhaps more subtle responses to perceived risks, such as denial or forms of adaptation rather than overt concern. One important general factor in determining risk perceptions which this research school identifies was the extent to which people feel in control of their life, as opposed to feeling controlled by other, perhaps obscure forces.

The Stallen and co-workers research begins to indicate an important divergence between the psychometric paradigm and social psychological research described in the next section. One can see that whereas the former is focused entirely upon risks and their attributes, the latter, towards which the Stallen work leans, is more sensitive to the broader attitudes of people towards their environment. Its explanations are rooted in these factors rather than in the narrower notion (which draws together psychometric approaches with decision analysis; see Fischoff, Hope and Watson, 1984) of what disutility people assign to different attributes of a risk.

3. SOCIAL PSYCHOLOGY

3.1 From Attributes to Belief Systems

Social psychological approaches concentrate upon the broader belief and value-systems into which people assimilate perceived properties of risky activities. Thus, it focusses more upon the interactions between social environment (values, beliefs and norms) and individual judgements, and less upon individual maximisation of "utilities" faced with supposedly universal attributes of specific risks. The main research tradition here has been that of Otway and co-workers at IIASA/IAEA in Vienna starting in the early 1970s. This was focused primarily upon understanding public attitudes to nuclear power, but it generated insights of wider relevance.

The central element of the approach of Otway and colleagues was attitude theory, using the expectancy-value theory of Fischbein (Otway and Fishbein, 1976; Thomas, Swatson, Fischbein and Otway, 1980). The determinants of attitude are the beliefs which people hold about a technology or risky activity, and the wider values with which they give operational meaning to those beliefs about a technology or risk, in other words, which aspects of those beliefs are most salient in forming attitudes or judgements about the acceptability of the technology. Thus, for example someone may believe that nuclear energy is cheap and abundant, and also a threat to decentralised forms of society. If they regard energy use as extravagant and the cost unimportant, but are strongly concerned about civil liberties, then nuclear energy will be judged unacceptable. Thus, there are essentially three steps - what clusters of beliefs do people hold about a technology? what are the social values which shape the saliency of these beliefs such that some are insignificant in influencing judgements, others are crucial? and how do those values translate salient beliefs into specific judgements of acceptability or otherwise?

Unlike the psychometric approach (and Starr's approach) the social psychology framework automatically highlights questions about how individuals maintain and deploy wider systems of values and beliefs, because it sees their attitudes to specific risks and technologies as integral parts of such wider patterns of individual and social identity. Risk perceptions cannot be understood without understanding something of these wider belief and value systems. This also naturally recognises the importance of social and individual differentiation into divergent attitude patterns. The other approaches take the supposedly universal risk attributes, or (in Starr's case) societally defined benefits as the key determinants of risk perceptions. Social variation in perceptions for these approaches thus tends to equate to problematic deviance rather a natural state of affairs. Psychometry handles such variance within the far narrower framework of different subjective (dis)utilities of the different attributes of risks - see Fischoff, Hope and Watson, 1984.

3.2 Saliency

Otway's group performed studies of nuclear experts as well as public groups, showing that the prime differences in risk perceptions of nuclear energy were to do with which beliefs about the technology were most salient to the different groups - that is what were their different value-systems which picked out different aspects as positive or negative signifiers. Although at first sight this might appear to be identical to the psychometric weighting of different attributes, it is wholly different because the latter assumes the respondent to be operating a utility-maximising strategy within a fixed and socially universal belief-value system. The range of recognisable social values is thus severely compressed by the psychometric approach into those which are expressed in the form of utilitarian weighting. The Otway approach relates the elements of belief about a technology to integral broader value-systems with which social groups maintain their identity consistently across the full range of social and technological issues and experiences.

Otway's work showed some superficial resemblances to the psychometric results, for example a dimension similar to the dread risk of Slovic and co-workers was found. However, the social psychology approach found that these dimensions were not universal at all; when they compared perceived risks across different energy technologies and across social groups such as those most in favour and those against nuclear power different dimensions waxed and waned in salience, not only in (dis)utility. Furthermore, when beliefs about several different energy sources were analysed together, five dimensions emerged (two relating to benefits, three to risks) as opposed to the two (both relating to risks) identified by psychometry; yet when nuclear energy was considered on its own a different factor structure was found. This and related results caused Otway's group to conclude that generalisations such as those offered by psychometry about the factors involved in all risk perceptions were misleading (Otway and Thomas, 1982) because they did not pay attention to the specific social and political contexts in which people experience risks and technologies.

3.3 Implicit Social Dimensions

One aspect of the Otway work which has been used to question the general validity of his distinctive interpreation of risk perceptions, is that it was focused on the nuclear power controversy. It is sometimes suggested that this particular technology may exhibit social and political implications and responses which are untypical of other technologies. Yet although this work was performed in the early and mid 1970s, well before any systematic research from a more sociological and anthropological orientation, the essentially sociological elements of Otway's work have been confirmed and extended by a substantial body of such research in the 1980s. As Otway himself pointed out (Otway and Thomas, 1982; see also Wynne, 1980) the psychometric work itself exposes sociological questions which have simply not been recognised. Otway and Thomas, for example, noted that Vlek and Stallen (1981) had discovered systematic individual differences with respect to the risk dimensions identified. For each of the respondents' judgements of riskiness, beneficiality, and acceptability, there was substantial agreement on one dimension and disagreement on the other. But in each case it was the dimension with most obvious institutional or political implications over which people disagreed: for example, roughly half of the respondents perceived riskiness to decrease with degree of organised safety (nuclear power and vaccination are more safe than drunken driving) whilst the other half believed the opposite (nuclear power and vaccination are higher risk than drunken driving, hence the need for organised safety measures). In one case organised safety is a signifier of high risk and institutional arrangements are presumably considered inadequate, whilst in the other case the existence of organised safety institutions is taken as a guarantee of safety. Completely contradictory social judgements or values are indicated here as the underlying determinants of risk perceptions. These are judgements of the social properties of the technologies in question - the structure of its controlling institutions.

3.4 Individual Values

Another strand of research which has emphasised the importance of values in determining risk perceptions is that of Green and colleagues

(Green and Brown, 1985; 1982). In some respects this work is similar to the psychometric approach discussed earlier, for example in identifying the significance of the distinction between "personal safety" and "threats to society" in people's risk perceptions. However, Green found these to be entirely independent dimensions of perceived risk, whereas Fischoff and colleagues found this difference to reduce to variables on the same dimension. Asking questions about comparisons revealed the difficulty which respondents had in treating these kinds of risk in the same framework. Green also found other discontinuities where the Oregon group found a continuum along a single dimension; for example, with respect to the difference between immediate deaths and delayed deaths. The reported lack of comparability between such qualitatively different risks led Green to emphasise the basic value-determination of risk perceptions, discarding the notion of risk attributes and subjective utilities. Unlike Otway, however, Green highlighted the individual as the unit of significance, and not the webs of social interaction in which individuals define their values and identity.

3.5 Composite Models

Lee's work (Lee, 1981; 1983; 1986) has also reflected the accumulating sense that for all its important contributions, psychometry has been hindered by the limitations of following a conventional reductionist scientific approach in which the item of interest, risk, has been treated as an experimentally isolable object, then decomposed into its supposedly internally related parts. Like all the approaches described in this section, and many others not reviewed (see, for example, Covello et al., 1985 for a comprehensive collection) there is dissatisfaction with the presumption that activities can or should be considered for their riskiness (and even then a very narrow definition of risk) in isolation from all their other qualities. Lee has used analytical techniques which attempt to identify "similarity space" between diverse perceived qualities of activities, with riskiness only one of several qualities plotted together to see mutual influences and interrelations between the different qualities considered. Although there is no quantification of statistically significant correlations between qualities, it is possible using this method to identify structural differences between social groups. A problem is that the qualities considered tend to be those assumed to be worth investigating by the analyst, although attempts to overcome this inherent difficulty have been made by extensive fieldwork in the case of radioactive waste risk perceptions (Brown and White, 1987).

4. SOCIOLOGICAL APPROACHES

4.1 Risks to Social Identities

There are now many and diverse bodies of risk perception research from a sociological perspective (Brown, 1989; Covello and Johnston, 1987; Douglas, 1986; Short, 1984). There are also countless case-studies of public responses to risks and accidents which offer rich insights into the complex influences upon perceptions and behaviour towards abstract and concrete threats. All these pieces of work, whether single case-studies or evolving theoretical programmes, share a basic analytical thread: that the fundamental categories of meaning and value to which people relate new experiences are their cherished and trusted social environments and identities. Thus, risks are defined primarily according to their perceived threat to familiar social relationships and practices, and not by numerical magnitudes of physical harm (nor even numerical magnitudes systematically reshaped by particular risk attributes such as catastrophic potential). Some of the case-studies of the effects of risks upon coherent communities, such as Levine's (1981) study of Love Canal, Edelstein's (1986) study of the effects of toxic waste pollution on US communities, or MacGill's (1987) study of the community around the Sellafield nuclear reprocessing plant, show most graphically the complex rationalities of ordinary people trying to maintain their own coherent networks of social support and identity in the face of fragmentary but often obscure forces. These frames of meaning and value subordinate the calculus of purely physical risk as seen by the scientist. This general approach thus takes further steps away from those centred around physical risk and risk attributes, and attempts instead to get inside the subtle "rationalities of everyday life" from where risks, risk-generating activities, and risk-managing institutions are identified and experienced. Furthermore, it claims to be able to return from such adventures with structured analysis of the basic factors at work. However, this is not to say that it claims to be able to make generalisations from which normative conclusions about acceptable risks and standards can be derived. The kinds of generalisation and normative conclusions that are drawn from this work usually focus on the process and social structure of risk and technology management rather than upon the specific content of such decisions.

4.2 Cultural Theory: Grid-Group Analysis

Perhaps the most well-developed theoretical framework in this area is the anthropological approach of Douglas and co-workers (Douglas, 1986; Thompson, 1989; Rayner, 1986a and b). Based originally upon ethnographic fieldwork in Africa (that is, participant observation in small-scale tribal societies, thus composing an account of the structure of belief-systems and social norms) this approach advances the theory that beliefs about nature, and about risks, are socially constructed so as to sustain established patterns of social relations. This is not seen as a deliberate process but a more fundamental interaction between experience of those social patterns and inferences about the ultimate structure of natural processes. Like psychology, this approach assumes that people attempt to create cognitive consistency between their experiences of different realms.

Thus, for example if the predominant experience of society is disorderly and amorphous, similar metaphors will arise in the attempt to conceptualise nature. Likewise, if society is experienced as tightly ordered, hierarchical and with little room for dissent or uncertainty, nature too will be seen in similar hierarchical and deterministic terms. Equally, if society is organised so as to manipulate nature as if purely a series of atomised resources for exploitation, such metaphors will also pervade our thinking about concepts of nature, and basic social relations or values. In order to capture this fundamental unity between beliefs and social norms, anthropologists use the concept of culture, in which belief systems are regarded as a tacit expression of dominant social values, and in which beliefs about nature are structured so as to present dominant social structures as if natural and inevitable.

Douglas first applied this basic approach to industrial society risk perceptions in 1974 (Douglas, 1974), though the work which was widely taken up was in 1982 (Douglas and Wildaysky, 1982). She noticed that in the raging environmental controversies and debates during the 1970s, different types of social group were emphasising not just different attributes of risks, but different kinds of risk altogether, even when they appeared to be arguing about the same problem. She argued that these differences in risk-selection could be ascribed to the different forms of social organisation of the groups involved, and their modes of interaction in the wider political culture. For example, environmental groups tended to emphasise catastrophic risks and long-term consequences, whilst entrepreneurs focused on risks to energy supply, continuing consumption and economic advance. Bureaucrats in regulatory agencies tended to recognise risks that could be quantified and subjected to management control programmes. Even where there was vast ignorance of effects, for example of the thousands of toxic chemicals in the environment, artificial scientific models were often created which in effect gave the pretence of hierarchical progression and orderly priorities in risk reduction. Douglas argued that these basic differences were systematic, and could be explained by reference to the kinds of social relationships familiar to each kind of group. Thus, environmentalists groups were observed to be insecure and egalitarian, with chronic concerns about membership and loyalty, and about maintaining democratic internal structures. Such groups, therefore, tend to be internally amorphous and fluid, but with very strong boundary protection with the outside world. There is a tendency towards schism and discontinuity. Douglas argued that this social structure could be found encoded in their beliefs about nature, uncertainties and risks. For example, without strong internal sanctions, external threats are used to maintain group solidarity and identity. Thus, risks with strong discontinuities such as catastrophic risks were emphasised as more important, and external threats from "big business", or scientific ignorance, emphasised. Also stressing long-term risks emphasises the importance of social continuity, and risks which depend upon the long-term maintenance of institutional vigilance and discipline (such as radioactive waste) are particularly threatening. Likewise, bureaucrats, within hierarchical, well ordered social contexts tended to recognise risks that could be quantified and managed by science, progressively reduced via incremental schemes, and which allowed comparisons for purposes of managerial allocation of society's resources.

These and many related observations from this school were not ad-hoc but structured by a deceptively simple analytical scheme, known as grid-group analysis (for a useful discussion see Johnson, in Johnson and Covello, 1987). The elements of social structure which determined the risk-cosmologies were just two orthogonal dimensions, which allowed just four basic types of social context, thus four basic cosmologies. The group dimension reflects the strength of group identity and boundary with the outside social world; the grid dimension reflects the extent to which any individual is constrained by surrounding "objective" social norms and sanctions. Only "high" and "low" values on each of these dimensions are described, reflecting the inherent empirical dfficulties with any more precise designations. Plotting the properties of groups in these independent thus orthogonal dimensions yields just four basic types, in each quadrant (Thompson, 1982, has proposed a fifth, at the intersect). High grid high group yields a hierarchical type of organisation with strong social boundary but also strong internal structure (high grid) in which the relationships between actors is rigid and precisely defined. Only risks which fit these categories can be recognised without tension. High group-low grid yields the sect-like structure described above in which a strong boundary is maintained by invoking external and cataclismic threats to keep an unruly, "ungridded" internal social flux together. Low group low grid is best typified by the individualist entrepreneur, whose concept of social relations is egocentric and opportunistic, with uncertainties and risks in nature an entirely positive resource, there to be exploited for further gain. Finally, the cosmology of the high grid-low group is that of people who feel no group solidarity or support, but who nevertheless feel highly constrained by external forces beyond their control or even understanding. This is, therefore, an amorphous and largely passive social type, perhaps the "silent majority" who simply feel disoriented and pushed about, and for whom risks are felt to be not personally controllable - they simply live in hope, but with considerable anxiety and very brittle faith - that someone else is indeed in charge and will manaǵe the risks in their interest. This social type is generally least organised and expressive, but a restless ambivalence and insecurity nevertheless results in occasional, fitful and unpredictable expressions of more concerted feeling, usually as a result of their structural power lessness, opposition to interventions in their lives.

The grid-group analysis thus offers a cultural model as to why societies exhibit such deep differentiation in the definition and perception of risks. It also gives a framework for understanding why even scientific knowledge is not sovereign over such conflicts, because it too has different idioms which are selectively picked out according to these underlying structures. There are questions about the analysis: for example on what level of social organisation, from small-scale groups to whole political cultures, is it supposed to apply? and are risks being selected according to a version of social interests structured in the way proposed, or are they defined in a more deterministic and involuntary way by the differentially structured nature of social experience? Also one can ask whether this leads to exclusive recognition of only the risks with the characteristics consistent with each social type, or the attributes consistent with each social type, of risks that are still universally recognised. In terms of methods too, anthropology and sociology are inherently at a disadvantage because it is impossible to offer controlled data from laboratory settings. On the other hand, the context-embeddedness of the data gives a richness and potential sublety which is unmatched by other methods. Equally, there will always be an element of analyst's interpretation - the data will never speak entirely for themselves. But at least this kind of approach explicitly recognises this problem. Whatever its shortcomings, the Douglas cultural theory has been the most widely taken up of all the sociological approaches to risk perceptions, perhaps because it appears to offer a quasi-scientific model of observed intractable elements of social differentiation and conflict in risk perceptions.

4.3 Tacit Perceptions of Social Control

Other sociological approaches are consistent with the fundamental point that the essential factor in public definition and perception of risks is maintenance of cherished social fabrics (Short, 1984; Brown, 1989). However, whether the essential units are generalisable in quite such a clear way as Douglas' approach proposes is open to question. Despite the ambitious attempt at conceptual simplification of grid-group analysis, it may be that the elements of social identity which operate as the fundamental determinants of differential risk problem-definitions are simply too complex and fluid in real societies to reduce into such a general scheme.

Nevertheless, sociological studies support the general conclusion that it is rational to define a risk according to the social structure of its control rather than its physical magnitude alone. This could be derived also from the social psychology tradition (Otway and von Winterfeldt, 1982) with its later emphasis upon the socio-political implications of technologies as the basic focus of perceptions, not risks as conceived by experts. From participant observation of the Windscale nuclear inquiry in 1977 Wynne (1980, 1982) noted how opponents defined the risk analysis issue before the inquiry as much larger than the narrow technical issue of the immediate effects of the proposed plant, because of systematically different judgements of the controlling institutions from those being made by the established experts (who were loyal members of those very same institutions). The scope of this physical risk problem was much greater than the environmental and public health effects of the single plant proposed. They saw this as leading "inevitably" to more fuel cycle facilities, and expanded nuclear power programmes, hence a much bigger physical risk problem. This larger physical risk boundary was based on an institutional judgement of the inevitability of further expansion should the plant be allowed. The experts assumed this to be open to further decisions, so adopted a limited physical risk problem. Wynne pointed out that each side, including the "purely" technical approach, was inevitably founded upon such social assumptions or judgements; there was an essentially symmetrical structure of rationality to each framework, "expert" and "public". He thus proposed the now widely accepted point that people judge a risk according to their perception of its controlling agents -

if they have a social track-record of secrecy, arrogance or incompetence, if they appear to dominate supposedly independent regulatory bodies and the policy making process, it is hardly surprising if people treat the risks as greater than those recognised in quantification of physical magnitudes based upon the tacit assertion that the institutional context does not matter. Indeed this could be regarded as a general statement of the "law" of risk perception described earlier about the more relaxed attitude towards self-controlled risks. Rightly or wrongly, people may feel that all else being equal, the most trustworthy agents are themselves.

4.4 Institutional Models

In a later piece of work Wynne (1989) attempted to reinterpret the findings of psychometric research in such sociological terms. He proposed that the two basic dimensions of risk attributes found in that tradition, unfamiliarity and dread, could both plausibly be redefined as sociological rationalities rather than cognitive or emotional factors. He also argued that expert risk perceptions of the same issue were equally socially based.

The hardest challenge in this respect is the public dread association of nuclear power with the atomic bomb, which is dismissed by experts as empirically unfounded, therefore irrational. Wynne argued that the association is based upon the tacit institutional assertion that the connection is real though different from the simple physical connection upon which the technical experts focus. This is because the institutions supposedly maintaining a strict separation between nuclear power and nuclear weapons, namely the IAEA under the Nuclear Non-proliferation Treaty, are judged not to be completely successful, and indeed could not be. Thus, the connection between nuclear power development and nuclear weapons proliferation is arguably real, even if the IAEA safeguards programme has kept it to less than it might have been. This social logic is matched by the tacit social logic of the "experts" which underpins the expert assertion of the lack of connection that the institutions have been successful. Each cognitive position is based upon different institutional assertions, or judgements. The expert assertion of the irrationality of the public dread association can be seen to be simply an authoritarian imposition of one set of social judgements or values over another. This also begins to explain why the repetition of the language of objective expert risk understanding against the irrationality of public risk perceptions, is socially provocative and - in terms of gaining public confidence - self-defeating.

This social perspective can be generalised to the broader understanding of risk perceptions and conflicts. It is not a sociological reductionist stance in that it can still acknowledge the force of technical analysis and evidence. But it emphasises that technical knowledge is only a part of the understanding of real risk systems, and technical experts also have to make assumptions and omit parts of the full picture in order to achieve controlled knowledge. Ordinary people may have relevant experience to contribute, or at least hold equally legitimate, but different founding assumptions about the issue from those of the experts. A deep problem for present decision making processes is that they have no regular and constructive ways of bringing these buried assumptions from either expert or lay public sides into the open for measured scrutiny and debate.

5. CONCLUDING REMARKS

Although the field surveyed all too scantily above looks extremely diverse, as the latter section indicates, there is perhaps more unity than appears at first sight. The wider issues exposed by the sociological approaches are identifiable but implicit and undeveloped in the other, more controlled but by the same token more reductionist approaches. This field demonstrates once again the truism that the responses one receives from (human) nature are largely determined by the questions one poses. To the extent that one shifts from examining in laboratory psychological experiments the "biasing" factors in public risk perceptions to analysing the social framing of both expert and public risk-definitions, one inevitably treads more overtly into political issues; but it is important to note that this is only exposing a dimension which is already there, albeit unstated and perhaps unintended, in the more "scientific" risk perception work. This is a dilemma confronting all those who work in the field, but also those who use the work in decision making.

It is resoundingly clear from the way that all the research understanding has developed, that there are no universal laws of risk perception which can be used to establish decision rules for defining socially acceptable levels of technical risk. Acceptability depends at least as much upon procedural and institutional structures of decision making in technological systems, as it does upon levels of physical risk - and this can be defended as perfectly rational, not merely an irresistible public demand that has to be tolerated.

Another accepted conclusion is that comparison of risk magnitude is not comparing like with like, and in ignoring many factors which people clearly judge to be important, it is an inadequate framework for normative thinking except in very limited and carefully defined situations.

One of the main ways in which understanding of risk perceptions has been used in decision making is simply to add new factors to the balancing of cost and benefit in classic cost-benefit analysis; if the public appears to weigth the disutility of a unit of radiation risk much more than the "same" unit of risk from other things, then for the same amount of assumed benefit one simply increases cost of the residual risk beneath a proposed limiting standard, thus driving the limit to a lower level of accepted risk. The multipliers for "risk aversion" in this kind of exercise are set intuitively, there being no warrant for anything more finely tuned. In routine cost-benefit optimisation language, the detriment curve is set more steeply, so that it cuts the curve of cost per increment of discharge-reduction (that is, risk-reduction) at a lower level of discharge, or risk.

This crude first-order method may be viable for routine decisions say in designing and regulating X-ray medical diagnosis or industrial inspection systems. Here there tend to be no controversies, and usually a single organisation wanting to get on with a legitimate job. In other settings, such as those which are typical of nuclear waste site decisions, the many dimensions of people's perceptions and responses have to be differentiated and weighted for significance and then somehow recombined into a decision. In this endeavour the psychometric paradigm of risks and attributes is tailor-made to fit into decision analysis using multiattribute utility analysis (MAUA). The various dimensions of risk, ranging from physical death and injury to concern, can be weighted for importance, then the disutility attached by respondents to each one can be elicited. The composite index of risk for the respondent can then be obtained in semi-quantitative form by simple integration, assuming independence between the preferences expressed (and of course a commensurable scale of utility - which means in practice, money - for all forms of human experience).

As the exponents of MAUA incorporating risk perception research have noted (Fischoff, Watson and Hope, 1984), although its traditional use is to clarify decisions for individual decision makers, this approach could in principle be used in public decision making, to render the structures of concern and interest entirely transparent and more democratic. Thus, even such advocates of MAUA recognise that its contribution is not to provide a sovereign consensual answer above all the contending social perspectives, to the question of where is the acceptable level of risk. The potential contribution lies in clarifying and thus improving the decision process and the depth of interaction between the groups involved.

Yet even with this enlightened conclusion there are further questions and problems. The MAUA attempt to incorporate risk perceptions in decisions still imposes the assumption that the problem recognised by every party is the same, and is risk-centred. Thus, it excludes a full recognition of the depth and diversity of problem-agendas and rationalities within which the qualities of risk recognised by psychometrics may or may not be recognised by ordinary people. They may set the experiences whose meaning these "scientific" methods of risk perception research try distantly to capture, in an entirely different frame of reference.

Thus, the dominant response to risk perception research in decision making, apart from just ignoring it, has been to try to enlarge the scope of existing decision methods, to try to capture the weights and "disutilities" to people, of an enlarged agenda of risk-attributes. This seems to act almost as a diversion from reviewing the need and opportunities to restructure decision making institutions and relationships so as to respond to perceptions that may not even recognise the supposed risk-centred nature of the decision problem. These interpretations suggest the need for enlarged mechanisms of genuine political negotiation rather than enlarged techniques for deluding decision makers into thinking they have tapped the full depth of meaning in public risk perceptions.

To give a concrete example from radioactive waste management, a repeated "attribute" of the risk which people appear to express, is the felt lack of independence of central government regulatory bodies from the nuclear industry. Another is the belief that if one allows a proposed plant of, say, 10,000 ton capacity, it will lead to expansion plans not long afterwards. One could respond to these institutional bases of risk perceptions in their own terms, rather than distort and emasculate them into the framework of "attributes" and "disutilities". It would be an option to offer a community its own independent research and monitoring capability, responsible to a local authority rather than to central government. It would also be feasible to offer a community an absolute right of veto on future expansion plans should it accept the proposed plant (in experience, the veto right not actually be exercised in the event).

A thorough scientific treatment of the question of public risk perceptions implies a thoroughgoing politics of technology. Risk perception research has at least given some indications of how that politics might be more constructively structured.

REFERENCES

Brown J. (1989) ed., "Environmental Treats: analysis, perception and management". Belhaven, London.
Brown J. and White J. (1987), "The Public's Understanding of Radiation and Nuclear Waste", Journal of the Society for Radiological Protection, vol. 7 pp. 61-70.
Covello V.T. (1987), "Decision Analysis and Risk Management Decision Making", Risk Analysis, vol. 7, pp. 131-139.
Douglas M. (1974), "Environment at Risk", in M. Douglas, Implicit Meanings, Routledge and Kegan Paul, London, 1974, pp. 64-81.
Douglas M. and Wildavsky A. (1982), "Risk and Culture", University of California Press, Berkeley.
Douglas M. (1986), "Risk Acceptability According to the Social Sciences", Russel Sage, Basic Books, New York.
Edelstein M. (1986), "Psychological Impacts of Toxic Exposure", in H. Becker and A. Porter (eds). Impact Assessment Today, vol. 2, Jan Van Arkel, Utrecht, pp. 761-776.
Green C.H. and Brown R.A. (1980), Through a glass darkly: perceiving perceived risk to health and safety, paper to workshop on Perceived Risk, Eugene, Oregon.
Green C.H. and Brown R.A. (1982), It all depends on what you mean by "acceptable" and "risk", Risk Analysis, vol. 2, pp. 94-113.
Johnson B. (1987), "The Environmentalist Movement and Grid Group Analysis: a modest critique", in V. Covello and B. Johnson, eds. (1987).
Johnson B. and Covello V.T. (1987), The Social and Cultural Construction of Risk, Kluwer, Boston.
Lee T.R. (1983), "The Perception of Risks in Risk Assessment". Royal Society, London.
Lee T.R. (1986), "Public Attitudes Towards Chemical Hazards", Science of the Total Environment, vol. 51, pp. 125-147.

Levine A. (1982), "Love Canal: science, politics and people", Lexington Books, Cambridge Mass.

MacGill S. (1987), "The Politics of Anxiety: Sellafield's Cancer-link Controversy". Pion, London.

Otway H. and Fishbein M. (1976), "The Determinants of Attitude formation: an application to nuclear power", RM-76-80, International Institute for Applied Systems Analysis, Austria.

Otway H. and Fishbein M. (1977), "Public Attitudes and Decision Making", RM-77-54, International Institute for Applied Systems Analysis, Austria.

Otway H. and Cohen J., "Revealed Preferences: Comments on the Starr Benefit-Risk Relationship, IIAA Research Memorandum, RM-75-5 International Institute for Applied Systems Analysis.

Otway H. and Thomas K. (1982), "Reflections on Risk Analysis and Policy", Risk Analysis, vol. 2, pp. 69-82.

Otway H. and von Winterfeldt D. (1982), "Beyond Acceptable Risk: the social implications of technology". Policy Sciences.

Rayner S. (1986), "Management of Radiation Hazards in Hospitals: Plural Rationalities in a Single Institution". Social Studies of Science, vol. 16, pp 573-591.

Rayner S. and Cantor R. (1987), "How fair is safe enough? the Cultural Approach to Societal Technology Choice", Risk Analysis, vol. 7, pp. 3-9.

Royal Society of London (1983), Risk Assessment: a Study Group Report, London, Royal Society.

Renn O. (1987), "Technology, Risk and Public Perception", Angewandte Systemanalyse, vol. 4(2), pp. 50-65.

Short J. (1984), "The Social Fabric at Risk: towards the social transformation or risk analysis". American Sociological Review, vol. 49, pp. 711-725.

Slovic P. and Fishchoff B. (1980), Perceived Risk, in R.C. Schwing and W.A. Albers, "Societal Risk Assessment: How Safe is Safe Enough?" New York Plenum Press, pp. 47-61.

Slovic P., Fischoff B. and Lichtenstein S. (1985), Characterising Risk in R.W. Kates et al. (eds) "Perilous Progress: Technology as Hazard". Westview Press, Boulder, pp. 91-123.

Slovic P. (1987a), Informing and educating the public about risks. In: H. Kunreuther and P. Kleindorfer (eds.), "Manging and Insuring against Hazardous Risks: from Seveso to Bhopal and Beyond". Springer, Berlin & London, pp. 307-324.

Slovic P. (1987b), "Perception or Risk". Science, vol. 236, pp. 280-285.

Sjoborg L. (ed) (1987), "Risk and Society: Studies of risk generation and reactions to risk". Allen and Unwin, London.

Stallen P.J. and Tomas A. (1985), "Public Concern about Industrial Hazards", paper to the Society for Risk Analysis meeting. October 1985. Washington D.C.

Starr C. (1969), "Social Benefit versus Technological Risk", Science vol. 165, pp. 1232-1238.

U.K. House of Commons Select Committee (1986), 1st Report, Radioactive Waste, London, House of Commons, pp. 78.

Vlek C. and Stallen P.J. (1980), "Rational and Personal Aspects of Risk", Actor Psychologica, 45, pp. 275-300.

Wynne B. (1980), "Technology, risk and participation: on the social treatment of uncertainty" in J. Conrad (ed) Society, Technology and Risk, Academic Press, London and New York, pp. 87-103.

Wynne B. (1982), "Rationality and Ritual: the Windscale Inquiry and Nuclear Decisions in Britain", Chalfont St. Cilca, U.K. British Society for the History of Science.

Wynne B. (1989a), "Frameworks of Rationality in Risk Debates: Towards the Testing of Naive Sociology", in J. Brown (ed) Environmental Threats: analysis, perception and management, Belhaven, London, pp. 33-47.

Wynne B. (1986b), "Building Public Concern into Risk Management", in J. Brown (ed) (1989), pp. 118-132.

DECISION-MAKING BASED ON RISK ANALYSIS:
COPING WITH UNCERTAINTY AND INCOMPLETE DATA

SAIFUL ISLAM
Research Unit Gottstein
Max Planck Society
P.O.Box 40 06 48
8000 Munich 40
Federal Republic of Germany

ABSTRACT. A number of scientific, ethical and sociological
questions relevant to radioactive waste disposal is discussed. In
the safety assessment of waste repositories, most of the data is
known only approximately. This gives the opponents and supporters
of the nuclear power programme an opportunity to arrive at
completely different conclusions regarding the safety of the
repositories. Scientists are urged not to give up the quest for
making the risk assessment as objective as possible.

1. INTRODUCTION

In the decision theory, a decision-maker is said to have a decision
problem when he or she has to choose one out of a number of possible
actions with an estimate to its consequences. If the consequences
of each possible action are known, the decision problem is one <u>under
certainty</u>. If they are unknown, the decision problem is one <u>under
uncertainty.</u>

Even if consequences are imperfectly known rather than unknown,
we may still be said to be taking decisions under uncertainty, if we
cannot assign probabilities to the different possible consequences.
If, on the other hand, probabilities can be assigned to the various
imperfectly known consequences, the decision becomes one <u>under
risk</u>.

Because of the significant time scale over which the
consequences of choice regarding radioactive waste repositories
apply, it would seem that the decision is, in principle, one under
uncertainty. But in calculations regarding the long-term isolation
of radioactive waste from the biosphere, it is assumed that
probabilities <u>can</u> be assigned to all possible consequences and
thereby the decision problem is changed from one under uncertainty
to one under risk.

A. Saltelli et al. (eds.), Risk Analysis in Nuclear Waste Management, 25–34.
© 1989 ECSC, EEC, EAEC, Brussels and Luxembourg.

2. UNCERTAINTIES IN THE TRANSPORT OF RADIONUCLIDES FROM THE REPOSITORY TO THE BIOSPHERE

A one-dimensional transport of a radionuclide through a uniformly porous geological medium can be modelled with the following equation(1):

$$\frac{\partial C}{\partial t} = \frac{\partial}{\partial x} \frac{\alpha u}{R} \frac{\partial C}{\partial x} - \frac{u}{R} \frac{\partial C}{\partial x} - \lambda C + q \tag{1}$$

where

C = concentration of radionuclide in the transporting water (kg/m^3)

u = ground water velocity (m/s)

α = longitudinal dispersion length (m)

R = retardation factor, which accounts for material absorbed in the rock

λ = decay constant of the radionuclide (1/s)

q = some source of C, including decay product from parent nuclide ($kg/m^3 s$)

Even this very simple equation can be solved analytically only under special boundary conditions (delta or point source and constant release of radionuclides). More often than not rather complicated equations are used to model radionuclide transport through the geosphere. Yet none of the some thirty numerial codes (12) now in use for modelling radionuclide transport is completely reliable.

Even if one could have a completely debugged numerical code, the problem of uncertainty in the input data would remain. The only parameter known with certainty in eq (1) is the decay constant (λ) of the radionuclides. The concentration (C) of the radionuclides in the transporting water is hard to estimate because the ground water velocity (u) is difficult to measure. The longitudinal dispersion length (α) cannot be known exactly for large scale systems -- the length of the geological pathways being of the order of kilometres. The pathway itself, i.e. the preferential path through which contaminants reach the biosphere, is one of the most uncertain parameters, due to the difficulties inherent in the investigation of fracture networks in rocks (in crystalline formations). The groundwater circulation over geological time scales is also uncertain.

Those who believe that the situation is risky will always claim that the retardation factor (R) is low. In fact, it is extremely complicated to measure R, because R is determined by a variety of processes collectively referred to as "sorption", which can cause radionuclides to move more slowly than the water which carries them. These processes include irreversible precipitation as well as ion-exchange and surface adsorption, diffusion in dead end pores etc. On the other hand, those who believe that the situation is uncertain can plausibly argue that the value of R cannot be known even approximately (within orders of magnitude) for long geological pathways.

3. RISK AND JUSTICE

We cannot be aware of most dangers, i.e. we cannot calculate even approximately the total expected risk we face. But still we have to decide which risks to take and which to ignore. At the level of public debate, we observe that people have differing levels of concern towards various dangers. People who are for 'law and order' are not necessarily concerned that crime might result from economic inequality. People concerned about the rapid increase in population are not necessarily concerned that economic well-being might reduce the population growth.

The key term in the debate over nuclear technology is justice. The technostructure (a term coined by John Kenneth Galbraith to denote top people in the administration, business, military, party, science and technology) has come up a number of times with the suggestion that radioactive and chemical waste can be reposed in poor countries - especially in China. Purely allocative economic efficiency would approve the disposal of radioactive and chemical waste in this way. Industrialized countries would rid themselves of the dangerous waste, and poor countries would be better-off economically.

Our sense of justice makes us suspicious of purely allocative economic efficiency - whatever the rhetoric or esoteric mathematics applied by some economists to 'prove' the virtue of pure competition. We know that the ruling class in poor countries will not ask their citizens whether they should import (at a negative price) the radioactive and chemical waste generated by the rich countries.

The philosopher John Rawls (15) puts it as the first principle of justice that everyone has the right to as much liberty as is consistent with a like liberty for all. The citizens in rich countries who are concerned about the disposal of radioactive waste feel a sense of responsibility towards future generations. They are worried about distributive justice, not allocative efficiency.

According to the Rawlsian principle of justice reasons for curtailing liberty must be derived from liberty preserving considerations. One generation or one country should not put other generations or other countries in a situation that is not acceptable to itself. So, as a first principle one should require that the radioactive waste must remain in the country which generated it and as a consequence benefited from it.

4. SUBJECTIVE IMMUNITY AND NUCLEAR TECHNOLOGY

The nuclear establishment is disconcerted that the public is not prepared to accept "reasonable risk". By comparing fatalities resulting from coal-burning plants (4 deaths per year per million kilowatt year) Chauncey Starr(18) comes to the conclusion that the public should accept one catastrophic reactor accident - of the Three Miles Island (TMI) type - per 100 reactor years. He assumes that catastrophic reactor accidents will claim only 10 lives.

The USA has about 100 nuclear reactors with about 1,200 years
of operating experience(11). According to Starr, 11 more TMI-type of
accidents should be acceptable to the populace of the USA if only
they would look at the matter in terms of probabilities. The whole
world has about 400 reactors with about 5,000 years of operating
experience(11). We know that 50 TMI-type accidents worldwide would
mean a certain termination of nuclear power programme. So are people
behaving irrationally? Not necessarily. Questions about acceptable
level of risk can never be answered merely by looking at the
probabilities. People are not convinced, probably cannot be
convinced, that the risk associated with car driving, smoking or
with the failure of a radioactive waste repository are of the same
kind simply because they can be expressed in numbers. Smoking might
give pleasure, car driving can be a necessity. Such factors
influence most people's attitudes rather than bald figures.

An accident in a coal-fired plant has only local effect both
in time and space. But the Chernobyl accident directly affected
people 2,000 km away and indirectly - due to the export of
contaminated food - people almost everywhere on the globe. Its
effect on the nearby region will continue for hundreds, if not
thousands of years.

People do not accept a risky situation when they cannot by
their action -- about where to live and what occupation to choose --
influence the effects of an accident. People take the risks
associated with natural disasters as a fact of life. Technological
disasters, on the other hand, if they are thought to be inflicted on
a large number of people by a powerful minority - in this context
the nuclear establishment -- make people very angry. Risk perception
may not be the issue at all, but indignation at the injustice done
Natural disaters are taken with less complaint, with less of a sense
of injustice.

We must also take into consideration that people grow used to
familiar risks, whether natural or technological, and they do not
always follow risk minimizing strategies. Mary Douglas points out(4):

"The best established results of risk research show that
individuals have a strong but unjustified sense of subjective
immunity. In very familiar activities there is a tendency to
minimize the probability of bad outcomes. Apparently, people
underestimate risks which are supposed to be under control. They
reckon they can cope with familiar situations. They also
underestimate risks of events rarely expected to happen ... For a
species well adapted to survive, neglecting low frequency events
seems an eminently reasonable strategy. To attend equally to all
the low probabilities of disaster would diffuse attention and even
produce a dangerous lack of focus. From the point of view of
species survival, the sense of subjective immunity is also adaptive
if it allows humans to keep cool in the midst of dangers, to dare to
experiment, not to be thrown off-balance by evidence of failures".

So public antipathy to nuclear technology may not be a
straightforward matter of fear; it may result from lack of
familiarity and a cosequent inability to invoke subjective immunity.

5. TRUST AND RISK PERCEPTION

In some societies, people believe that they are capable of
influencing political decisions. Whether or not the confidence is
justified, it is a necessary ingredient of a democratic society.
Erosion of trust in institutions is dangerous for democracy. Trust
was breached when the public discovered that the nuclear
establishment tended to exaggerate the technology's safety.

The designers of nuclear reactors claim that the chance of
reactor accidents should be less than one in one million. The U.S.
Rasmussen report(16) and the Federal German safety study(3) found
that there is a fifty-fifty chance that such accidents can occur
every 23,000 and 10,000 reactor years respectively. Two major
accidents, Three Miles Island and Chernobyl (three, if we count
Windscale), suggest that the standard has not been met in practice.
Using a statistical method suggested by S. Islam and K.
Lindgren(13), it can be shown that the fifty-fifty chance of a
reactor accident is about one in every 2,000 reactor years · or, at
present, about 5 calendar years (400 reactors x 5 years).

The public has always been told that the nuclear fuel cycle is
closed, i.e. no radioactivity ·· especially plutonium · can ever leak
out of the cycle. Understandably, the public became enraged and
felt cheated over a bribery scandal involving a Federal German firm
dealing with nuclear fuel and radioactive waste. Apparently, a
large quantity of plutonium was sold to a third party in
contravention of the non proliferation treaty, or was disposed
illegally. Under section 334 of the Federal German Criminal Penal
Code bribing becomes an offence only if a civil servant is
bribed(19). Most of the persons involved cannot even be
prosecuted. Although the firm gets all the contracts from companies
owned by municipalities and communities, it is a private company,
and its executives or employees can bribe, without fear of
prosecution, executives of publicly owned but privately incorporated
electrical utilities. This "taking commercial advantage" does not
constitute a criminal offence. People can rightly ask why such
important duties, such as the handling of radioactive waste, are in
the private sector. The public has good cause to fear that they are
not in control, not even through their legal institutions.

6. RISK PERCEPTION IN HIERARCHICAL ORGANIZATIONS

Most of the risk assessment work on radioactive waste management is
performed in hierarchical organizations. Risk is perceived
differently in different types of organizations. A hierarchy is
characterized by inertia and by its avoidance of completely new
ideas. It cannot perceive the unexpected risk. Douglas and
Wildavsky(5) observe: "Leaving aside government, what do those
organizations whose survival depends on correct calculation
capitalist firms · do about risk? Do they attempt comprehensive
calculation of the risk in their environment or do they rely on

different sources? The evidence against comprehensive calculation is overwhelming ... If there has been any finding that has been amply documented in studies of decision-making by large scale organizations of the hierarchical kind, it is that most alternatives, most of the time, are not considered as candidates for adoption. Only a few ideas - those best known and closest to existing programs - are given attention".

Risk tends to be underestimated in radioactive waste management because many of the risk assessors form part of a bigger hierarchy. Political decision makers want to know the judgement of scientists on the safety of waste repositories. In principle, the judgement should be objective. The goal of the political decision makers is, however, known to scientists. They want safe repositories. A bias is thus introduced in the risk assessment problem. It is difficult for scientists, especially for the junior scientists who actually do the calculations, to avoid the bias because their rank in the hierarchy is rather low.

The late great physicist Richard Feynman, who as a member of the Space Shuttle Commission investigated the Challenger disaster, found out on the very first briefing that O-ring scorching was responsible for the accident(8). It was clear to him immediately because he was not a part of the establishment - the big hierarchy. In his own words(8): "One of the ways I was unique was that I was not connected to any organization -- I had no weakness from that point of view. I was, of course, connected with Caltech, but that's not a weakness! For example, General Kutyna was in the Air Force, so he couldn't say everything exactly the way he wanted, because he might get in trouble with the Air Force. Sally Ride still had a job at NASA. Everyone on the commission had some kind of connection and therefore some kind of weakness, but I was apparently invincible". The danger in a hierarchy is that "in hierarchies the goals are multiple and vague, their multiplicity making it easier to satisfy different elements of the firm and to retrospectively rationalize whichever ones happen to be accomplished."(5)

Feynman also found out that the information flow was not functioning properly at NASA(8): "By now we had found out that the flight failed because one of the seals had broken, and the higher-ups had told us they didn't know anything about seals problems -- even though I was able to find out about it right away at JPL (Jet Propulsion Laboratory) before I even went to Washington. We saw that NASA had no system for fixing the problem, even though engineers were writing letters like 'HELP!' and 'This is a RED ALERT'. Nothing was happening." If extreme care is not taken to guarantee the free flow of information, a hierarchy filters and selcts information by a process which is normally inefficient.

The next finding of Feynman is even more startling. It demonstrates simultaneously the systematic underestimation of risk and the arbitrary selection of risk by a hierarchy: "Mr. Ulian had to decide whether to put explosive charges on the side so ground control could destroy the shuttle in case it was falling onto a city. The big cheeses at NASA said: 'Don't put any explosives on, because the shuttle is so safe. It'll never fall onto a city.'

"Mr. Ulian tried to argue that there was danger. One out of every 25 rockets had failed previously, so Mr. Ulian estimated the probability of danger to be 1 in 100 - enough to justify the explosive charges. But the higher-ups at NASA said the probability of failure was 1 in 100 000. That means if you flew the shuttle everyday, the average time before your accident would be 300 years - everyday, one flight, for 300 years - which is obviously crazy! Mr. Ulian also told us about the problems he had with big cheeses - how they didn't come to the meetings sometimes and all kinds of other details."(8)

Douglas and Wildavsky have an explanation for this phenomenon(5): "The selection of risks worth taking and avoiding (in a hierarchy) is made by a process, not by a person ... its information is so heavily filtered that it easily misses a point".

This is, of course, all highly relevant to the question of risk assessment as regards radioactive waste management. The work must inevitably be done in large if not hierarchical organizations - risk assessment performed by small firms is unreliable due to their lack of wide-ranging expertise; private organizations may be biased because of their financial dependence on commissioning authorities. We cannot and should not avoid big organizations' participation in nuclear risk assessment, but what we must do is reduce the inherent dangers in the big organizations.

The group of risk assessors must be independent, i.e. its dependence on outside expertise must be a minimum and it must be publicly funded. The group should have as many different types of scientists as possible: physicists, geologists, chemists, engineers, climatologists, oceanographers, mathematicians, statisticians, biologists, physiologists, soft-ware engineers etc. And the problem of coping with uncertain data and incomplete information must be squarely tackled.

7. COPING WITH UNCERTAINTY AND INCOMPLETE DATA

The publication of the scientific correspondence concerning the frequency of catastrophic reactor accidents(13) was followed by a number of critical scientific correspondences(2,6,7,9,10,17). But there was also a consensus of opinion amongst them that the main conclusion of the original correspondence(13) nevertheless remained valid: failure rate estimates of one accident per million reactor year are much too optimistic in view of past experience.

It is heartening that this particular debate on catastrophic reactor accidents at least remained objective among the scientists. Different correspondents had different attitudes towards the nuclear power programme. But they did not let their personal bias influence their scientific judgement. Scientists involved in the assessment of risk should try not to be involved in the political debate. They should not be concerned whether their findings help the environmentalists, conservatives or socialists.

Of course, scientists cannot and do not operate in a vacuum, they work within a society. But when they are asked to give their opinion on radioactive waste management, they should not be concerned with unemployment, economic decline, carbon dioxide pollution or the ozone hole. They should only give evidence on the environmental impact of radioactive waste management. Experts might disagree but they must know why they disagree - this requires a high degree of objectivity. In the end, society will take a political decision, but scientists should not be involved in the politics in their scientific work.

The main responsibility of the scientists is to be objective. This responsibility results from their social responsibility of not putting society in a dangerous position by giving a biased opinion. Sometimes suggestions like(14): "Site-specific studies should concentrate on obtaining information to satisfy regulatory agency requirements" are seen in the literature. This is certainly not the best approach. The responsibilities of the scientists are far more wide-ranging than a "duty" to the regulatory agencies. They should undoubtedly concentrate on certain important aspects of the problem - dealing with all minor dangers will diffuse their attention and unnecessarily complicate the calculation, and may render resulting recommendations useless for a political decision maker. But if in the process of their investigation, they do find some information which is relevant for risk assessment, but was overlooked by the regulating agency, they must include the information in their calculation. Their credibility will be seriously damaged if they neglect any relevant data which their scientific judgement dictates them to consider.

The best course of action for scientists is to assemble all possible and relevant information pertaining to the geological system in which the waste will be reposed. If the data is incomplete, they must say so. One should not try to reduce the error bars in the data by "plausible" arguments. Objectivity should also be the criterion for choosing the method of analysis, because the controlling factor in many decision problems is how the problem is defined.

Because of the uncertainties in the data, statistical methods must be used in risk assessment. Sensitivity analysis, which is aimed at the identification of the most important system parameters, must be used with extreme caution. A method should not be chosen just because of its speed and economic efficiency. More consideration must be given to its robustness and numerical stability.

In physics, statistical methods are used to set limits on different possibilities. In radioactive waste management, statistical methods are still mainly used for analyzing data. Serious consideration should be given to the possibility of using statistical methods in conjunction with available theory to make the field of investigation more precise. The dialogue about risk is conducted in two languages: traditional rhetoric and mathematical statistics. The statistical method, since it can be improved objectively, must be improved.

Finally there is the need for openness. If the risk analysis is carried out with public money, the public surely has a right to know how the analysis was done. The brief of the risk assessors is not to make the nuclear repositories safe or unsafe, but to be as objective as possible. They must be able to convince the public that the parameters used in the safety analysis, although not exhaustive, are the most important ones. If they hide anything from the public, the public would rightly feel offended and this will increase the public mistrust in expert opinion.

It is difficult to be objective on such a complicated issue. But the quest for finding an objective method should not be given up. Mathematicians did not give up axiomatization just because axiomatization cannot be complete.

8. EPILOGUE

The extent to which the data pertaining to waste repositories is uncertain is the main issue in the debate over nuclear power generation. There are two types of opponent to the nuclear power programme. The first believe that highly risky decisions were taken by pursuing such a programme; the second argue - plausibly enough that probabilities cannot be assigned to the imperfectly known consequences of the decisions taken. Since everyone agrees that radioactive waste is extremely hazardous, the second group demands that nuclear power generation be stopped immediately.

So the decision problem is complicated. If there were a consensus that nuclear power is needed, as in the aftermath of the launching of the 'Atoms for Peace' programme, the decision problem could be approached by commissioning more research for gathering new information. The new information might asses probabilities more accurately. But the nuclear power programme is severely contested. Each side in the debate is thought by the other to be biased. Each takes the evidence of the other to be self-serving and therefore false. It is important that this fact is recognized, and that the debate is political. It is highly unlikely that a solution would be acceptable simultaneously to the nuclear establishment and the general public, let alone to the dissenters.

The best way to deal with uncertainty is to avoid taking decisions which have very long-term consequences. Technologies connected with harmful chemicals and radioactive material which must be excluded from the biosphere for a million years should be selected only in dire emergency. A million years is after all an eternity for a species that abandoned gathering and hunting no more than ten thousand years ago.

But a large amount of radioactive waste has already been generated, and it must be reposed somewhere. As a scientist, my duty, if required to calculate the saftety of a repository, is unquestionably to forget these personal views and to proceed fully objectively.

ACKNOWLEDGEMENT

This paper is based on the works of Professor Mary Douglas. I am
sincerely thankful to her for encouraging discussions. I am
thankful to Professor Klaus Gottstein for giving me an opportunity
to work at his research unit, for fruitful discussions, for his
constant encouragement and for arranging financial support in the
form of a Max Planck Society scholarship. I thank Amanda and Paul
Lomas for helpful discussions and Ingrid Hajjar for typing various
versions of the paper.

REFERENCES

1 Bear, J. (1979) Hydraulics of Ground Water, McGraw-Hill, New York.
2 Chow, T.C. & Oliver, R.M. (1987) 'How many reactor accidents will
   there be?', Nature 327, 20-21
3 Deutsche Risikostudie Kernkraftwerke (1979), Federal Ministry of
   Research & Technology, Bonn and Verlag TÜV Rheinland, Köln.
4 Douglas, M. (1986) Risk Acceptability According to the Social
   Sciences, Routledge & Kegan Paul, London.
5 Douglas, M. & Wildavsky, A. (1982) Risk and Culture, University
   of California Press, Berkeley.
6 Edwards, A.W.F. (1986) 'How many reactor accidents?', Nature 324,
   417-418
7 Edwards, A.W.F. (1987) 'How many nuclear reactor accidents?',
   Nature 327, 663.
8 Feynman, R.P. (1988) 'An Outsider's Inside View of the Challenger
   Inquiry', Physics Today 41:2, 26-37
9 Fröhner, F.H. (1987) 'Once again, how many ractor accidents?',
   Nature 326, 834
10 Hsü, K. (1987) 'Nuclear risk evaluation', Nature 328, 22
11 IAEA Bulletin (1988) 30:2, 67.
12 INTRACOIN (International Nuclide Transport Code Intercomparison)
   (1984), Swedish Nuclear Power Inspectorate, Stockholm.
13 Islam, S. & Lindgren, K. (1986) 'How many reactor accidents will
   there be?', Nature 322, 691-692.
14 Keeney, R.L. & Nair, K. (1977) 'Selecting Nuclear Power Plant
   Sites in the Pacific Northwest using Decision Analysis', in E.E.
   Bell, R.L. Keeney & H. Raiffa (eds.), Conflicting Objectives in
   Decisions. International Institute for Applied Systems Analysis,
   Vienna and John Wiley, Chichster, 298-322.
15 Rawls, J. (1972) A Theory of Justice, Harvard University Press,
   Cambridge, Mass.
16 Reactor Safety Study (1975) Nuclear Regulatory Commission,
   Washington, D.C.
17 Schwartz, J. (1986) 'How many reactor accidents?', Nature 324,
   622
18 Starr, C. (1969) 'Social Benefit versus Technological Risk',
   Science 165, 1232-1238
19 Strafgesetzbuch (1981), Deutscher Taschenbuch Verlag, München

Section 2 – From reactor safety to disposal safety

The contribution from Dr. Garrick summarizes the content of two separate lectures which he gave at the seminar. The Probabilistic Risk Assessment (PRA) approach is presented, with some examples taken from the reactor safety field. The use of event trees and decision trees is illustrated and an extension of these methodologies to waste transportation and waste disposal is presented.

PROBABILISTIC RISK ASSESSMENT IN THE NUCLEAR SECTOR: FROM REACTOR SAFETY TO DISPOSAL SAFETY

B. JOHN GARRICK
Pickard, Lowe and Garrick, Inc.
2260 University Drive
Newport Beach, CA 92660

ABSTRACT. A primary concern of those responsible for the planning and implementation of programs for the hazardous waste disposal process is the identification of the risks impacting the safety of the process. A methodology for probabilistic risk assessment (PRA) is presented that will enable decision makers to identify those risks most important to both ensuring safety and reducing resources allocated for hazardous waste disposal. Examples of PRA applications will demonstrate both the type of information produced by the PRA and the use of this information by decision makers for prudent risk management. Based on the extensive experience base in performing PRAs in the nuclear power and chemical process industries, it is proposed that the PRA thought process be adopted so that the stages of hazardous waste disposal from storage at the reactor site to emplacement at a geological repository can be analyzed.

1. Introduction

Disposal of high level nuclear waste has been a concern since the atomic age began in the late 1940s. Deep-mined, underground repositories for waste were proposed in the United States in 1957 by a National Academy of Sciences committee. Currently, the United States Department of Energy (DOE) has the goal of siting, licensing, constructing, and operating a geologic repository by 1998. To prevent wastes from contaminating the biosphere, issues of the safety and reliability for such a facility are of chief concern.

Due to the public fear of anything related to nuclear power and due to the resulting stringent criteria involving power generation, nuclear power plants have been required to demonstrate their safety and reliability of operation. To accommodate these requirements, owners and operators of nuclear power plants have turned to probabilistic risk assessment (PRA) techniques. In the dozen or so years that PRAs have been performed, they are now gaining wider acceptance from both utility and government officials. Managers now recognize that PRAs enable system owners and operators to make better decisions for resource allocations because of an increased knowledge of plant weak spots and operations.

A. Saltelli et al. (eds.), Risk Analysis in Nuclear Waste Management, 37–53.
© *1989 ECSC, EEC, EAEC, Brussels and Luxembourg.*

The first part of this paper will highlight the purpose of, methodology for, and experience in performing PRAs in the nuclear power industry. In the second part, discussions will focus on the application of PRA techniques to transportation, onsite storage, and geological storage of nuclear waste.

2. What is a PRA?

A PRA is an integrated model of the response of an engineered system to disturbances during operations [1]. The purpose of the PRA is to help owners and operators make decisions about system operations. This is accomplished through the PRA's addressing of three basic questions:

1. What can go wrong during operation?
2. How likely is it to go wrong?
3. What are the consequences when it goes wrong?

The first question is answered in terms of scenarios; the second, by quantifying the knowledge of the likelihood of each scenario; and, the third, by quantifying our knowledge of the response of the plant, its systems, and operations in terms of release states, damage states, and scenario consequences. Answers to these three questions together form the risk triplet, $R_i = <S_i, p_i, x_i>$. Each scenario is associated with a triplet. Figure 1 illustrates that many scenarios typically exist, and that the likelihood and damage for each independent scenario are explicitly stated in the triplet. The set of all scenarios results in a specification of the risk associated with system operations.

ANSWERS TO: (1) WHAT CAN GO WRONG?
(2) WHAT IS THE LIKELIHOOD?
(3) WHAT IS THE DAMAGE

SCENARIO	PROBABILITY	DAMAGE
s_1	p_1	x_1
s_2	p_2	x_2
s_3	p_3	x_3
.	.	.
.	.	.
.	.	.
.	.	.
s_N	p_N	x_N

$$\text{RISK} = \{ <s_i, p_i, x_i> \}$$

Figure 1. Quantitative Definition of Risk

Answers to the likelihood and consequence questions can be either qualitative or quantitative; however, the true power of PRA is revealed through the quantitative response to these questions [1,2].

If the scenarios are ranked in order of increasing damage and the likelihoods are cumulated from the bottom up, the result, when plotted on a log-log scale, is the so-called risk curve, as depicted in Figure 2. We express our uncertainty with a family of curves, as in Figure 3. Each curve has a probability associated with it and represents information about the likelihood and consequence. Consider the curve, P3, in Figure 3; this curve expresses our confidence that there is P3 probability that a damage level of x1 or greater will be observed with a frequency of Φ_1. As our knowledge of the system increases, our confidence in the ability of the system to operate safely also increases; thus, we expect that the curves in Figure 3 will move closer together as more information becomes available.

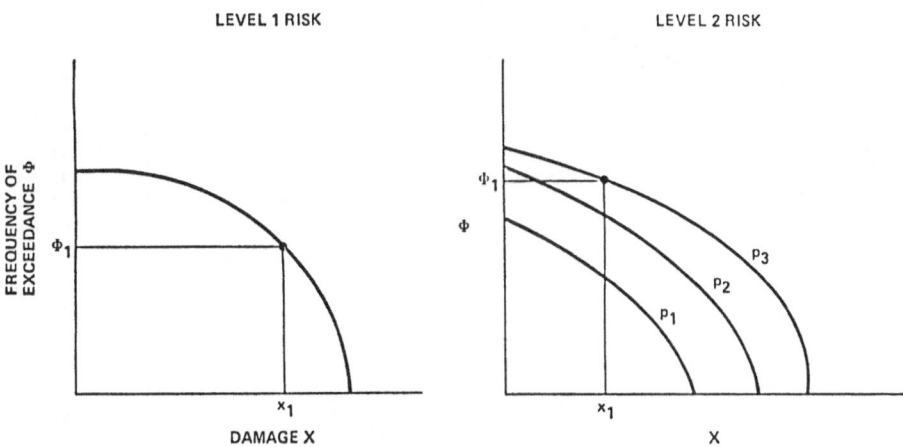

Figure 2. Form of the Results   Figure 3. Form of the Results

To describe our confidence that a specific damage level, such as X_1 = core melt, will occur, we can view the results in an equivalent way to Figure 3, as in Figure 4.

The methodology used to construct these risk curves is not complex [1]. Modeling tools include the use of master logic diagrams, event sequence diagrams, and event trees. The master logic diagram is a top-down view of the entire system. Through this diagram, as in Figure 5, initiating events are identified. Initiating events are perturbations to steady-state operations. For a geologic repository, a master logic diagram would identify all possible perturbations.

Consider the initiating event "Groundwater enters repository site." A set of events characteristic of the initiating event and the interaction with the waste repository can be identified and displayed in an event sequence diagram, as in Figure 6. Next, a set of questions regarding the initiating event can be asked; for

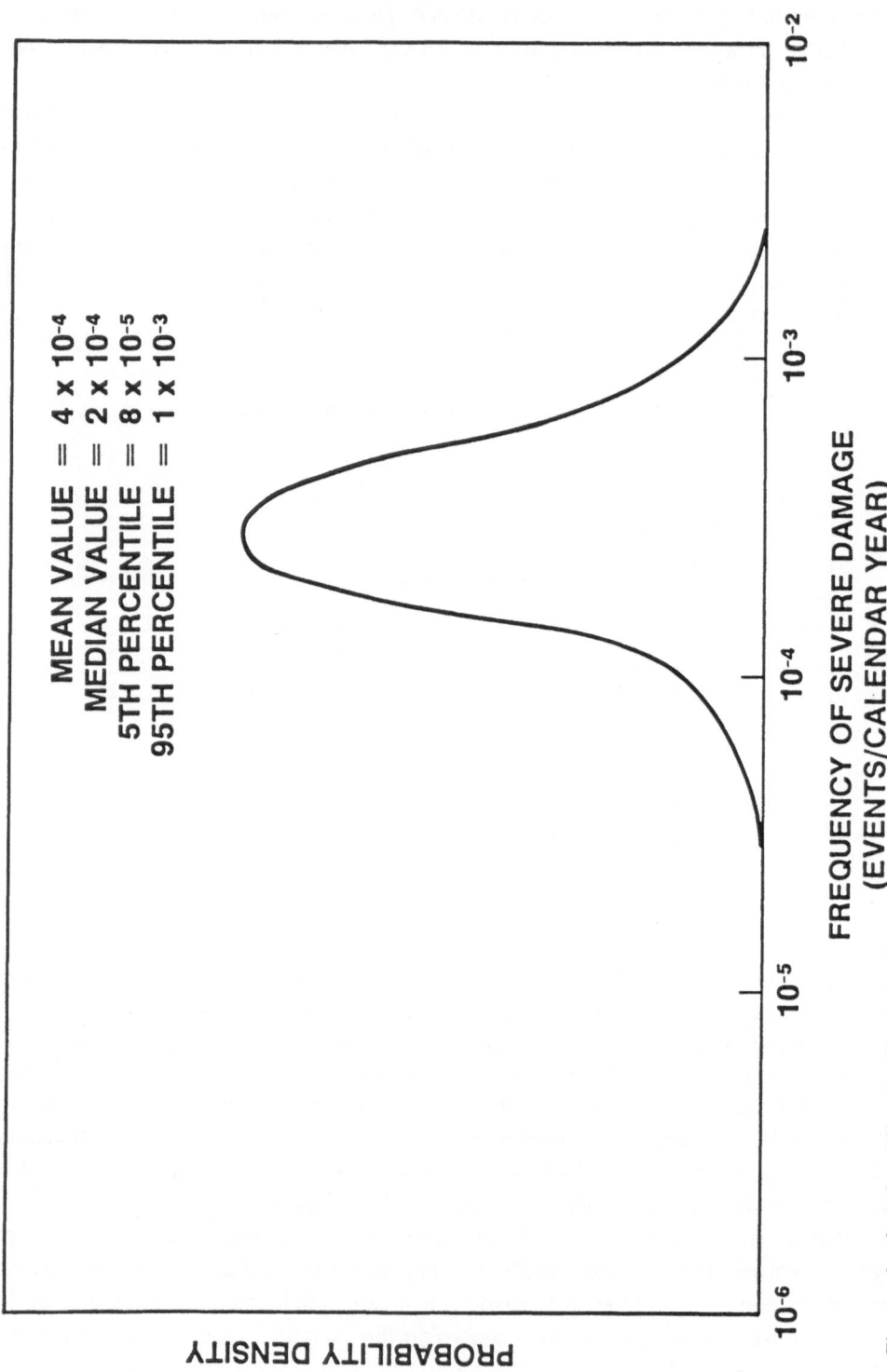

MEAN VALUE = 4×10^{-4}
MEDIAN VALUE = 2×10^{-4}
5TH PERCENTILE = 8×10^{-5}
95TH PERCENTILE = 1×10^{-3}

PROBABILITY DENSITY

10^{-6} 10^{-5} 10^{-4} 10^{-3} 10^{-2}

FREQUENCY OF SEVERE DAMAGE
(EVENTS/CALENDAR YEAR)

Figure 4. Another Way to View the Results

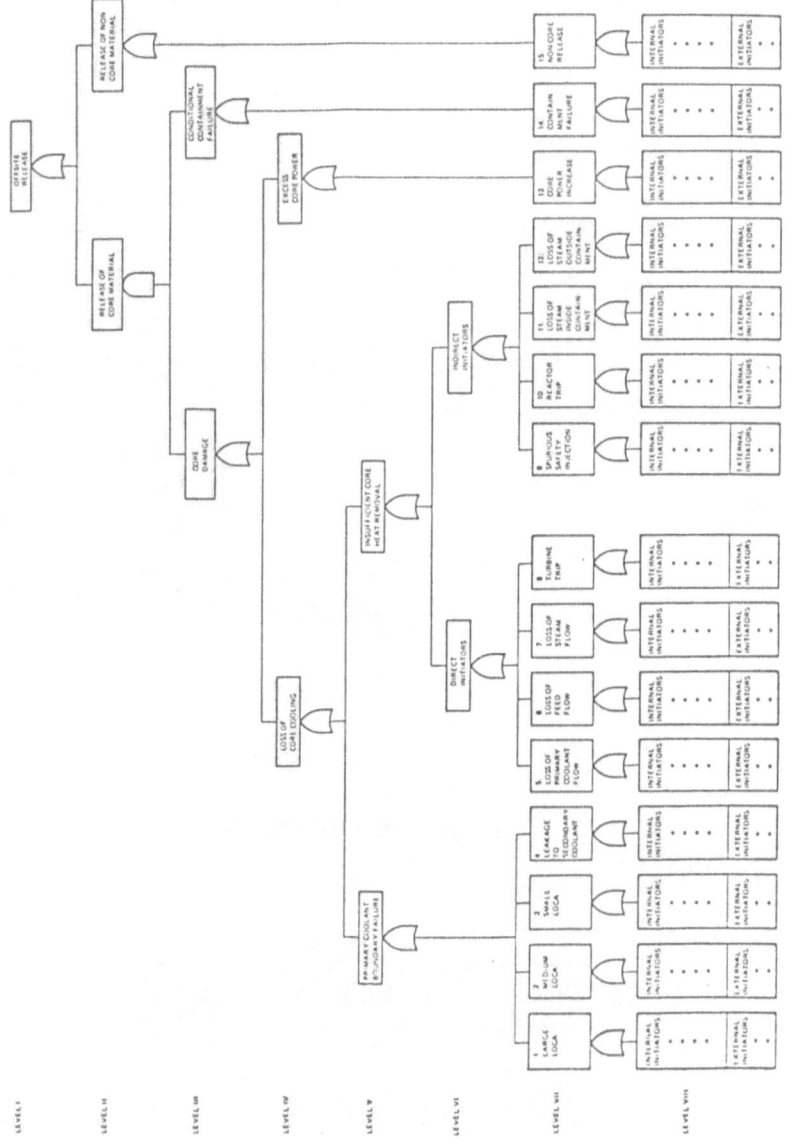

Figure 5. Master Logic Tree for a Typical PWR

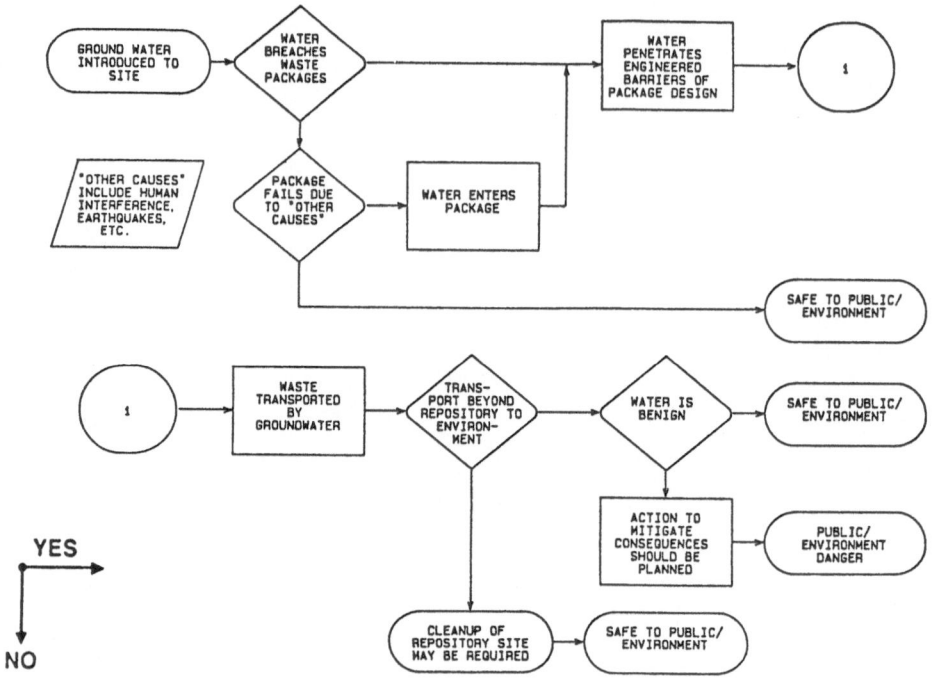

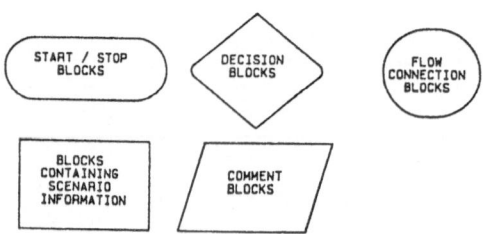

SYMBOL DEFINITIONS

Figure 6. Event Sequence Development for Repository
Operating Stage

example, "Will the water breach the waste packages?" The consideration of all questions pertinent to risk enables the identification of all scenarios. Constructed scenarios are then cast into a form called the event tree, which is suitable for quantification. The event tree, shown in Figure 7, contains all questions that are asked pertinent to scenario description. This event tree is entered with the initiating event and is "walked through," left to right, by answering a series of questions that reveal each scenario of the tree. For example, the highlighted portion in Figure 7 is one scenario: groundwater introduced to the site; water does not remain outside waste package; water transports waste beyond repository boundaries; waste has become benign due to time elements; the result is no hazard to public/environment.

Quantification of scenario likelihood is accomplished by determining the probabilities of the various branch points in the event tree. Propagation through event trees, for each initiating event and scenario results in a set of scenarios tabularized in the format of Figure 1, which, as discussed earlier, is converted into the form of Figures 3 and 4.

2.1 LESSONS LEARNED FROM PRAS

The PRA methodology highlighted above has been applied to a number of engineered systems [3]. In addition to performing PRAs on nuclear power plants, PRAs have been performed on chemical process facilities and space systems. Although developed to address issues in nuclear power generation, the fundamental PRA tools and the output form of a PRA are applicable to any engineered system.

From the performance of over 25 PRAs, the Pickard, Lowe and Garrick, Inc. (PLG), team has identified output characteristic of any PRA. Essentially, all PRAs identify those initiating events, scenarios, systems, and causes that contribute to the risk of engineered system operations.

The set of initiating events for each PRA is different; this is primarily because each plant or engineered system has a unique design or function. However, a general lesson that is learned is that different initiators are important for different damage states. For example, initiators often important for core melt are often not important contributors to offsite health effects. The principal reason for this difference is rooted in the effectiveness of the containment systems. For PRAs performed on waste repositories, effective waste package containment systems might yield similar results.

The PRA identifies those sequences of events most important to risk. Such information puts risk contributors in context with how things go wrong and in what sequence. Often, a few scenarios dominate risk. Similarly, a few systems or causes often contribute significantly to risk; Figure 8 shows that the reactor scram systems and diesel generators together contribute over 90% to the risk of one particular plant's damage.

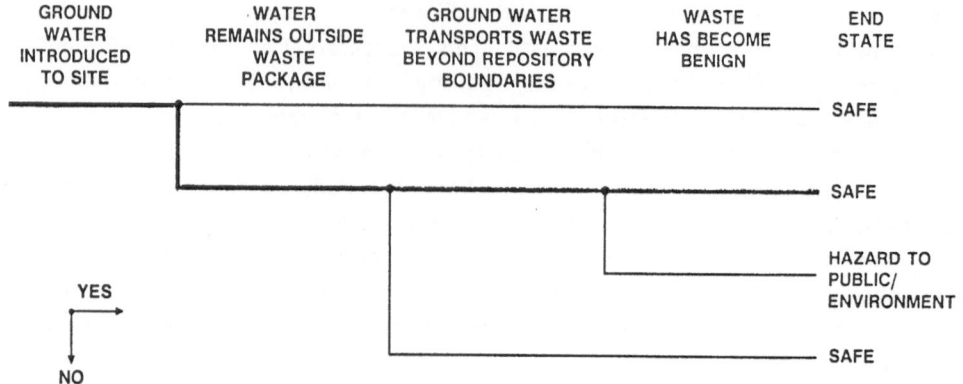

Figure 7. Event Tree Development for Repository Operating Stage
Initiating Event: Ground Water Introduced to Site

RANKING	SYSTEM FUNCTION	FREQUENCY (EVENTS/CALENDAR YEAR)	PERCENT
1	REACTOR SCRAM	4.1×10^{-4}	52
2	DIESEL GENERATORS	3.2×10^{-4}	40
3	SAFETY VALVES CLOSE	8.2×10^{-5}	10
4	RELIEF VALVES CLOSE	6.8×10^{-5}	8
5	MANUAL PRESSURE RELIEF	4.2×10^{-5}	5
6	LIQUID POISON INJECTION	3.6×10^{-5}	4
7	RELIEF VALVES OPEN	2.5×10^{-5}	3
8	CORE SPRAY INJECTION	1.7×10^{-5}	2
9	CONTAINMENT SPRAY	1.0×10^{-5}	1
10	EMERGENCY CONDENSER OPERATION	5.6×10^{-6}	< 1

Figure 8. Systems Contributing to Plant Damage

An effective risk management program will use PRA-generated information to best allocate resources to implement design changes, procedural changes, or, possibly, additional system or human action modeling. Figure 9 illustrates the PRA iterative process. In this example, after design and procedural changes and a subsequent restructuring of the risk model, the overall core melt frequency was reduced by a factor of almost 20, and the percentage contributions to core damage were spread more evenly over many systems. Assuming that the last iteration core melt frequency was acceptable, such a distribution of system contribution to risk indicates that the plant design and risk model have converged on a set of acceptable risk contributors.

Typical examples of cost-effective risk management and risk-reduction opportunities identified by the PRA process are:

- At one plant, the dominant contributor to risk was the seismic interaction between the Unit 1 and Unit 2 control rooms. During earthquakes near 0.27g, the control room roofs were envisioned to impact each other and to likely result in Unit 2 control room roof collapse. The $200,000 repair cost included the installation of energy absorbers between the roofs. This installation essentially eliminated seismic risk as being significant, and reduced core melt frequency by more than a factor of 10.

- The service water system at one particular plant had two sources of cooling water: a cooling pond and a cooling tower for use during very hot weather. The PRA found that an actuation signal could close the cooling tower valves to the train when valves to the pond failed to open. Without water, the service water pumps would soon burn up, and the system would cease functioning. An interlock was proposed that would not allow the cooling tower valves to close before the pond valves had opened. The costs of this modification were compared to the reduction in the frequency of severe core damage, and it was found that an inexpensive change would reduce the service water system unavailability by a factor of about 10. The service water interlock was installed for only $5,000; as a result, core melt frequency was reduced by 30%.

  For this same plant, an improvement to the safeguards chilled water system was found to yield an additional 10% reduction in core damage frequency at an estimated cost of only $15,000, and two improvements to the auxiliary feedwater system were identified as yielding large gains in plant safety at a total cost of only about $2 million each.

Many examples similar to these can provide further evidence of prudent decisions based on the results of a PRA. It is believed that the PRA thought process that has been successfully applied to other technologies is equally

SYSTEM(s) OR OPERATOR ACTION	PERCENT REDUCTION IN CORE DAMAGE FREQUENCY IF THE INDIVIDUAL SYSTEM (OR OPERATOR ACTION) FAILURE FREQUENCY COULD BE REDUCED TO ZERO			
	FIRST ITERATION	SECOND ITERATION	THIRD ITERATION	FOURTH ITERATION
1. ELECTRIC POWER	11	65	43	52
2. AUXILIARY FEEDWATER	9	11	11	31
3. TWO TRAINS OF ELECTRIC POWER RECOVERED				21
4. LOW PRESSURE INJECTION/DECAY HEAT REMOVAL	4	3	8	19
5. FAILURE TO RECLOSE PORV/PSVs		5	20	17
6. ESFAS/ECCAS			14	15
7. HIGH PRESSURE INJECTION SYSTEMS .	3	9	15	14
8. OPERATOR RECOVERY OF ELECTRIC POWER DURING STATION BLACKOUT		50	14	14
9. SUMP RECIRCULATION WATER SOURCE				11
10. COMPONENT COOLING WATER			3	8
11. THROTTLE HPI FLOW (OPERATOR ACTION)			1	4
12. FAILURE OF MAIN STEAM SAFETY VALVE TO RECLOSE			1	4
13. SERVICE WATER	32	<1	10	4
14. SAFEGUARDS CHILLED WATER	20	8	13	1
15. BWST SUCTION VALVE				1
16. CONTAINMENT ISOLATION			1	
17. STEAM GENERATOR TUBE RUPTURE	7	<1		
18. HIGH PRESSURE RECIRCULATION (OPERATOR ACTION)	5	2	11	
19. MAIN FEEDWATER			1	
20. FAST TRANSFER OF ELECTRIC BUS		4	7	
21. OTHER	4	6	6	< <1
RELATIVE CORE MELT FREQUENCY	1.00	0.30	0.10	0.06

Figure 9. Contributors to Core Damage for Four Phases of Risk Management

applicable to the handling of high level nuclear waste. From storage at the reactor site, to transportation to and from a monitored retrievable storage (MRS) facility, to permanent emplacement in a geologic repository, the PRA process will be effective in helping decision makers be aware of primary contributors to risk and therefore enable the proper allocation of resources.

2.2 TRANSPORTATION RISK

Transportation of high level radioactive wastes has been of concern since the 1970s. PRA techniques have been used to help to determine the optimal types of transportation vehicles to transport spent fuel. In particular, a 1977 PRA and cost-benefit study considered the advantages of using special trains rather than regular trains to transport spent fuel [4]. Special trains have strict limitations on speed (30 mph), passing conditions, and train size. Because of these limitations, in comparison with regular trains, special trains have an increased cost $20,000 (1977) per each one-way, 1,000-mile shipment. With respect to regular trains, accident data from U.S. government studies on rail transportation of spent fuel were used, and, from a cost-benefit perspective, a value of $1,000/man-rem was used as a utility function for radiation based on [5].

The risk analysis was structured into a decision tree format (Figure 10) with the decision options of either using a special train or a regular train. Accident scenarios were identified that included the consequences of release. Emphasis was placed on probabilistic quantification of the consequence of a decision with uncertainty intervals. For each decision, an expected utility was determined and presented in the form of risk curves and risk surfaces. The format of such a risk curve is shown in Figure 11. This figure illustrates that the risk analysis has the goal of determining the probability of attaining or exceeding a certain damage level, d, where damage is measured in terms of health effects on people.

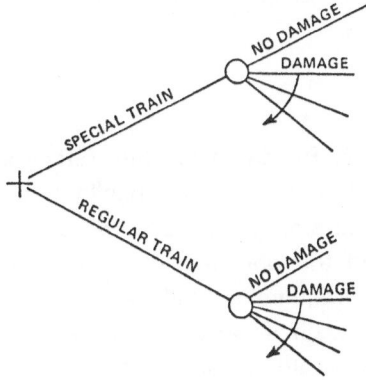

Figure 10. Special Train Decision Tree

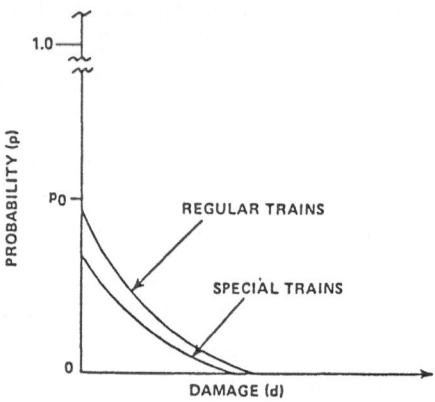

Figure 11. Effect of Special Trains on Risk Curve

Quantification of the risk model yielded the number of people affected from an accident, the radiation dose received, and the cost-benefit of operating special trains. The resulting constructed risk surface provided an upper bound on the risk reduction possible if a special train were- used. Numerical integration of the probability density surface and allowance for uncertainty associated with discretization resulted in a range of risk reduction. In this example, special trains resulted in a cost reduction of at least $269 per shipment and no more than $2,690 per shipment. However, the increased cost of a special train is $20,000 per shipment; therefore, special trains are not justified from a cost-benefit perspective.

It is interesting to note that the DOE is planning to use special trains for shipment of high level wastes. Although the cost of special trains is not justified from an economic-health risk reduction point of view, the great public and legislative fear of nuclear waste has prompted officials to expend excessive resources to convince nuclear energy skeptics that transportation of nuclear waste will pose no threat to the public or environment.

2.3 ONSITE STORAGE RISK

In 1983, the Electric Power Research Institute (EPRI) sponsored a limited-scope PRA study to review four alternative types of onsite storage programs [6]. Generic pool, cask, cassion, and vault storage concepts were quantitatively analyzed during this study. The study had the objective of identifying a spectrum of accidents including considerations of equipment failure rates, onsite transportation risks, external events, and human actions. Accidents were to be quantified and ranked by a relative hazard index to illustrate distinctions among the four storage concepts.

Storage designs were generic since specific storage containers, transport vehicles, facility layouts, and operational procedures were not specified. Thus, considerable lack of precision in frequency and consequence estimates resulted. In addition, external events (earthquakes, tornados, and aircraft crashes) are site-dependent; therefore, there is uncertainty in their application to the generic results. However, even with these limitations, an upper bound was used for order of magnitude conclusions.

Essentially, classical PRA modeling techniques were employed. A master logic diagram was constructed; initiating events and scenarios were quantified. Upon completion of quantification, results were presented in the form of discreditized risk curves. Details can be found in [6].

The study concluded that handling, onsite transportation, and placement operations yielded the highest risk. The pool concept had the greatest risk of all of the types of storage because of high consequences associated with rare earthquake events. None of the dry storage concepts exhibited any significant advantage over another. Finally, the study concluded that the absolute public risk of any onsite storage concept is negligible compared to that from an operating reactor at the same site.

2.4 GEOLOGICAL REPOSITORY RISK

Of the risk assessment studies performed to determine the safety and reliability of geological repositories, the most ambitious has been the Sandia National Laboratory/NRC Study of August 1987 [7]. This study had the objective of developing a comprehensive risk assessment methodology for use in evaluating potential disposal sites. The methodology was to contain procedures and tools that were necessary for performing a detailed assessment of postclosure risk (including uncertainties). Once developed, the methodology was to be applied to a hypothetical site representative of actual physical sites in the United States.

The study performed by Sandia considered 12 accident scenarios constructed from a large number of hypothetical repository-breaching events and radionuclide transport mechanisms. The quantity of radionuclide discharge rates from each scenario was predicted, and risk curves illustrating human risk to cancer were developed. The risk curves were conditional on the occurrence of the scenario; i.e., the probability of the scenario occurring was not considered.

The primary results from the study were that three variables are most important in determining health risk: (1) time of onset of migration of radionuclides, (2) strength of radionuclide source, and (3) migration time to the environment.

The weakness of these results is they are too generic due to study scope limitations. This study did not consider all initiating events and scenarios that are important to disposal facility, environment, or human risk. This fundamental weakness results in an analysis that cannot ensure that those initiating events or scenarios most important to risk have been identified or prioritized. Unlike classical PRAs, this study did not demonstrate an attempt to ensure completeness of the model. The second weakness of the study is that no account was made for the likelihood of occurrence of initiating events or accident scenarios. With respect to the risk triplet, $R_i = <s_i, p_i, x_i>$, discussed earlier, the lack of scenario-likelihood information makes the analysis of little practical use for risk management and decision-making purposes.

One explanation of these weaknesses was proposed by Bernard Cohen [8]. Cohen argues that PRAs, in the classical sense, are not suitable for analysis of geologic repositories because many variables have great uncertainties in each proposed accident scenario. It is acknowledged that Cohen's argument has some physical basis since the mechanisms for geologic and hydrologic natural alterations, as well as transport of waste materials, are poorly understood. However, the fundamental purpose in performing a PRA is to identify and quantify our lack of knowledge. Only through the systematic identification of system weaknesses can we hope to properly allocate resources in order to reduce our uncertainty and increase the likelihood of providing a safe and reliable system.

2.5 MERITS OF PRA FOR GEOLOGICAL WASTE REPOSITORIES

The PLG team strongly believes that classical PRA techniques can be used to perform meaningful analyses on geological waste repositories. Essentially, the stages from spent fuel storage at the reactor site to ultimate emplacement in a geological repository can be analyzed through a phased PRA process. In phased PRAs, the engineered system being analyzed is characterized by changes in system configuration and environment conditions. Phased analyses have been successfully applied to chemical process facilities and space systems. Analyses have provided valuable information about the integrated operational phases of the system; such information would not have been revealed had the entire phased PRA process not been completed. During an analysis on the hydraulic power units for the U.S. space shuttle, component failure modes were identified that had different impacts depending on which phase of the mission they occurred [9]. Only via a phased-PRA analysis were changing system configurations and nonsteady-state operations effectively analyzed. With respect to disposal of high level waste, a phased approach will enable the best control over the evolution, analysis, and assured completeness of the risk model. An example of the possible stages of model development is shown in Figure 12. Essentially, at each stage of the model, initiating events and scenarios are identified that could lead to those end states most important to risk. To illustrate application to waste disposal at a geologic repository, consider the initiating event, "groundwater introduced to site," shown in Figure 6. A basic and yet complete set of events can occur that will result either in the transport of waste from the site to the environment or in the containment of the waste at the site. The event tree in Figure 7 lists along the top row those "top events" that, when quantified, will enable quantification of all tree scenarios. For

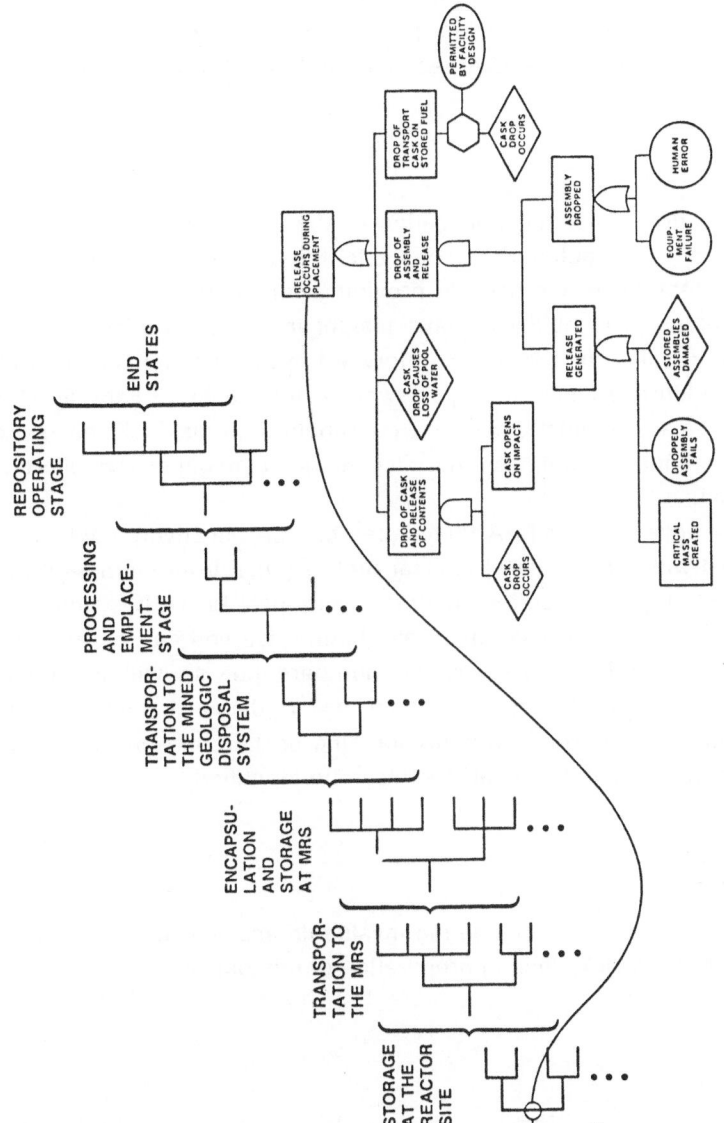

Figure 12. States to Event Tree Linking

each scenario, completeness of the model is attempted through the construction of detailed logic models. See [1] for details. During the quantification of the model, explicit statements are made about the uncertainty in both the model and the input data. Finally, as in the case of nuclear power risk, our complete state of knowledge about the repository site will be expressed in the form of probabilistic risk curves.

3. Conclusion

There is little doubt that the application of the PRA thought process to nuclear power plants has contributed enormously to the understanding of nuclear plant safety. Its greatest value has been in providing perspective and context to what is important to safety. The ability to have insight into how to allocate resources to ensure continued safety is a major step forward in the risk management of nuclear power plants. Figure 9 illustrates the power of the process. Each iteration of the risk model reveals the most important contributors to risk. Fixes or corrective actions can be implemented, and the risk model can guide the effort toward a balanced design.

It is strongly believed that PRA techniques can be effectively applied to analyze risks associated with all phases of nuclear fuel disposal from storage at the reactor to final emplacement at geologic repository. From past PRAs, it is well documented that important accident scenarios that have been observed can be anticipated with quantitative risk models. However, the numbers per se will not be the most important result from performing PRAs for waste disposal systems. The most important result will be a realistic understanding of the entire nuclear fuel disposal process so that ways to fix the weak spots can be identified.

4. Acknowledgment

The author wishes to acknowledge the invaluable assistance of Brian A. Fagan of Pickard, Lowe and Garrick, Inc., in preparation of this paper.

5. References

1. Pickard, Lowe and Garrick, Inc., "Methodology for Probabilistic Risk Assessment of Nuclear Power Plants," June 1981.

2. Garrick, B.J., and S. Kaplan, "On the Quantitative Definition of Risk Analysis," *Risk Analysis*, Vol. 1, No. 1, 1981.

Section 3 - The analyst's answer: methodologies and methods

This is the largest section of the course, and begins with an introduction to
System Variability Analysis (Dr. Goodwin). Uncertainty Analysis and
Sensitivity Analysis are described (Dr. Saltelli), with some relevant examples
being given. Special emphasis is given to the activities of the Probabilistic
System Assessment Code (PSAC) Users group, coordinated by the OECD/NEA, which
acts as a forum for the discussion of the PSA methodologies, and assists a
large number of national and international organisations in their code
bench-marking activities. A paper from Dr. Mobbs demonstrates the applications
of the ICRP principles in risk computation.

AN INTRODUCTION TO SYSTEMS VARIABILITY ANALYSIS

Bruce W. Goodwin
Atomic Energy of Canada Limited
Whiteshell Nuclear Research Establishment
Pinawa, Manitoba, Canada
ROE 1L0

ABSTRACT. The disposal of nuclear wastes has challenged environmental analysts to deal with two important issues:
- the need to assess complex engineered and natural systems, and
- the need to consider very long time frames.

The challenge is being met with an assessment approach known variously as systems variability analysis, probabilistic systems analysis and probabilistic risk assessment.

This report describes the systems variability analysis (SVA) approach. It also shows the applicability and requirements of SVA in assessments of complex environmental systems over long time frames.

1. INTRODUCTION

The scope of many environmental impact assessments has changed significantly over the past few decades. A driving force for change can be traced to national and international programs aimed at the disposal of nuclear wastes. "Disposal" implies the permanent and safe isolation of wastes, without the need for long-term institutional controls. (In contrast, "storage" implies temporary isolation, an ongoing need for monitoring and the possibility of corrective activities.) In the discussion below, we shall see how consideration of "disposal" has influenced the scope of environmental assessments.

Some assessment studies focus on one or more important components of a larger system. For example, we might consider only the potential impacts associated with the contamination of a stream near a waste management facility, if we know that contaminants could significantly affect only that stream. In other cases, many components of the environment may interact significantly, and we must examine numerous pathways within a larger system. The disposal of high-level nuclear wastes is an example of a complex interaction of many components.

Assessments of the disposal of nuclear wastes must take into consideration lifetimes of radioactive constituents. Typical high-level wastes from nuclear reactors contain isotopes of uranium, activation products, fission products and decay products, some having

A. Saltelli et al. (eds.), Risk Analysis in Nuclear Waste Management, 57–68.
© 1989 ECSC, EEC, EAEC, Brussels and Luxembourg.

lifetimes in excess of one million years. For such assessments, the time frame of concern extends far into the future.

The need to treat complex environmental systems over long time frames places an extraordinary demand on our assessment techniques. Most of the available data and knowledge of the environment is restricted to studies of short duration and limited ranges of experimental conditions. We may need to extrapolate the data and understanding to chemical and physical conditions and to time scales that are not amenable to laboratory and field studies. Clearly our assessment must deal with the uncertainty and variability inherent in making such extrapolations.

These challenges can be met with the systems variability analysis (SVA) approach to environmental assessments. In particular, the goals of SVA include the following:

(1) Development of an overall "systems" approach to account for the major components of a complex environmental system. The systems approach allows us to examine the relative importance and impact of different components of the system. Preliminary assessment studies can help focus detailed research effort on those components predicted to have the greatest influence on estimated impacts. Future assessment studies can then help in optimizing the system in terms of safety and cost.

(2) Development of the capability to account for the effects of uncertainty and variability in the system. This capability is required to deal with extrapolations of current data and understanding into the distant future and to conditions not directly accessible in laboratory and field studies.

In the following sections, we will investigate how the SVA approach meets these goals. We will also examine the special requirements and limitations placed on the data and models used by SVA codes. We start by outlining a typical assessment process, and then describe the role played by SVA.

2. DESCRIPTION OF THE ASSESSMENT PROCESS

Figure 1 represents the assessment process as a series of seven steps.

(1) The process starts with describing the system to be assessed and the criteria to judge its safety and performance. Some of the examples described below relate to the Canadian concept for disposal of nuclear fuel waste in plutonic rock of the Canadian Shield, but similar examples could be drawn from other systems.

(2) The second step is to systematically identify the scenarios relevant to the assessment (Cranwell et al., 1982; 1987). A scenario is a combination of phenomena that could affect the performance of the disposal facility. The "central" scenario consists of phenomena that have a high probability of occurrence. An example could be contaminant transport in ground-

water from a deep geological disposal site to man's environment. There may also be alternative scenarios, consisting of phenomena less likely to occur. An example might be a meteorite impact of sufficient magnitude to result in direct release of contaminants to man's environment. For each scenario, we describe the factors that could affect the performance of the facility. We also identify what types of impact are important, such as radiological dose to man due to contamination of foodchains by radioactive wastes.

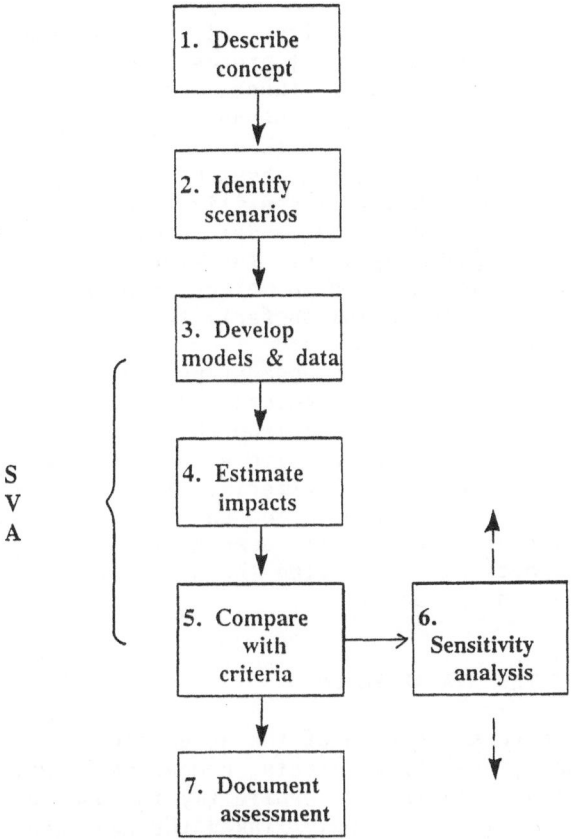

Figure 1. A simplified schematic of the assessment process.

The next four steps form the core of most quantitative assessments. These steps might be repeated for each scenario, depending on the constraints and requirements that have been identified in step 2.
   (3) The important components of the system are usually described using mathematical models and associated data. Frequently there are both complex models for use in detailed research studies and simpler models for use in assessment studies (Dormuth and Lyon, 1985).

(4) The assessment analyst selects a suitable combination of models and data to provide the required estimates of impact. For quantitative assessments of complex systems, the models and data are assembled as a computer code.

(5) Impacts are estimated and then compared with available criteria, to provide information on whether the facility represents an acceptable risk to man and the environment. Radiological criteria for the Canadian concept are set by the Atomic Energy Control Board (AECB, 1987). (Regulatory agencies may use estimated impacts to help derive suitable acceptance criteria.)

(6) The next step is sensitivity analysis, which plays an important role in the assessment process. We use sensitivity analysis to help identify those components and parameters that are most important in producing unacceptable or acceptable impacts. We then use this information to guide future efforts. In preliminary assessment studies, we use the understanding gained from sensitivity analyses to direct further development of our models and data, and/or to focus on the underlying research. In later assessment studies, we use sensitivity analysis to select among several disposal options and/or to establish facility siting and design constraints.

(7) The final step in the assessment process is documentation. The assessment document typically describes the entire assessment process, starting from the description of the facility or concept and ending with a comparison of estimated impacts with regulatory criteria. It also includes details on model and data verification and validation, and the identification of important components of the system.

In the next section, we describe the SVA approach, including its role in the assessment process.

3. DESCRIPTION OF THE SVA APPROACH

The SVA approach covers the core of the assessment process: developing models and data, analyzing impacts, comparing impacts with criteria and sensitivity analysis. We explicitly include sensitivity analysis because it can be strongly linked through feedback loops to the other steps.

Figure 2 demonstrates how these steps are accommodated in a computer code. The figure is a schematic overview of SYVAC (SYstems Variability Analysis Code), one of several computer codes that use the SVA approach. The models and data required by SYVAC are developed using information from experimental and field research, detailed analyses and expert opinion. These models take into account the basic objectives of the assessment, as well as the constraints imposed by the availability and quality of data and the availability of computer resources. Figure 2 identifies three submodels (vault, geosphere and biosphere) that make up the system model required to assess the

Canadian concept for nuclear fuel waste disposal. These submodels
estimate a unique set of impacts or consequences given a unique set of
input parameters.

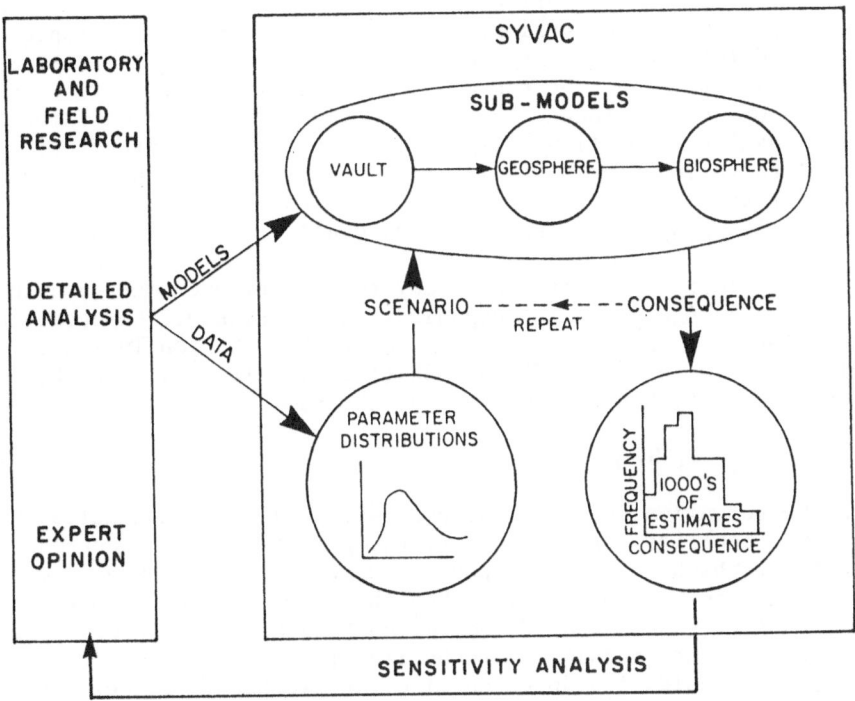

Figure 2. Overview of SYVAC, typical of many computer codes that implement systems variability analysis.

In the analysis of impacts, the SVA approach quantifies the uncertainty and variability in the system description by using probability distributions to define parameter values. (Codes not using the
SVA approach typically use one or more unique sets of values, such as
"best estimate" or "worst case" sets. See also ERL (1985) for other
approaches.) SYVAC starts by selecting a possible realization of the
system by sampling a value for each parameter from its specified distribution. The parameter values are then passed to the system model
and used to simulate the behaviour of the system and calculate the
corresponding impact (consequence) for that realization of the system.
SYVAC repeats the selection and calculation steps, typically a thousand times or more, to build up a histogram of the magnitude of the
estimated consequences versus the frequency with which they occurred
in the set of simulations.

In common with other SVA codes, SYVAC produces a frequency distribution of estimates of impact, rather than just a single estimate.
This distribution reflects the uncertainties and variabilities associated with the modelling of the system.

Each of the thousands of impact values can be individually compared with regulatory criteria. More importantly, however, the impacts can be evaluated as a set to provide additional information on the degree of confidence that the criteria are met. Suppose, for example, that one of our criteria is based on the travel time of a radionuclide through a barrier. We can examine the set of impact values to determine the fraction of estimates in which the travel-time criterion is exceeded. We can also use statistical tests to calculate a measure of our confidence in this fraction. This way of interpreting the results is particularily useful when the regulatory criteria are applied to impacts having inherent uncertainties. Most regulatory agencies responsible for regulation of nuclear wastes demand that uncertainties, when important, must be dealt within the assessment process.

For nuclear waste disposal, many regulatory agencies (and the Nuclear Energy Agency of the Organization for Economic Co-operation and Development) include a radiological risk criterion in their regulations (see AECB (1987) and NEA (1984)). The net risk, R, of a disposal facility has been defined as the sum:

$$R = \Sigma_i \ P_i \ H_i \ g \qquad\qquad\qquad [1]$$

where the summation index i runs over all applicable and mutually exclusive scenarios. P_i represents the probability of occurrence of scenario i, which has a corresponding dose H_i. The parameter g is a factor that converts dose into a health impact on man. (Note that this use of g assumes that H_i is in the "stochastic" range, where health effects are considered to be linearly related to dose. Equation [1] also holds for doses outside the stochastic range, except that the product (H_i g) is then equal to unity, since nonstochastic doses are assumed to produce a health impact.)

Deterministic assessment codes produce a single estimate of H_i for each scenario. Codes following the SVA approach provide many estimates of H_i, and an estimate of the expected or average value of dose taking into account uncertainty and variability. Codes following the SVA approach produce a measure of our statistical confidence in the estimated average dose. We can also use Equation [1] to obtain a corresponding measure of confidence in the estimated risk. (Note that this discussion refers only to our confidence in the precision of the estimates. It does not refer to our confidence in the accuracy of the estimates, which could be much more dependent on the quality of our models and data (ERL, 1985).)

We can improve the precision of the estimates by performing more simulations to reduce the standard error associated with mean dose. (The standard error is approximately proportional to the reciprocal of the square root of the number of simulations.) We can also reduce standard deviations by using different sampling strategies, such as Latin hypercube and importance sampling (Johnson and Lucas, 1986). Latin hypercube sampling, for example, can produce smaller standard errors than Monte Carlo sampling for the same number of simulations (McKay et al., 1979; Iman et al., 1980). However, this result has been proven only when the calculated consequence displays a monotonic dependence on all of the model parameters. For many assessment models, the dependence may be much more complex (Andres, 1987).

We can also obtain increased confidence in our estimates by reducing the uncertainty in the input parameters, since the spread in output estimates is a reflection of the spread in input parameters. The number of input parameters does not, in principle, affect the confidence associated with our estimates. This result is more obvious if all parameters have similar effects on the impact: the "noise" tends to average out (or the standard error is reduced) if we take many samples of a few parameters or a few samples of many parameters. (However, the number of parameters does directly affect our ability to perform sensitivity analysis.)

We can improve our understanding of results from a SVA calculation with sensitivity analysis studies, to identify important submodels, components of a submodels, parameters and even parameter ranges. Within the context of sensitivity analysis, an "important" factor is one that tends to give rise to high (or low) consequences. This information may lead to improved risk estimates, by directing research efforts towards more productive and significant topics. We can also use sensitivity analysis to optimize design or siting constraints, which could reduce risks and improve facility performance. While sensitivity analysis can improve our understanding, and thus our confidence in the precision of our risk estimates, it does not yield a direct measure of confidence.

We can investigate the measure of confidence using "uncertainty" analysis. (Uncertainty analysis is sometimes used with the same meaning as SVA; however, its meaning here is defined below.) Uncertainty analysis is directed at deducing the important factors (submodels, components of a submodel and/or parameters) that contribute most to the uncertainty in the consequences. Within the context of uncertainty analysis, an "important" factor is one that gives rise to a large (or small) spread in the consequences. Results from uncertainty analysis can then be used, much like the results from sensitivity analysis, to guide further research efforts and/or to establish optimal siting and design constraints. Unlike sensitivity analysis, however, uncertainty analysis can yield an increased confidence in the estimate of risk.

4. MODEL AND DATA REQUIREMENTS

The development of suitable models and data is a key element of the SVA approach. Some SVA requirements are shared by other assessment tools, including the deterministic approach. In particular, two common requirements are as follows:
  (1) Extrapolation. The assessment models and data must be capable of extrapolation.
      This requirement can place severe constraints on our selection of models and data. In general, we should avoid empirical models in favor of models with strong theoretical

backing, so that extrapolations are more defensible. Nevertheless, there are instances where we must use empirical models because no other suitable models are available. One well-known example is the use of the distribution coefficient (K_d) to describe the interaction of dissolved contaminants with surrounding solid media. We can find support for the K_d model from thermodynamic arguments, but the support is limited to trace concentrations of contaminants. The K_d model does not apply to chemical reactions such as precipitation and formation of colloids. Despite these limitations, the K_d model is sometimes extrapolated to cases where concentrations may be high and/or where precipitation may occur.

Frequently we choose to use models and data that are known to provide "conservative" extrapolations of our understanding. A conservative extrapolation is one which will over-estimate negative impacts (or underestimate positive impacts).

(2) <u>Verification and validation</u>. The credibility of our assessments will depend strongly on how well the models and data have been verified and validated. Verification consists of demonstrating that the models and data are accurately represented within a calculation or associated computer code. Validation involves showing that the underlying phenomena are correctly modelled, so that the essential physical, chemical, geological and biological features of the disposal system are properly represented.

Verification and validation present a difficult challenge in nuclear waste disposal, particularily because of the need to extrapolate our models and data in time and space. Verification can be accomplished by performing detailed comparisons of different computer codes that were developed to assess similar problems. Validation requires comparison of our estimates with observations. Significant support for the assessment can be achieved through comparisons of code predictions with observations from natural analogues. Both of these topics are the subject of two separate sessions at this seminar, and are not discussed further here.

There are also several major requirements of our models and data that are, for the most part, unique to the SVA approach. They include the following:

(3) <u>Uncertainty and variability</u>. Some assessment studies must take into account the uncertainties and variabilities in the system, which arise from extrapolations of models and data.

One of the primary purposes of SVA is to account for uncertainty and variability in data, which we accommodate through the use of probability distributions. We start by determining which of the parameters used in a model may not be well characterized. One such parameter is the solubility of uranium, which frequently appears in models of the leaching of nuclear fuel wastes. We can estimate the solubility of uranium by drawing on an extensive thermodynamic database. Nevertheless, we can never remove all sources of uncertainty and variability. We must contend with inaccuracies and deficiencies in the database. We must also contend with the fact that uranium solubility is a sensitive function of Eh and pH. We may find it very difficult to predict the values of these two chemical parameters in complex geological media, and in the presence of radiolysis and hydrothermal reactions between the media and nearby groundwaters. We may also need to consider the variabilities in pH and Eh that could occur at different locations within the host media. We must therefore carry out a detailed analysis to determine what the probability of different uranium solubilities could be, and quantify our uncertainty through the use of a probability distribution.

A few general guidelines are available that cover the derivation of probability distributions from existing data and information (for examples, see PSAC (1987)). For the most part, derivation of probability distributions must rely heavily on field and laboratory investigations, coupled with the analysis and judgement of experts who have a good understanding of the quality and meaning of the data.

Normally we associate uncertainty with data rather than models. However, we can also deal with uncertainties in models within the SVA framework. As a simple example, suppose that we are uncertain whether doses from ingestion of water would be higher if the source of the water were a lake or a well. In this situation, we could include models of a lake and a well in our SVA code, and compare the doses from both models after each simulation. For an overall conservative calculation, we then select the highest dose for each simulation.

Another type of model uncertainty is dealt with through the use of a simplified model, with appropriate modification of parameter probability distributions. A common example is the use of the linear K_d model to describe contaminant sorption. Laboratory results may indicate that some contaminants follow nonlinear Freundlich or Langmuir sorption isotherms, but the data may be limited to substantially different concentrations, to experimental conditions that are not directly relevant or to measurements taken over a short time. It may be very difficult to extrapolate and apply these sorption models in a satisfactory fashion. We might therefore decide to reinterpret the available data in terms of the K_d model. In our reinterpretation, we might also assign parameter distributions that have a larger range to accommodate the modelling uncertainty. (We may also skew the distributions so as to yield estimated impacts that are more conservative. It may be very difficult, however, to prove that the resulting distribution is truly conservative, particularily for nonlinear models.) This method lacks mathematical rigor, but it provides a pragmatic solution to a formidable problem.

(4) <u>Model complexity</u>. The models used in the assessment must be relatively simple (computationally efficient) and robust (numerically stable).

The reason for these requirements arises because our models may be exercised thousands of times or more in a SVA calculation. We therefore require models that are efficiently computed, unless we have substantial computer resources at our disposal. Our models must also be robust because thousands of different combinations of parameter values must be handled. We must therefore develop general numerical algorithms instead of optimizing them to handle a few special combinations of parameter values.

For purposes of sensitivity analysis, simple models can be more desirable than complex models. Assume that we may choose between an accurate but very complex model, and a less accurate but simple model. The simple model should contain all of the essential and important features, but it will not accurately describe the fine details. Sensitivity analysis will be easier for the simple model because we have removed some of the "noise" generated by random perturbations of unimportant parameters. This factor may be very important when we come to defend models that are used to extrapolate our understanding. We may also find it easier to obtain support for simple models through comparisons with natural analogues, because frequently we can find analogous behaviour only for some of the more important features.

(5) <u>Parameter correlations</u>. The model parameters must take into account correlations. This requirement is unique to the SVA approach, where we could sample related parameters thousands of times. We must ensure that, in every simulation, we have not chosen a set of parameter values that is inconsistent with physical reality. We should also ensure that, for a group of simulations, we have not introduced artificial or "spurious" correlations.

Parameter correlations can be accommodated in at least two ways. The preferred approach is to include correlations through explicit equations. We know, for example, that the solubilities of uranium and thorium are closely related under some chemical conditions. We could account for this correlation by using thermodynamic relationships that relate thorium and uranium solubilities to basic chemical parameters such as pH and groundwater composition. We could then sample these chemical parameters and calculate the required solubilities, so that the resultant solubilities are effectively correlated. The second approach is to include correlation coefficients in our sampling scheme. In this case, we sample an "independent" parameter, and use the sampled value and a correlation coefficient to calculate the value of the "dependent" parameter.

Spurious correlations occur because of chance when samples
of unrelated parameters show an unexpected correlation in one
set of simulations. Another set of simulations may not have
the same unplanned correlations. We can avoid this problem
by choosing large sets of simulations, or by making a judi-
cious selection of samples (Iman and Conover, 1982).

Note that the number of possible correlations is of the
order n^2, where n is the number of input parameters. We
expect that a much smaller number of correlations is likely
to be important and/or known. We can test whether or not a
correlation is significant by carrying out two sets of simulations - one where the two parameters are varied with full
correlation, and the other where they are independently
sampled. If we observe that the correlation has a significant influence on the result, then we must incorporate an
appropriate correlation in our sampling procedure.

5. SUMMARY

Systems variability analysis is a valuable tool for environmental
assessments. Its major strength lies in its ability to quantify the
uncertainty in complex environmental systems.

The SVA approach includes developing models and data, analyzing
impacts, comparing impacts with criteria and sensitivity analysis. In
common with conventional assessments, the SVA approach produces estimates of environmental impact. However, SVA goes beyond the conventional assessments by quantifying the uncertainty in its
estimates of impact. In addition, SVA provides a measure of our confidence in the estimated impact values.

Application of SVA requires the following:
- derivation of probability distributions that capture the
  relevant uncertainties in model parameters,
- development of models (and associated data) that are efficiently computed and numerically stable, and
- description of correlations between parameters that are
  dependent on one another.

These issues have not yet been fully resolved, but we have a substantial set of rules and guidelines available.

The SVA approach is still being improved. There has been a great
deal of activity over the past ten years, much involving assessment
studies of the disposal of high-level nuclear waste. Considerable
progress is being made by members of international groups such as the
OECD/NEA Probabilistic System Assessment Code User Group. Although
SVA is still evolving as an assessment tool, its capabilities have
been demonstrated (and are detailed in another session at this seminar). SVA is uniquely poised to meet the challenges of assessing the
long-term environmental impacts of complex systems.

References

AECB (1987). Regulatory objectives, requirements and guidelines for
the disposal of radioactive wastes - long-term aspects. Atomic
Energy Control Board (of Canada) Regulatory Document R-104,
P.O. Box 1046, Ottawa, ON K1P 5S9.

Andres, T.H. (1987). Statistical sampling strategies. In proceedings
of an NEA Workshop on Uncertainty Analysis for Performance
Assessments of Radioactive Waste Disposal Systems, Seattle, WA, 1987
February 24-26; OECD/NEA, Paris, pp.70-84.

Cranwell, R.M., Campbell, J.E., Helton, J.C., Iman, R.L.,
Longsine, D.E., Ortiz, N.R., Runkle, G.E. and Shortencarier, M.J.
(1987). Risk methodology for geologic disposal of radioactive
waste: final report. NUREG/CR-2452; SAND81-2573, Sandia
Laboratories, Albuquerque, NM.

Cranwell, R.M., Guzowski, R.V., Campbell, J.E. and Ortiz, N.R.,
(1982). Risk methodology for geologic disposal of radioactive
waste: Scenario selection procedure. NUREG/CR-1667; SAND80-1429,
Sandia Laboratories, Albuquerque, NM.

Dormuth, K.W. and Lyon, R.B. (1985). The link between detailed
process models and simplified models. In proceedings of the Second
NEA Workshop on System Performance Assessments for Radioactive Waste
Disposal, Paris, 1985 October 22-24; OECD/NEA, Paris, pp. 81-87.

ERL (1985). Handling uncertainty in environmental impact assessment.
Report (volume 18) prepared for the Ministry of Public Housing,
Physical Planning and Environmental Protection by Environmental
Resources Limited, 79 Baker Street, London.

Iman, R.L. and Conover, W.J. (1982). A distribution-free approach to
inducing rank correlation among input variables. Communications in
Statistics B11 (3), 311-334.

Iman, R.L., Davenport, J.M. and Zeigler, D.K. (1980). Latin hypercube
sampling (program user's guide). SAN79-1473, Sandia Laboratories,
Albuquerque, NM.

Johnson, K. and Lucas, R. (1986). The use of importance sampling in a
trial assessment to obtain converged estimates of radiological risk.
Report No. DOE/RW/87.080. Prepared for the Department of the
Environment by CAP Scientific Limited.

McKay, M.D., Conover, W.J. and Beckman, R.J. (1979). A comparison of
three methods for selecting values of input variables in the
analysis of output from a computer code. Technometrics 21 (2).

NEA (1984). Long-term radiation protection objectives for radioactive
waste disposal. OECD/NEA, Paris.

PSAC (1987). Probability density function justification. Report of a
Topical Meeting of the Probabilistic Systems Assessment Code User
Group. Radioactive Waste Management Committee, OECD/NEA, Paris.

TECHNIQUES FOR UNCERTAINTY AND SENSITIVITY ANALYSES

Andrea Saltelli
Commission of the European Communities
Joint Research Centre - Ispra Establishment
Institute of the Environment
21020 Ispra (Varese) - Italy

1. INTRODUCTION

1.1. Uncertainty and Sensitivity Analysis

Uncertainty and sensitivity analyses constitute an essential element of a Probabilistic Safety Assessment (PSA). PSA is a complex multi-task, multi-disciplinary procedure including:
1) identification and description of the different scenarios involving radionuclide release from the repository for the site under investigation;
2) assessment of the probabilities of occurrence of the various events capable of triggering or perturbing the release scenarios;
3) identification of the relevant release, transport and dispersion mechanisms in the various barriers;
3) modelling of the radionuclide release and transport through the various system components, following the scenarios considered; modelling of the radionuclide dispersion or reconcentration in the environment, assessment of their concentration in food chains and related intakes to man;
4) assessment of the radiological consequences to individuals;
5) evaluation of the range of variability of the relevant input parameters;
6) analysis of the uncertainty in the consequences linked to the uncertainty in the model input parameters;
7) analysis of the sensitivity of the output with respect to input parameters, choice of model, model assumptions and boundary conditions;
8) optionally, conversion of doses to risk;
8) Comparison against criteria

The above scheme is only given for illustrative purposes; it does not include, for instance, the links and the feed-backs between the various stages of the analysis and the process of model verification and validation which remains one of the most difficult tasks of the assessment. Uncertainty and sensitivity analysis are part of this

A. Saltelli et al. (eds.), Risk Analysis in Nuclear Waste Management, 69–95.
© 1989 ECSC, EEC, EAEC, Brussels and Luxembourg,

complex procedure. They play a crucial role because of the long time span which is to be covered by the analysis and of the difficulty to make precise, unique predictions of the evolution of the system over such a range. Prediction uncertainty is an intrinsic feature of the problem, which can be reduced - not eliminated - by laboratory and field research. It is the purpose of UA to quantify this uncertainty and to make its evaluation as transparent as possible in order to provide a scientific basis for the performance evaluation.

In the following discussion of UA only the uncertainty due to the input data is addressed. It is implicitly assumed that conservative assumptions have been made in the selection of scenarios and in the choice of the models and boundary conditions, thus biasing the effect of uncertainty in these latters toward the pessimistic end.

2. SPECIFIC DIFFICULTIES OF THE UNCERTAINTY AND SENSITIVITY ANALYSES FOR THE RADWASTE MODELS

Before we describe the specific techniques used in uncertainty analysis and sensitivity analysis (UA and SA in the following) it is useful to underline the factors which complicate this task for the models under consideration.

2.1. Range of Variability

When UA was originally developed, in the context of the reactor safety studies, it was supposed to assess the impact of the uncertainty of the input parameters on some output variable, for instance the lifetime of a physical component, and constituted part of a system reliability analysis. In that context the input parameter uncertainty was moderate, let us say within one order of magnitude. In the radwaste context, however, such an uncertainty can cover many orders of magnitude. This is due to a variety of factors such as the long time span covered by the analysis, the difficulty to define the geochemistry of the system, the impossibility to predict future human habits and diets. Classical examples of very uncertain parameters (covering several orders of magnitude) are the spent fuel solubility in groundwater, the time of occurrence of the next glaciation episode, the water flow rate in the geosphere, etc. As a result the distribution function of the output, for instance a dose rate, varies over several orders of magnitude. Furthermore the mean of this quantity, which is relevant for risk calculation, is quite unstable, being determined by the few values in the upper tail of the distribution. This implies that a high number of runs (computer simulations) is needed to achieve a mean with a pre-established level of confidence.

2.2. Model Nonlinearity

Models used to compute the radionuclide migrations through the various compartments of the system (near field, far field, geosphere) are strongly nonlinear. Often many sets of input values result in no dose (knot), and the distribution of the non-zero outputs is also skewed. To a first approximation the transfer functions are described by steps or pulses which are smoothed by dispersion and or diffusion. This behaviour makes sensitivity analysis particularly difficult. Model non-monotonicity is an other element of concern for sensitivity analysis.

2.3. Time Dependence

The models' output is a time dependent function of the input variables. Consequently the relative influence of the input variables on the output uncertainty is function of time.

2.4. Computer Cost

The computer codes where the model has been implemented is time consuming and expensive to run. Consequently the number of simulations must be optimised.

The constraints and the difficulties imposed by the above problem on the different stages of the analysis will be discussed in the following sections.

3. SOME PRELIMINARY ELEMENTS OF MONTE CARLO COMPUTATIONS

3.1. Data Distribution Assignment

Uncertainty in the value of an input parameter may arise from one or more sources. Uncertainty might be due to a limited knowledge which can eventually be resolved by further experimental research. The uncertainty on the dose factor of Neptunium 237, for instance, has been reduced by laboratory investigation. There also exist an intrinsic uncertainty, non-susceptible of experimantal resolution. An example is provided by the time of occurrence of a geological event whose probability is - in principle - constant, such as the occurrence of a fault crossing the repository. Value distributions are also used for "effective", or "lumped" parameter, where the indetermination arising from an incomplete understanding of a complex phenomenon is resolved by the choice of a conservative model in conjunction with a number of effective parameter. For example, the complex geochemistry of the nuclide/medium interaction is generally simplified to a linear sorption isotherm, where the relevant parameter (a sorption constant) is allowed a wide range of values. Uncertainty is also due to the extrapolation of parameter value to longer time spans, or to the process of averaging its spacial and temporal variability to a single value to be fed into the model. For instance, rock permeability can vary considerably within a given formation, and will also vary as result of seasonal or geological alteration. Often in this case the model considers a constant permeability, whose value is sampled from a distribution. The above remarks indicate tight links which exist between models and input data distribution assignment. In fact the choice of the distribution is made by the modellists on the basis of the available information. For the Canadian concept assessment a number of guidelines have been established to help in the selection (Stephens, Goodwin and Andres,1987). We quote from there a selected list of often used distributions together with the conditions for their applicability:

1) <u>Constant</u>. Used for parameters with well defined,fixed value. Example: radionuclide decay constant.
2) <u>Uniform</u>. For parameters with limits which are known or çan be approximated; all values are equally probable. Example: porosity of the vault backfilling.
3) <u>Piecewise uniform</u>. For parameter of unknown distribution, or with two or more disjoint sets of possible values, or weighing different probability density functions describing alternative situations. Example: probability of selecting different regional biospheres.
4) <u>Log-Uniform</u>. For intrinsecally positive parameters, with limits which are known or can be approximated; relative weights poorly known (but lower values are more likely. Example: hydraulic conductivity of buffer and backfilling.
5) <u>Normal</u>. For parameters which are the sum of values or variables. Example: consumption rates for food.
6) <u>Log-Normal</u>. For parameters which are the the product of values or variables. Example: sorption constant for nuclide.

Log-Normal distributions have been widely used in probabilistic risk assessment, mainly because many environmental quantities appear to be log-normally distributed (see, for example Shaeffer and Hoffman, 1979). Weibull distributions have recently been used to fit lumped biosphere parameters (Marivoet and Van Bosstraeten, 1988).

Normal distributions are representative of the scatter in a repeated measurement of the same datum.

Uniform and Log-Uniform distributions constitute a last recourse when no information on relative frequencies is available.

3.2. Sampling

The purpose of the sampling is to generate an input data matrix of the form

$$\underline{X} = \begin{array}{ccccc} x_{11} & x_{12} & \cdots & x_{1K} \\ x_{21} & x_{22} & \cdots & x_{2K} \\ & & \vdots & \\ x_{N1} & x_{N2} & \cdots & x_{NK} \end{array}$$

containing the values assigned to the K variables in the N runs. For each variable X_j the N values X_{ji}, i=1,2,...N must represent a sample from the distribution of the variable X , or - in other words - once plotted in a histogram, the X_{ji} must resemble the input distribution of the variable X_j. The simpler method to obtain such a sample is to generate a random number r in the range (0,1) and to solve for x the integral equation

$$r = \int_{-\infty}^{x} f(u)*du$$

where f(u) is the distribution of the variable under consideration. Repeating the above procedure N times for each variable results in a "purely random" sample of the input vector. Whenever the barrier submodels are relatively simple and the computer code is not particularly time consuming, random sampling is probably the best technique to use. Often - however - the computer program involves time consuming steps, such as numerical solution of differential equations, series approximation of analytical solutions. Then a rationalisation of the sampling procedure is sought in order to reduce the number of runs. In fact other sampling strategies are available which are more effective than the purely random sampling, in the sense that the convergence of the output distribution is more rapid.

One of the most widely used techniques is the Latin Hypercube Sampling (LHS; Mc Kay et al., 1979). According to LHS the range of variability of each variable is divided into a number N of non-overlapping intervals, where N equals the number of model runs. Once this partition is done for all the K variables the sample space is divided into N^K cells (Figure 1). When using LHS N such cells are selected in the sample space in such a way that each interval is sampled exactly once, so that the entire range of each variable X_j is explored. LHS can be considered as a method of selecting, out of the possible N^K cells, a subset of N cells in such a way that the scanning of the sample space is optimised (Figure 1).

Because of the random pairing in the input values, spurious correlations among the input variables may be introduced. Correlations are particularly undesirable when the sample has also to be used for sensitivity analysis purposes. A techique is available (Iman and Conover, 1980) to eliminate the spurious correlations. This technique allows any kind of pre-specified correlation to be imposed on the input variables, so that the sample matrix $\underline{X}$ can be elaborated either to eliminate all the undesired correlations or to impose correlations among two or more variables when this is desired. This technique is implemented in LISA and can be used with either random or hypercube sampling. LHS has a shortcaming, however, in that it gives a biased estimate of the output variance. Nevertheless the bias can be proved to be negligible when the input-output relationship is monotonic for all the variables (Iman and Helton, 1985).

Other sampling algorithms are also available, such as stratified sampling and importance sampling. The latter is particularly useful to explore the tail of the output distribution (the upper tail, generally), as the risk is very often determined by the few high dose outcomes. With this method the sampling of the input variables is forced in a sub-interval of the input distribution to produce higher doses, so that the upper tail of the output distribution is better explored. This can only be made for those variables whose monotonical relationship with the output has been ascertained.

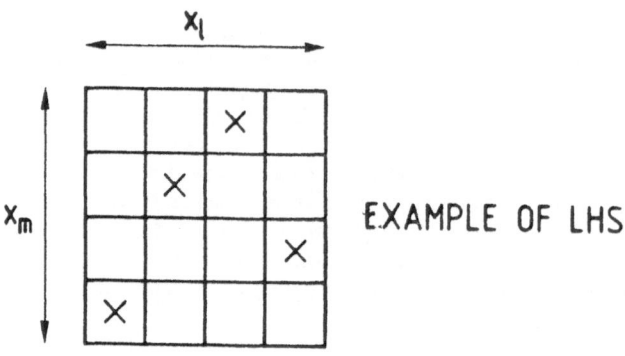

Figure 1.   Example of Latin hypercube sampling for K=2 variables and N=4 runs.

3.3. Post-processing

It is common practice in PSA codes to divide the computation in (at least) two stages: in the first one the model is executed a pre-established number of times in order to produce a statistically significant set of dose curves as a function of time (Figure 2). These curves can either represent individual nuclide contribution or total dose for each run, and are stored as an intermediate file. In the second stage these curves are elaborated in order to determine:
* means and confidence bounds and
* histogram and cumulative distribution (uncertainty analysis)
* ranking of the contributing nuclides and of the influential parameters (sensitivity analysis).

The above are generally computed at different time points. It is also possible to divide the computation into three stages, performing the preparation of the sampling matrix separately.
    When using LISA, for example, both uncertainty and sensitivity analysis are performed using the Statitistical Post-Processor SPOP (Saltelli, 1987). Having the post-processor separate from the main com-

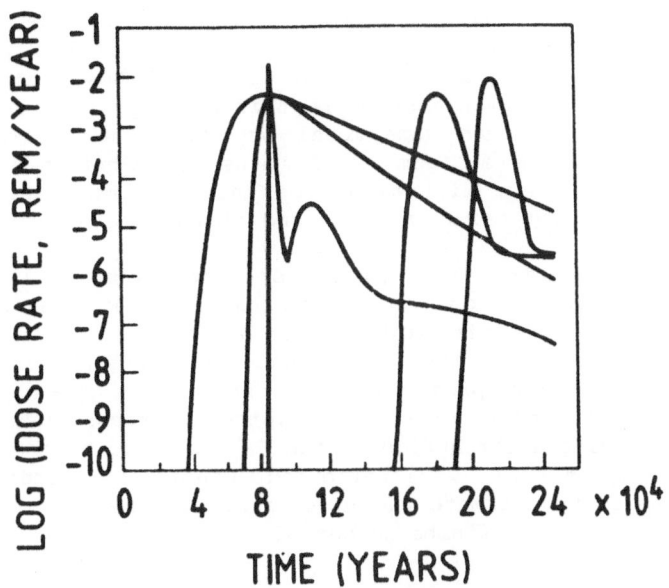

Figure 2. Total dose rate versus time curve for five runs from a Monte Carlo simulation.

puter programs has a number of obvious advantages, such as, for instance, the possibility of exchanging data files among different organisations in order to intercompare the performances of the various UA, SA techniques.

4. UNCERTAINTY ANALYSIS

4.1. Limits of the Analysis

Different kinds of uncertainties are may exist in a probabilistic risk assessment, which are linked to

Scenarios. Is the considered scenario an adequate picture of reality? Have all the relevant scenarios been identified?...

Events and processes. Hasn't any potentially deleterious event been overlooked? Are the relevant transport processes understood? Is the extrapolation to long time span correct?...

Models. Are there errors in the formalisations of the process? Are the boundary conditions representative of reality? Are the numerical methods accurate?...

Input data. Which is the parameter value? Which the influence of parameter uncertainty on the output?...

The uncertainty analysis described in this section, which is essentially based on a Monte Carlo approach, is particularly suited to handle the latter class of uncertainties, i.e. those linked to the input data. This limitation must be kept in mind, if the result are to be used properly. Although some model or process uncertainty can be dealt with through the assignment of value distribution to relevant lumped parameters, there will always be uncertainties which are not resolved by the Monte Carlo approch and which are to be dealt with separately.

In this context then the purpose of uncertainty analysis is the characterisation of the output distribution. In Figure 3 one such distribution is shown. This histogram is an estimate of an empirical distribution function, and the problem here is to characterise the true underlying distribution, producing - for instance - confidence bounds on the distribution itself or estimating the error associated with its mean. The underlying distribution is that which would be obtained if an infinite number of runs were performed. Confidence bounds and error bounds are needed to estimate the uncertainty in the dose distribution f(H) which is due to the finite number of runs employed.

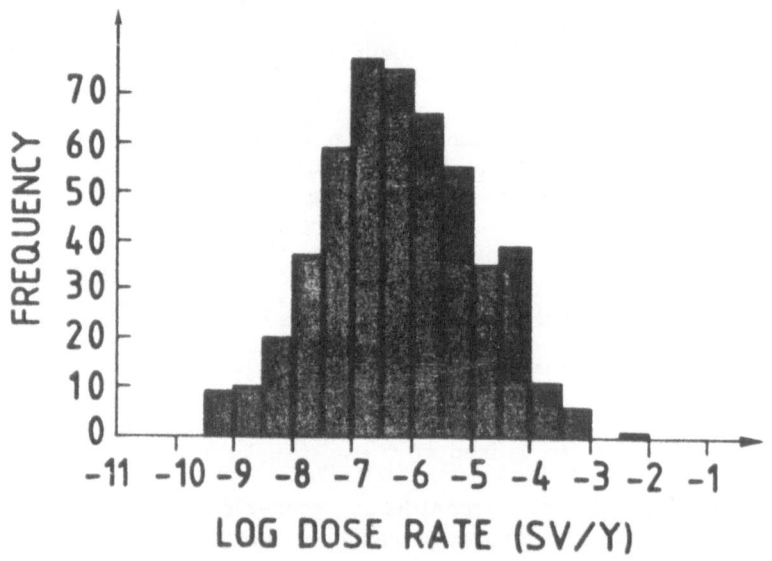

Figure 3. Example of dose histogram at a given time point.

4.2. Mean Estimation

The dose distribution functions are generally skewed (Figure 3). The mean is largely determined by the low percentage of results in the upper tail of the distribution (those between 10^{-3} and 10^{-2} Sv/a in the example of the Figure), and repeating the Monte Carlo simulation with different seed for the random number generation can result in dose means which are quite different from each other. In Figure 4 a different type of histogram is shown, which gives the incremental contribution to the dose mean for each bin of the horizontal axis. Figure 4 has been derived from Figure 3 and shows that the upper dose bin between 10^{-3} and 10^{-2} Sv/a contributes for about 40% to the dose mean. This gives a strong indication that convergence of the dose mean has not been reached in this simultation and that a higher number of runs is needed.

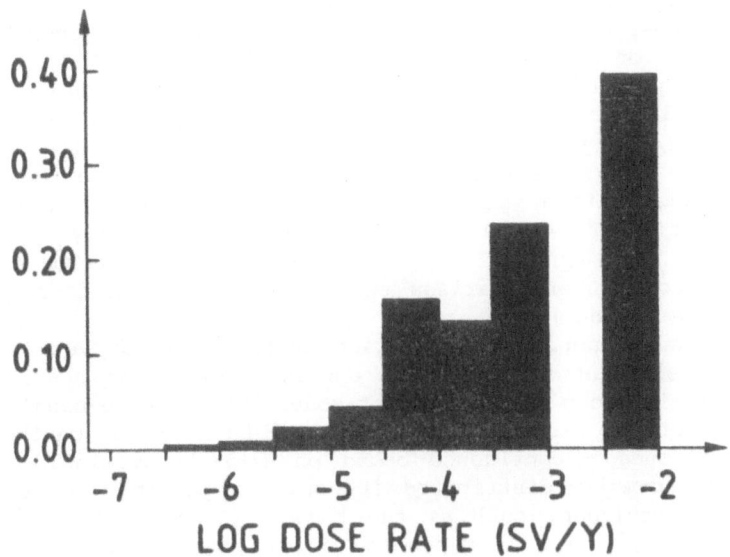

Figure 4.   Incremental contribution histogram. The ordinate give the incremental contribution of each bin of the histogram in Figure 3 to the value of the mean dose.

In fact a correct estimation of the mean is needed in order to compute the risk. Risk can be defined as

$$R = \int_{H_{min}}^{H_{max}} H\ p(H)\ r(H)\ dH$$

where

R   = risk
H   = annual dose
p(H) = probability of the annual dose
r(H) = dose to risk conversion factor

If H_{max} < 1 Sv/a then r(H) = 0.01 Sv^{-1} (constant) and

$$R = r \int_{H_{min}}^{H_{max}} H \, p(H) \, dH = r \, E(H)$$

where $E(H)$ is the expectation value or mean. It can be seen that the risk is linked to the arithmetic mean of the calculated dose. As shown above (Figures 3 and 4) it is difficult to estimate a stable value of the arithmetic mean for a parameter whose distribution is spread over several orders of magnitude.

Importance sampling may give a faster convergence of the mean for a smaller number of runs, because more outputs will be generated in the high dose tail of the output distribution.

Another possibility involves the analytical determination of the true underlying distribution. This is obtained by solving, in the Laplace space, all mass transfer equations constituting the model. Such solutions are then integrated over the various parameter distributions to get the expectation value of the dose in the Laplace space. Actual risk as function of time are then computed by numerically inverting the Laplace risk (Robinson and Hodgkinson, 1986). This approach has been taken in a code intercomparison exercise of the PSAC user group (see later in this seminar).

Although attractive, the Laplace approach is not yet at a mature stage of development for practical applications, as it has a number of important limitations in the type of model which can be handled.

For practical risk calculations, confidence bounds on the risk can be computed once a confidence bound on $E(H)$ is available. If the H values were normally distributed it would be possible to use standard parametric techniques, such as the t test (Conover, 1980) to compute confidence bounds on $E(H)$ depending on the sample size. Actually the hypothesis of sample normality is unrealistic for the dose distribution functions considered here. Experience from previous assessments has shown instead that the dose distribution can be roughly Log-normal (Wuschke et al., 1981; Bertozzi et al.,1984; De Marsily et al., 1988).

When the distribution function of a variable is unknown (or simply non-normal) a non-parametric test based on the Tchebycheff's theorem can be used to estimate confidence bounds on the mean (Saltelli and Marivoet, 1988). For instance, the 95% confidence bound on the expectation value can be shown to be:

$$E(H) = \langle x \rangle \pm \frac{4.472s}{\sqrt{N}}$$

where $\langle x \rangle$ is the sample mean obtained though a Monte Carlo simulation of N runs, s is the sample standard deviation and sqrt=square root.

This equation is very similar to the equivalent parametric formula for the confidence interval on the mean based on the Student's t test

$$E(H) = \langle x \rangle \pm \frac{1.96s}{\sqrt{N}}$$

The t test is commonly defined as parametric because the distribution function of the sample is assumed to be normal. When the assumption of normality is dropped (Tchebycheff's theorem) the confidence interval is more than doubled.

4.3. Use of Cumulative Distributions. Kolmogorov Statistic

As an alternative to the use of the risk concept, involving a dose to risk conversion factor, the consequences of the simulation can be analysed by replotting the dose histogram of Figure 3 in an inverse cumulative distribution plot (curve F(H) in Figure 5). For each dose value given on the abscissa, the ordinate gives the percentage of outputs exceeding that dose value. This can be considered as an estimate of the probability of exceeding that dose level. In this way an alternative definition of risk is possible, which is the probability of exceding a certain fraction of the natural background dose (Canadian Atomic Energy Control Board, 1985). This hypothetical fraction has been indicated as H* in Figure 5.

The L(H) and U(H) curves in Figure 5 represent the 95% confidence bounds on the inverse cumulative distribution, which have been calculated with the Kolmogorov statistic (Conover, 1980). The U(H) curve gives us, with 95% confidence, the maximum probability of exceeding the h* dose. U(H) is computed by adding to F(H) the appropriate quantile W(y) of the Kolmogorov test statistic, where y is the desired confidence level (e.g. y=0.95 for a 95% confidence bound). The W(y) value depends on the sample size, i.e. the bigger the sample size, the smaller the incertainty on F(H) (Conover, 1980; Saltelli and Marivoet, 1988).

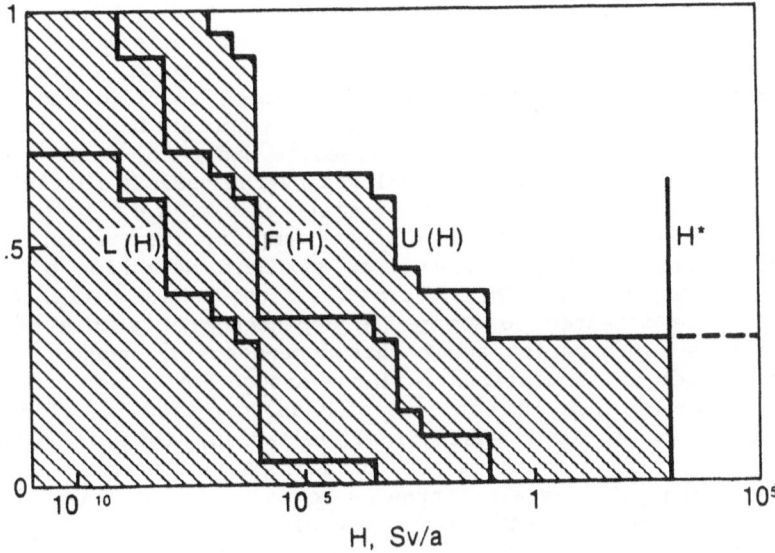

Figure 5. Downward cumulative distribution of dose with 95% Kolmogorov confidence bounds. H* represents an hypothetical dose limit.

4.4. A Worked Example: Convergence of Results as a Function of Sample Size for the Level 0 PSAC Intercomparison

Codes intercomparisons are an important element of the process of code verification and validation. Two such intercomparison have been promoted by the Probabilistic System Assessment Codes (PSAC) users group of the OECD/NEA, which shall be illustrated later in this seminar. We shall present here the results of the convergence analysis performed by Terry Andres (in Saltelli et al., 1987) on the results from the first intercomparison, named Level 0.

The Level 0 exercise was mainly intended to check the performances of the Monte Carlo framework of the codes using simplified barrier models. Twenty-one contributions were received in all for this inter-comparison, involving twelve different groups from eight countries and one international organisation; some submitted more than one solution in which different sampling methods or number of runs were used. Two contributions were withdrawn.

Each contribution is identified by the a fictitious name given in Table 1, which also indicates the sampling method used, the number of runs and some explanatory comments describing the particular features of the simulation. In the covergence analysis given in Figure 6 each participant is indicated by the letter given in Table 1.

The purpose of the convergence analysis is to verify the random nature of the residual error in the various contributions after correction of the systematic errors. This is needed because of the stochast-ic nature of the output; the codes to be compared are Monte Carlo in type, the input being sampled from density distributions of input variables and the output being provided in the same form (histograms). As a consequence, each output variable is affected by a random error. urthermore the standard error is not the same for all the codes, being dependent on the sample size employed and the various outputs cannot be compared against an exact (eg. analytical) solution, as it is not available. As a result the random errors associated with the apparently correct contributions must be analysed,in order to ensure their compatibility with the sample size employed by the various participants.

Mean estimates obtained from small samples tend to be more variable than mean estimates from large samples. More specifically, the standard error of a variable (i.e. standard deviation of its sample mean) varies as $1/sqrt(N)$, where N is the sample size. Convergence analysis was intended to study the dependence of sample mean dose on sample size. Information about this dependence is needed to determine whether a set of simulations is large enough for a particular application. For example, if an estimate of mean dose used in calculating risk is subject to a large statistical variation, then it might not be adequate for analyzing the safety of a disposal concept.

This analysis was also intended to identify outliers. In this case outliers are results that are unusually far from the centre of the data, after their sample size has been taken into account.

The three largest samples amongst the contributions were 3000 runs (participant Budhi in Table 1), 7865 runs (Bruce2), and one million

Table 1. List of Level 0 intercomparison participants. Each participant is identified by a fictitious name in the text and by a letter in Figure 6 (MC=purely random sampling; LHS=Latin hypercube sampling).

Participant	Identifier	Sampling Methods	Number of Runs	Comments
Steve	A	MC	1000	1000 time points
Jan	B	LHS	1000	500 runs for the mean of maximum dose equation
Budhi	C	MC	3000	
Andy1	D	MC	1000	1000 time points
Toshi	E	LHS	1000	
Timo1	F	MC	500	
Timo2	G	LHS	500	
Peter1	H	MC	1000000	Special code optimised for Level 0
Peter2	I	MC	1000	Special code optimised for Level 0
Withdrawn1	J			
Withdrawn2	K			
Ron	L	MC	2000	500 runs for the mean of maximum dose estimation
Bruce1	M	MC	1000	Automatic time step size
Bruce2	N	MC	7865	Automatic time step size
Shelley1	O	LHS	100	
Shelley2	P	LHS	1000	
Niels1	Q	MC	946	
Niels2	R	LHS	858	
Niels3	S	LHS	196	Spurious correlation removed
Andy2	T	LHS	1000	Spurious correlation removed
Alex	U	LHS	2000	

runs (Peter1). The means of these three sets of simulations are in agreement. This was taken as evidence that the largest set of runs was probably correct (ie. no systematic errors are present). With this assumption, the Peter1 results with a million simulations were used as a reference to compare against the rest for the results of the final iteration.

Figure 6 is a scatterplot of dose at time=10^7 years versus the logarithm of the number of simulations. Superimposed on the figure are

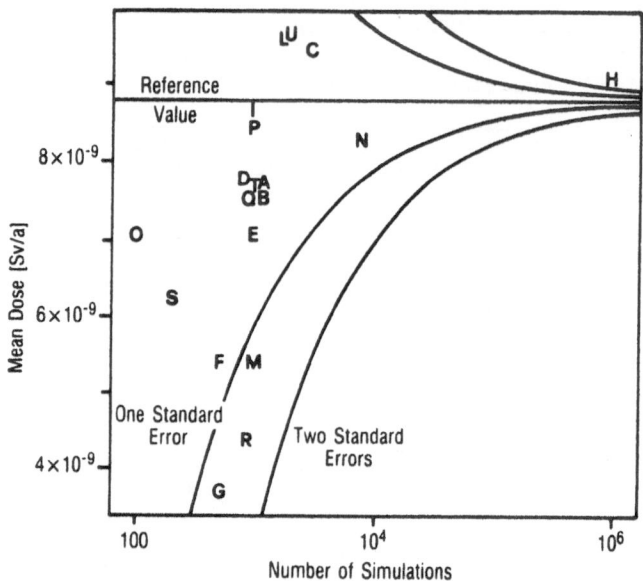

Figure 6. Scatter plot of mean dose vs number of simulations at $t=10^7$ years.

five reference lines. From top to bottom, these lines are:
- reference value plus two and one standard errors;
- reference value;
- reference value minus one and two standard errors;
    The standard error used in creating Figure 6 was calculated as:

STANDARD ERROR = STANDARD DEVIATION / sqrt(SAMPLE SIZE)

where STANDARD DEVIATION represents the million-run estimate of standard deviation of dose. The four reference lines depending on standard error are curved because of the dependence on sample size shown in this equation.

    Figure 6 has no outliers (i.e. no values beyond the two standard error lines). This means that there is no evidence to suggest that any of the points are in disagreement with the Peter1 million-run results. It confirms that the contributed values are in excellent agreement, as suggested by other analysis techniques employed in the analysis of the Level 0 results.

    A few of the plotted values lie between the one standard error and two standard error reference lines, and can be attributed to expected statistical variation alone.

    Figure 6 shows points that are not placed symmetrically about the central reference line, and there is a clear preponderance of values below the reference line. The source of this asymmetry is not known. It may be due to the skewness of the dose distribution. If mean dose is dominated by infrequent large dose values, then small samples will not collect enough observations with large doses to estimate mean dose accurately. They will give estimates of dose that are too low, as shown here. Evidence in support of the hypothesis that skewness causes asymmetry comes from two sources. First, in Figure 6, the only mean dose estimates that are at least as large as the reference value come from samples with at least 1000 simulations. Second, participant Peter has made the following remark concerning the dose distribution:

"From my million run case I find that the 10 million year case is indeed very skewed. Thirty five per cent of the average comes from just 0.19% of the runs, 80% of the average comes from 0.9% of the runs and 96% of the average from 1.6%. At the other times things are less skewed. At 100,000 years 52% of the average is from 1.4% of the runs and at 1 million years 50% of the average is from 1.6% of the runs."

Figure 6 can also be used to estimate how many simulations are needed for convergence of the results. For example, suppose we want one standard error to be no more than 10% of the mean dose value. Then rough measurements from the figures suggest that about 2,000 simulations are needed at 10^5 and 10^6 years, and about 10,000 simulations are needed at 10^7 years (using Peter's results, more precise numbers are 3.1, 2.4 and 11.0 thousand simulations). It is not necessary to use a million simulations to make this estimate. Based on the Bruce1 1,000 simulation results, the estimated numbers of simulations needed for this level of convergence are: 2.8, 2.0 and 10.3 thousand simulations. It therefore seems reasonable that, in another application, a scoping set of perhaps 1,000 simulations could be first of all carried out to estimate how many simulations would be needed for adequate convergence.

In conclusion, this technique gave no evidence of systematic errors in any of the results and showed one possible way to estimate the number of simulations needed to achieve a specified level of convergence in results.

5. SENSITIVITY ANALYSIS

5.1. Problems Linked to the Model

The analysis of the sensitivity of model output data to the value of input parameters is a crucial step in the analysis of model performance, especially when the model itself is complex and involves many variable parameters.

The complexity of Sensitivity Analysis (SA) can also be increased by the factors mentioned above in this lecture, which are:
- Model non-linearity and/or non-monotonicity.
- Time dependence.
- Large range of variation of the input parameters.
- Existence of "ties" in the output vector (a possibility is that many output data are zero).
- Cost of computer simulation.

5.2. Definition of the Methods

It is well known that nonparametric statistics based on the ranks of both input and output vectors are an appropriate tool for tackling sensitivity analysis problems (Iman et al., 1978). Intercomparisons have also been carried out on the relative performance of SA techniques (Iman and Helton, 1978; Saltelli and Marivoet 1988). Only a few techniques are described here which posses the following characteristics:
* They are applicable in conjunction with a Monte Carlo simulation performed using random or hypercube sampling. More precisely it is assumed that the same sample used for uncertainty analysis is used for the analysis of model sensitivity. Techniques based on fractional factorial design - for instance - are not discussed here.
* They are "global" analysis techniques, i.e. the impact on the output is studied over the entire range of variation of the input variables, and not just around some central point in the variables' space. For this reason the adjoint method and differential analysis are not discussed here.

On the basis of the previous experience four techniques have been selected. They are the Spearman rank correlation coefficients, the partial rank correlation coefficients, the standardised rank regression coefficients and the Smirnov test. These techniques are complemented by input/output scatterplots based on the ranks.

5.3. Techniques

5.3.1. Spearman coefficient. The Pearson product moment correlation coefficient (PEAR) is the usual linear correlation coefficient computed on the $x(i,j)$'s, $y(i)$'s, $(1=1,2...N)$ (see legenda in Table 2). For non-linear models the Spearman coefficient (SPEA) is preferred as a measure of correlation, which is essentially the same as PEAR but using the ranks of both the Y and $X(j)$ vectors instead of the raw values (Conover, 1980).

$$SPEA(Y,X(j)) = PEAR(R(Y), R(X(j)))$$

The basic assumptions underlying the Spearman test are:
a) both the $x(i,j)$'s and the $y(i)$'s are random samples from the respective populations;
b) the measurement scale of both variables is at least ordinal.

The numerical value of SPEA, commonly known as the Spearman "rho", can also be used for hypothesis testing, to quantify the confidence in the correlation itself. This is done by making first the following base hypothesis:

"no correlation exists between Y and X_j"

$SPEA(Y,X_j)$ is than computed from a simulation of a given number of runs N, and its value is compared with the quantiles of the Spearman

Table 2. Legenda for the SA symbols.

X(j)	vector of the x(i,j)'s
x(i,j)	value of variable number j in run number i
<x(j)>	mean of the x(i,j)'s over the N runs
x*(i,j)	standardised variable value
Y	vector of the y(i)'s
y(i)	value of the output at run number i
<y>	mean of the y(i)'s over the N runs
y#	value of the output computed with the
N	number of runs
K	number of variable parameters regression model based on the SRC's (or SRRC's)
R-squared	model coefficient of determination
F(X(j))	empirical cumulative distribution of the x(i,j)'s

test distribution. The comparison is made at a certain pre-established level of significance (alpha), and the hypothesis of no correlation is rejected if SPEA is either lower than W(alpha/2) or higher than W(1-alpha/2), where the W's are the quantiles of the test distribution. The level of significance alpha is the probability of erroneously rejecting the hypothesis, i.e. - in this context - the probability that the test indicates a correlation when Y and X_j are actually uncorrelated. To apply the test at a 0.05 significance level W(0.025) and W(0.975) must be computed or read on tables (Conover, 1980). Taking - for instance - N=500, the Spearman quantiles are:

W(0.025) = -0.088
W(0.975) = 0.088

The hypothesis of no correlation is rejected if SPEA, as computed from a 500 run simulation, falls outlisde the range (-0.088,0.088), and the probability of an erroneous rejection, when Y and X_j are actually uncorrelated, is 0.05.

5.3.2. Partial Correlation Coefficient and Standardised Regression Coefficient. The Partial Correlation Coefficients (PCC's) and Standardised Regression Coefficients (SR C's) are two very useful correlation estimators, which can also be used on the ranks of the (Y, X(j)) values (Partial Rank Correlation Coefficients PRCC's and Standardised Rank Regression Coefficients SRRC's) (Iman, Shortencarrier and Johnson, 1985).
The SRC(Y,X(j))'s are the coefficients of the regression model for Y; they may provide an approximation to Y in the form

$$y^{\#}(i) = \sum_{j=1}^{K} \left\{ SRC(Y,X(j)) \, x^*(i,j) \right\}$$

where the x*(i,j) are the standardised variables

$$x^*(i,j) = (x(i,j)-<x(j)>)/s(X(j))$$

and $<x(j)>$ and $s(X(j))$ are respectively the sample mean and standard deviation (see legenda in Table 2).

When using the SRC's it is also important to consider the model coefficient of determination R-squared(Y). This coefficient provides a measure of how well the linear regression model based on the SRC's can reproduce the actual output vector Y. In particular:

$$\text{R-squared}(Y) = \sum_{j=1}^{K} \left\{ (y(i)^\# -<y>)^2 \right\} / \sum_{j=1}^{K} \left\{ (y(i)-<y>)^2 \right\}$$

where $<y>$ is the mean of the output values y(i) and the $y^\#$'s are the model prediction based on the SRC's, so that R-squared(Y) represents the fraction of the variance of the output vector explained by the regression. The closer R-squared is to unity the better is the regression model performance. The coefficients SRC(Y,X(j))'s can themselves provide a very effective measure of the relative importance of the input variables. Of course the validity of the SRC's as a measure of sensitivity is conditional on the degree to which the regression model fits the data, i.e. to R-squared.

The PCC's can be considered as an extension of the usual correlation coefficient (PEAR) and to represent that part of the interdependence between two variables which is not due to the correlation between these two variables and the remaining ones. When the PCC's are used they can provide a ranking of the input variables by indicating the strength of the linear relationship between Y and any variable X(j). When the PRCC's are used the linear relatioship between the ranks of Y and X(j) is measured. This gives an effective estimation of sensitivity.

5.3.3. Smirnov test. The Smirnov test (SMIR(Y,X(j)) is a "two-sample" test designed to check the hypothesis that two different samples belong to the same population. The application of such "two-sample" tests to sensitivity analysis comes from the idea of partitioning the sample of the parameter X(j) under consideration into two sub-samples according to the quantiles of the output distribution. If the distribution of X(j) in the two sub-samples can be proved to be different then the parameter under consideration is recognized as influential.

For instance the values x(i,j)'s corresponding to output y(i)'s above the 90th quantile of the F(Y) distribution may constitute one sub-sample, and all the remaining x(i,j)'s the other sub-sample.

For this statistic to be applicable, a number of basic assumptions must be satisfied by the two sub-samples under consideation, viz.
a) the two sub-samples are random samples;
b) the two sub-sample are mutually independent;
c) the measurement scale of X(j) is at least ordinal;
d) the random variable must be continuous.

When using SMIR the empirical cumulative distribution F(X(j)) is computed on the two sub-samples, and the two distributions are compared with each other. If the two distributions are different, it can be said that the parameter influences the output, and that high outputs are preferentially associated with high,or low, parameter values.

More quantitatively the Smirnov statistic is defined as the maximum vertical distance between the empirical cumulative distribution functions of the two sub-samples (see Figure 7). SMIR can also be used for hypothesis testing (Conover, 1980).

5.3.4. Scatterplots on ranks. Figure 8 is an example of rank scatterplot arising from a 1000 run simulation for the variable FLOWV1. It refers to a particular time point (t=2.5 10^4 a) and has been built by replacing the value of both the dose rate and the variable FLOWV1 by their ranks. This Figure is useful in pointing out two important characteristics of the sample: <u>first</u>, the large number of ties, represented by a row of constant values of dose at approximatively rank=260; these values are actually about 520 zeros (no dose at all), which are given by the ranking algorithm the average value rank=260; <u>second</u>, the monotonical relationship between input and output. Scatterplot are frequently employed to verify the characteristics of the sample, as showed in the next section.

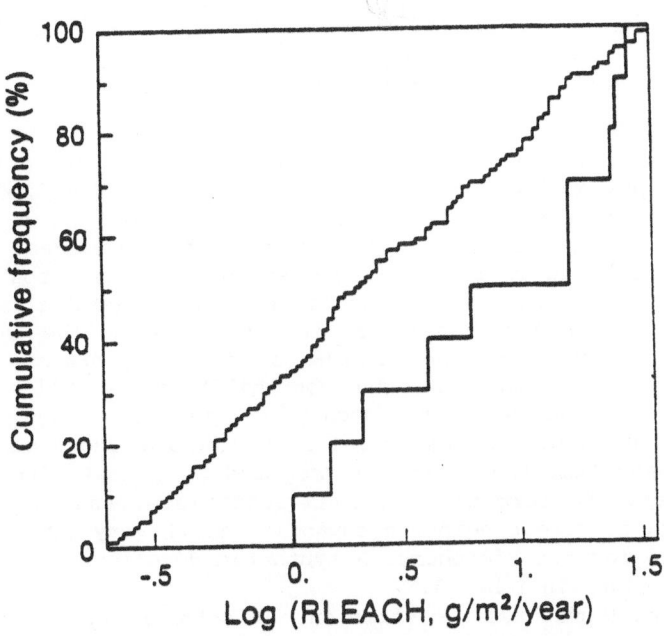

Figure 7. Example of plot for the Smirnow test. The curve with coarser steps is the cumulative distribution of the variable for the sub-sample containing 90% of the data; the curve with larger steps relates to the sub-sample containing the 10% of the data associated with the 10% highest outputs.

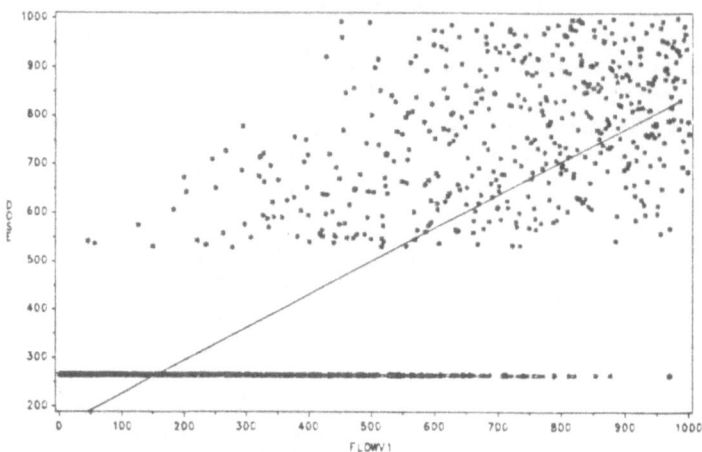

DOSE – FLOWV1 SCATTERPLOT AT T=2.5 E4

Figure 8. Scatterplots of rank of dose against rank of variable. The straight line is the linear regression over all the time points. The large number of ties (zero doses) is shown by the bottom row of constant ranks.

5.4. A Worked Example: Sensitivity Analysis for the Level E Intercomparison

We shall present here some results of a preliminary calculation made at the JRC of Ispra on the sensitivity analysis for the test model employed in the second exercise promoted within the PSAC group, which was named Level E (for exact) intercomparison after the exact analytical method which is used to check the convergence of the output distributions. The analysis has been performed on a set of data provided by Jim Sinclair, of the Harwell laboratory in the UK, using the version 4 of our SPOP post-processor. The finding of this exploratory study are quite interesting, as they show the possibilities, as well as the deficiencies, of the most frequently used SA estimators.

A description of the test model is beyond the scope of this note; we can just say that the system is composed by a source term, characterised by a waste form with a given leach rate, and by a geosphere with two with different sorption capabilities. A very simple diluition model is used for the biosphere. A synthetic description of the system variables is given in Table 3.

In Figure 9 the model coefficient of determination is plotted as function of time, together with the percentage of non-zero runs. This figure tells us that the regression model based on the SRC's has a drop of efficiency around the time point $t=10^5$ a. This point corresponds also to the maximum of the percentage of non-zero runs curve. The value of the statistics SPEA, PRCC SRRC and SMIR for the most important variables are given in Figures 10 to 13. Actually what is

Table 3. Variables of the Level E model.

FORTRAN NAME	DISTR. TYPE	DESCRIPTION	MODULE
CONTIME	Uniform	Containment Time	Source
RELRNP	Loguniform	Release rate	Source Nuclide, NP237
RELRI	Loguniform	Release rate	Source Nuclide I129
PATHL1	Uniform	Path Length	Layer 1
FLOWV1	Loguniform	Flow Velocity	Layer 1
RETF1	Uniform	Retardation Factor	Layer 1
PATHL2	Uniform	Path Length	Layer 2
FLOWV2	Loguniform	Flow Velocity	Layer 2
RETF2	Uniform	Retardation Factor	Layer 2
STFLOW	Loguniform	Stream Flowrate	Output

plotted in the fugure is not the raw value of the statistic, but the ank given by the statistic to the variable. Figure 10, for example, tells us that all the statistics consider (SPEA, PRCC, SRRC, SMIR) consider the variable FLOWV1 as the most important one (rank=1) for all the time points exept the strange $t=10^5$ a point. Something similar appears in Figures 11 and 12, where the variables PATHL1 and RETF1 loose part of their influence for the same time point. The variable STFLOW, on the contrary, becomes more influential around this time point (its rank passes from only 9 - the lowest - to 1 - the highest -, Figure 13)). What is happening?

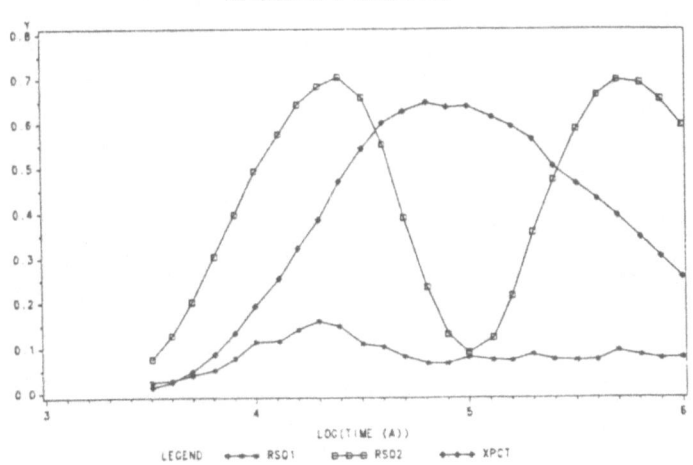

Figure 9. a) Percentage of non-zero runs (diamond); b) model coefficient of determination on raw value (star); c) model coefficient of determination on ranks (square).

SPEA PRCC SRRC AND SMIR STATISTICS AS FUNCTION OF TIME
FOR THE VARIABLE FLOW1

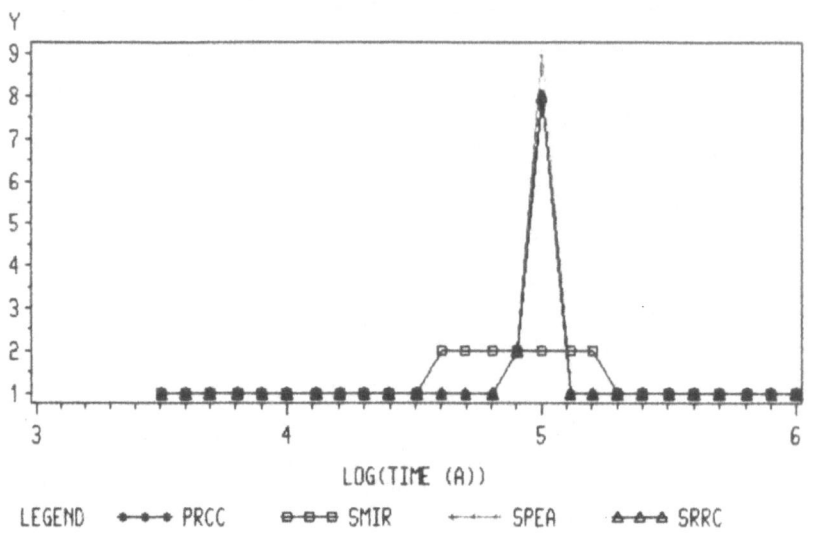

SPEA PRCC SRRC AND SMIR STATISTICS AS FUNCTION OF TIME
FOR THE VARIABLE PATHL1

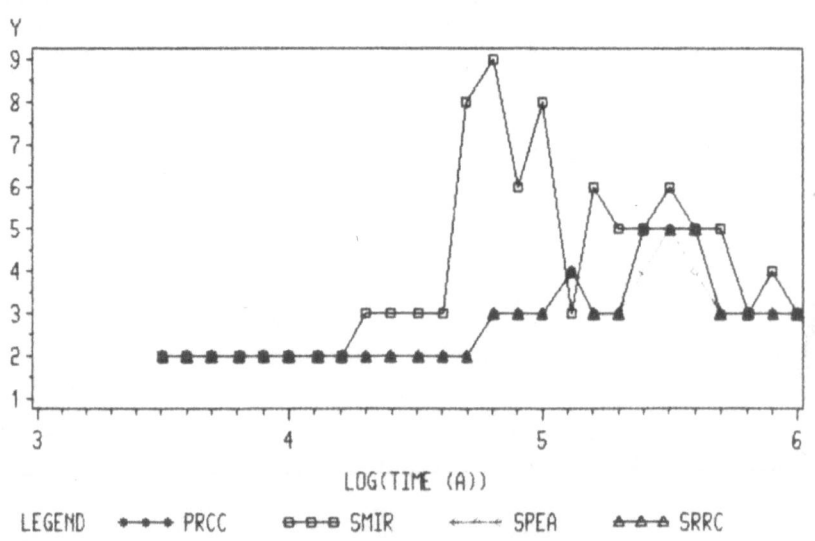

Figure 10 and 11. Plots of the ranks given by the sensitivity analysis estimators SMIR (square), SPEA (diamond) and PRCC, SRRC (star and triangle, always superimposed) to the 2 most important variables.
RANK=1 => most influential variable;
RANK=9 => least influential variable.

SPEA PRCC SRRC AND SMIR STATISTICS AS FUNCTION OF TIME
FOR THE VARIABLE RETF1

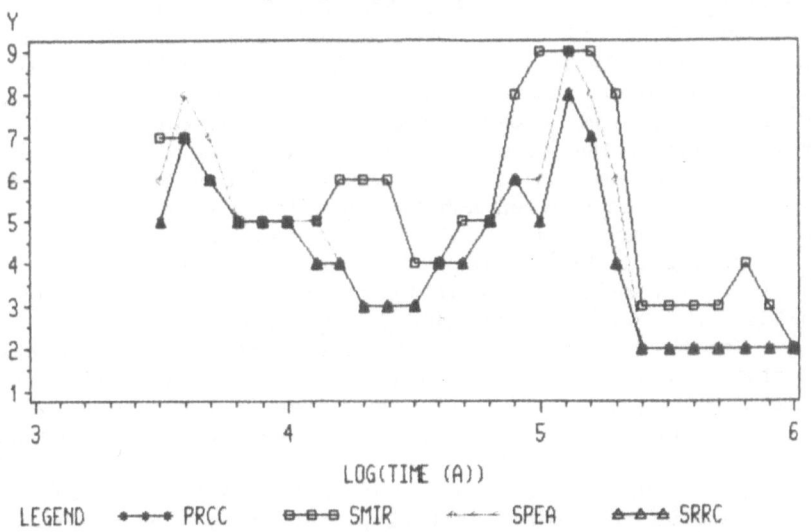

SPEA PRCC SRRC AND SMIR STATISTICS AS FUNCTION OF TIME
FOR THE VARIABLE STFLOW

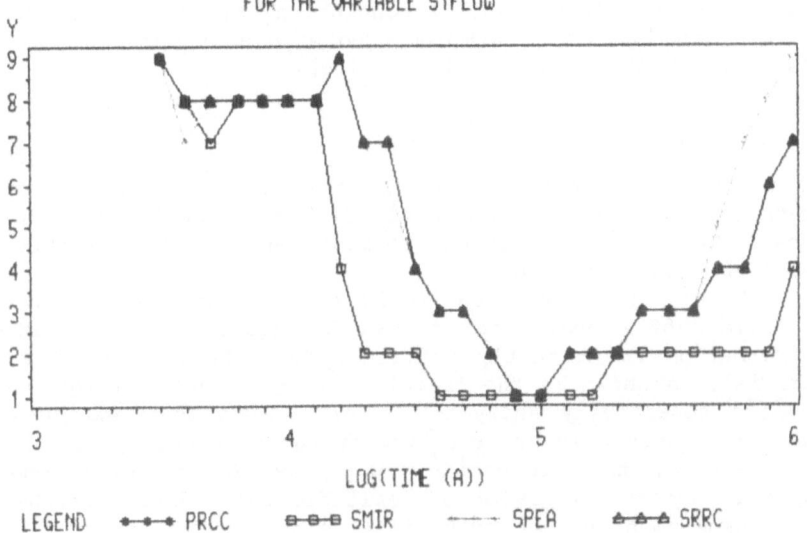

Figures 12 and 13. Plots of the ranks given by the sensitivity analysis estimators SMIR (square), SPEA (diamond) and PRCC, SRRC (star and triangle, always superimposed) to the 3rd and 4th most important variables.
RANK=1 => most influential variable;
RANK=9 => least influential variable.

The reason for this behaviour is the non-monotonicity of the model, shown by the dose/variable scatterplots. Figure 8 shows the scattergram for the variable FLOWV1 at t= 2.5 10^4 a. The straight line is the calculated linear regression on the ranks. There is a clear positive correlation, i.e. high doses are associated to high FLOWV1 values. At large times (e.g. 10^6, Figure 15) there is a clear negative correlation: high FLOWV1 values mostly correspond to zero doses at t=10^6 because the time pulse is produced at earlier time points. For intermediate times, such as the t=10^5 time point (Figure 14), the situation is in between and the relationship between dose and FLOWV1 is clearly non-monotonic . The linear regression on the points of the scatterplot in Figure 14 is an horizontal line. All the sensitivity analysis estimators fall in presence of such a non-monotonicity. This holds for the variables PATHL1 and RETF1 as well.

For this time point (t=10^5 years) the variable identified as most influential is STFLOW , which has a monotonic (and linear) influence on the output. Yet with just one identified influential variable the SRRC's cannot build an effective regression model and the R-squared value is very low.

The above example shows that, although the nonparametric tests are an effective tool for sensitivity analysis , they can lead to erroneous results. In the example above, in fact, the variable STFLOW is not the most important one, and it shows up as influential only because of the model non-monotonicity which hides the influence of the other variables. The SA techniques, in conclusion, are not to be used indiscriminately, and are to be complemented with an understanding of the system (see also Saltelli and Marivoet, 1988).

6. CONCLUSIONS

A short and necessarily incomplete presentation has been given of methods to be used in stochastic risk assessment for nuclear waste disposal. Two consideration can be made here, which should be kept in mind when applying these tools.

First, the statistical tools required to analyse the probabilistic behaviour of a geological system do exist, provided that the starting information is correctly utilised. Nevertheless the task can be non-trivial, as shown by the examples given on the convergence analysis and on the sensitivity analysis. Expert's judgement and knowledge of the system is needed in order to verify the results.

Second, the stochastic safety assessment is strongly dependent upon a correct definition of realistic input distributions and the correct assessment of the probabilities of occurrence for various types of rare events. These facets should not be hidden in presenting the results.

Experience from the reactor safety studies has shown that uncertainties is Risk Assessment can be much higher than risk assessors claim. The limitations of this kind of treatment should always be kept in mind, to avoid, for instance, that a confidence bound on a risk figure (computed on purely statistical basis) be taken or presented as the real uncertainty on such a risk.

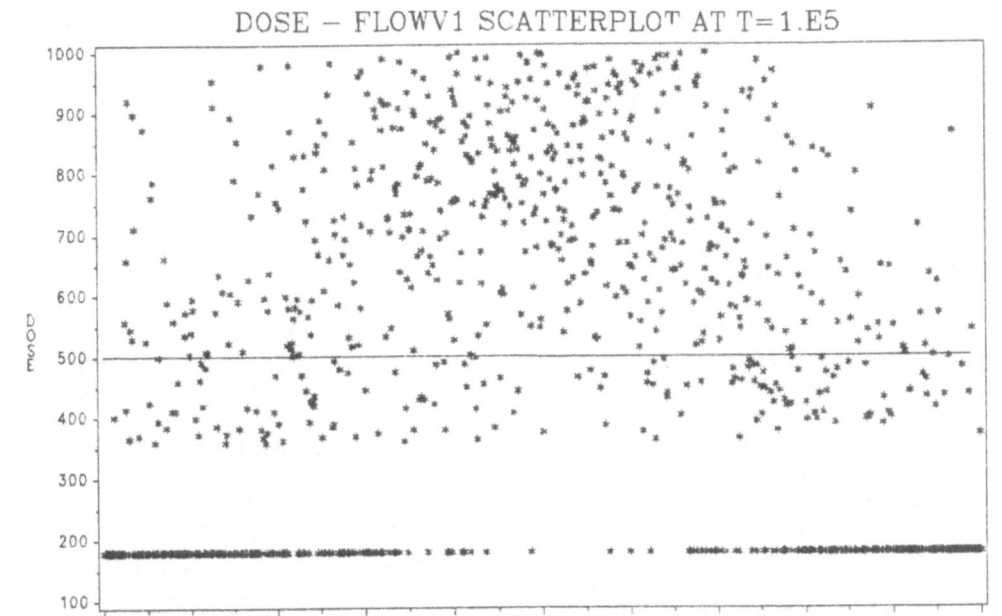

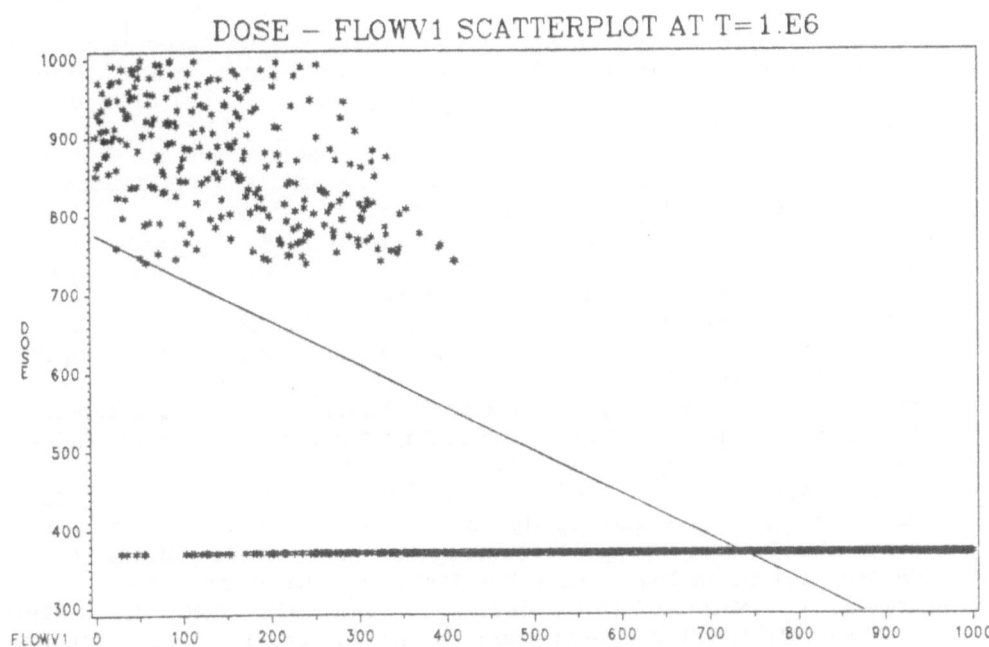

Figures 14 and 15. Scatterplots of rank of dose against rank of variable. The straight line is the linear regression over all the time points.

The huge amount of studies recently devoted to the safety of nuclear waste disposal should ensure a carefully considerd technology. At
the same time it must be aknowledged that objective difficulties exist
in assessing the risk for those systems where few accidents have occurred (eg. reactor system) or where accidents did not occurr at all because no such system has yet been created (the waste disposal system).

REFERENCES

Bertozzi, G. et Al.: Waste disposal into a plastic clay formation. Risk
    analysis for probabilistic scenarios. European Applied Research
    report 6,1985 (EUR 9641 EN).
Canadian Atomic Energy Control Board - regulatory document R-71: Deep
    geological disposal of nuclear fuel waste. Background information and
    regulatory requirement regarding the concept assessment phase. Ot-
    tawa, Ontario 1985.
Chatfield, C. and Collins, A.J.: Introduction to Multivariate Analysis;
    Editors: Chapman and Hall, London (1983).
Conover, W.J.: Practical Nonparametric Statistic, John Whiley & Sons,
    New York 1980.
Marsily, G. De, Behrendt, V., Ensmineer, D.A., Flebus, C., Hutchinson,
    B.L., Kane, P., Karpf, A., Klett, R.D., Mobbs, S., Poulin, M. and
    Stanners, D.A.: Subseabed disposal of high level radioactive waste,
    Volume 2, Radiological assessment, OECD/NEA, Paris 1988 (in prepara-
    tion).
Goodwin, B.W. and Saltelli, A., Probabilistic System Assessment Codes.
    Level 0 Intercomparison: Problem Specification Note. Atomic Energy of
    Canada Ltd. report (1986), (also as appendix in Saltelli et al.,
    1987).
Iman, R.L. and Helton, J.C.: A comparison of uncertainty and sensitivi-
    ty analysis techniques for compute models. Sandia Natl. Laboratories
    report NUREG/CR-3904, SAND 84-1461, 1985.
Marivoet, J. and Bosstraeten, C. Van: A simplified bisaphere model
    applicable in computer codes for probabilistic performance assessment
    for radioactive waste disposal (in preparation).
Robinson, P.C. and Hodgkinson, D.P.: Exact solutions for radionuclide
    transport in the presence of parameter uncertainty. AERE report R
    12125 ,1986 Harwell, UK).
Saltelli, A., Bertozzi, G. and Stanners, D.A.: LISA - A Code for Safety
    Assessment in Nuclear Waste Disposals; Program Description and User
    Guide, JRC Ispra Report EUR 9306 EN, Luxembourg (1985).
Saltelli, A.: PREP and SPOP Utilities. Two FORTRAN Programs for Sample
    Preparation, Uncertainty Analysis and Sensitivity Analysis in Monte
    Carlo Simulation. Programs Description and User's Guide. Nuclear
    Science and Technology report EUR 11034 EN , Luxemburg 1987.
Saltelli, A., Sartori, E., Andres, T.H., Goodwin, B.W. and Carlyle,
    S.G. PSACOIN Level 0 Intercomparison. An international Code Intercom-
    parison Exercise on a Hypothetical Safety Assessment Case Study for
    Radioactive Waste Disposal Systems. OECD - NEA publication, Paris
    (1987).

Saltelli, A. and Marivoet, J.: Safety assessment for nuclear waste disposal. Some observations about actual risk calculations. Radioactive Waste Management and the Nuclear Fuel Cycle, 1988 (in press).

Saltelli, A. and Marivoet, J.: Nonparametric statistics in sensitivity analysis for model output. A comparison of selected techniques. 1988, paper in progress.

SAS User's Guide: Statistics, Version 5 Edition ISBN 0-917382-66-8, SAS Institute Inc., Cary NC, USA (1985).

Shaeffer, D.L. and Hoffman, F.O.: Uncertainties in radiological assessment. A statistical analysis of radioiodine transport via the pasture-cow-milk pathway. Nuclear Technology 45 (1979).

Sherman, G.R., Donahue, D.C. and King, S.G.: SYVAC2 - A System Variability Analysis Code for Assessment of Nuclear Fuel Waste Disposal Concepts, TR-317 (1986).

Stephens, M.E., Goodwin, B.W. and Andres, T.H.: Guidelines for selecting probability density functions for SYVAC3-CC3 input parameters Atomic Energy of Canada Ltd. report (1987), (also in OECD NEA report PSAC/DOC(87)10, Paris 1987).

Wuschke, D.M. et Al.: Second interim assessment of the Canadian concept for nuclear fuel waste disposal Atomic Energy of Canada Ltd. report AECL 8373-1, 1986.

APPLICATIONS OF SYSTEMS VARIABILITY ANALYSIS CODES

Bruce W. Goodwin
Atomic Energy of Canada Limited
Whiteshell Nuclear Research Establishment
Pinawa, Manitoba, Canada
ROE 1L0

ABSTRACT. Several computer codes have been developed over the past decade that implement the systems variability analysis approach. These codes are of particular interest for environmental assessments where uncertainty and variability must be taken into account.
     To date, most applications of these codes have dealt with assessments of the disposal of nuclear wastes. The types of wastes range from low-level reactor wastes and uranium mine tailings to high-level used fuel and reprocessing wastes. This overview report examines two recent applications of the systems variability analysis approach. It also investigates other potential applications to more conventional types of wastes.

1. INTRODUCTION

Systems variability analysis (SVA) represents a relatively new approach to the preparation of environmental assessments. Most of the development for SVA has occurred over the past decade in response to international and national programs aimed at the disposal of nuclear fuel wastes.
     The concept of "disposal" has two significant characteristics that influence the assessment process:
- the disposal facility should provide for the <u>permanent</u> and safe isolation of wastes, and
- the disposal facility should not require subsequent monitoring and/or remedial action.

The first characteristic implies that an environmental assessment should extend until such time as the potential hazards of the waste are not significant. The second characteristic implies that our assessment must identify and quantify all potential hazards, including hazards with relatively low probabilities.
     Systems variability analysis deals with these requirements by providing a framework within which we can
- Build a complete description of the environmental system. We require a systems approach to handle the complex interplay of

97

processes that may lead to a potential hazard. The systems
approach can provide a structure for input of data and information from many different scientific disciplines.
- Accommodate our uncertainty and variability in the system.
  This issue could be very important in our assessments,
  particularly when we must extrapolate our current imprecise
  understanding in time and/or space.
This report describes the application of SVA to long-term environmental assessments. The bibliography section identifies many
relevant reports, but we shall focus on two specific applications to
give an in-depth flavor of systems variability analysis.

2. DEVELOPMENT OF SVA - THE PSAC USER GROUP

Table 1 identifies individuals and organizations that are actively
developing and applying the approach. It lists the members of the
Probabilistic System Assessment Code (PSAC) User Group.   The PSAC User
Group was organized in 1985 and is sponsored by the Nuclear Energy
Agency (NEA) of the Organization for Economic Co-operation and
Development.   Objectives of the Group center around the development and
application of the SVA approach, including the exchange of codes and
information, code comparisons and discussion of topical issues:
- The NEA Databank maintains copies and documentation of SVA
  codes, which are available to NEA members. In addition, the
  Databank has devised a common format for saving the output
  from SVA codes. This common format facilitates the exchange
  of raw results for subsequent analysis and plotting.
- The PSAC User Group has established plans for a set of code
  comparisons. The "Level 0" comparison has already been completed (Saltelli et al., 1987), and provides a verification
  test of the basic functions of an SVA code. The "Level E"
  comparison is in progress and has the attraction that SVA
  results can be directly compared with an "exact" solution
  (Robinson and Hodgkinson, 1987). Other planned comparisons
  will provide for tests of models for deep disposal concepts
  (Level 1a), shallow land burial (Level 1b) and sensitivity
  analysis techniques (Level ESA).
- At each of the semi-annual Group meetings, time is set aside
  for the discussion of topical issues, including data acquisi-
  tion (for probability density distributions), sensitivity
  analysis techniques, methods for the presentation of SVA
  results, sampling strategies, software quality assurance and
  the reduction of detailed research models to SVA models. More
  information on these meetings, including some written reports,
  are available from the NEA Secretariat.
The PSAC User Group has been active for three years, and its membership
has increased steadily. Members of the Group are responsible for much
of the recent development and application of SVA.

TABLE 1. Participants in the PSAC Users Group

Member	Organization	Code
Jan Marivoet Belgium	Centre d'Etude de l'Energie Nucleaire (SCK/CEN)	LISA-SCK
Bruce Goodwin Canada	Whiteshell Nuclear Research Establishment (AECL)	SYVAC
Steve Wilkinson Canada	Chalk River Nuclear Laboratories (AECL)	COSMOS
Ron Holmes Canada	National Uranium Tailing Program (CANMET, EMR)	UTAP
Timo Vieno Finland	Technical Research Center (VTT)	LHSSIM
Alex Nies Fed. Rep. of Germany	Gesellschaft fur Strahlen und Umweltforschung mbH (GSF)	EMOS
Toshinori Iijima Susumu Mitake Japan	Japan Atomic Research Institute (JAERI)	MIGRAT3-CIRCLE
Nils Kjellbert Sweden	Swedish Nuclear Fuel and Waste Management Company (SKB)	PROPER
Benny Sundstroem Sweden	Swedish Nuclear Power Inspectorate (SKI)	SYVAC/SU, PARVAR
Piet Zuidema Switzerland	National Cooperative for the Storage of Radioactive Waste (NAGRA)	SYVAC
Brian Thompson United Kingdom	Department of the Environment (DOE)	SYVAC/AC, TIME2
Shelly Mobbs United Kingdom	National Radiation Protection Board (NRPB)	ESP
Jim Sinclair United Kingdom	Atomic Energy Research Establishment (AERE, Harwell)	MASCOT

continued. . .

TABLE 1. Participants in the PSAC Users Group (concluded)

Budhi Sagar United States	Basalt Waste Isolation Project (BWIP, Hanford)	SPAM
Al Liebetrau United States	Battelle Pacific Northwest Laboratories (PNL)	AREST
Evaristo Bonano United States	Sandia National Laboratories (SNL)	various
Andrea Saltelli CEC	Joint Research Centre of Ispra (CEC)	LISA

3. APPLICATIONS OF SVA

The next two sections of this report concentrates on two specific
applications drawn from the author's experience. (Many other
applications may be found in the bibliography). Since these two
applications have been detailed in the literature, we will focus on
some of the highlights using figures and plots taken from appropriate
references. The two applications are
- assessment of the Canadian concept for nuclear fuel waste
  disposal, using the computer code SYVAC2 (Wuschke et al.,
  1985); and
- assessment of the closeout of uranium mill tailings, using the
  computer code UTAP3 (SENES, 1986a).
Note that other sessions in this seminar will also touch on applications of SVA. In particular, D.A. Stanners and S. Mobbs describe
assessments of sub-seabed disposal, performed using the systems variability analysis approach.

4. ASSESSMENT OF NUCLEAR FUEL WASTE DISPOSAL

Figure 1 illustrates the Canadian concept for nuclear fuel waste disposal. The major features of the concept are a series of engineered
barriers (the UO_2 waste form, containers, buffer and seals) and natural
barriers (500 to 1000 m of crystalline rock) (Goodwin et al., 1987a;
Wuschke et al., 1985). These barriers isolate the nuclear waste from
the accessible environment.

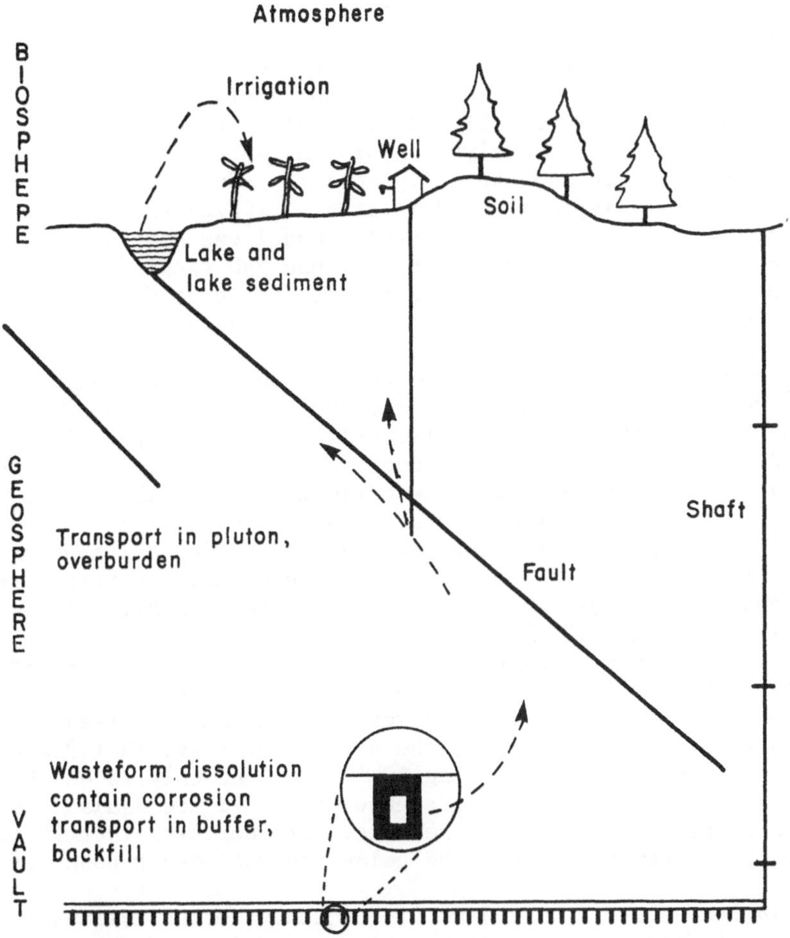

Figure 1. Schematic of the Canadian concept for nuclear fuel waste
disposal. The concept uses a series of barriers designed to
isolate nuclear fuel waste from the accessible environment.
Engineered barriers include the waste form, container,
buffer and seals. The major natural barrier is 500 to 1000 m
of crystalline rock.

The goals of the Canadian Nuclear Fuel Waste Management Program
are as follows:
- to demonstrate an assessment method and its capabilities;
- to identify and evaluate potential impacts;
- to assess the acceptability of the concept based on established
  criteria; and
- to deduce important factors affecting impacts, and develop
  siting and design constraints.

Criteria for the assessment are provided by the Atomic Energy Control
Board (AECB, 1987). These criteria include a time scale of 10^4 years,

a risk limit of 10^{-6} health effects per year, and consideration of chemical toxicity impacts.

Two preliminary assessments have been completed, and we shall examine some results from the most recent application, produced using the computer code SYVAC2. SYVAC2 is a computer code that implements the systems variability analysis approach (Sherman et al., 1986). It was used to evaluate radiotoxicity impacts (Wuschke et al., 1985) and chemical toxicity impacts (Goodwin et al., 1987b) from the disposal of high-level nuclear wastes. The system model was made up of vault, geosphere and biosphere submodels, and had 540 sampled parameters describing 59 radionuclides and 8 chemically toxic elements. Further details of the system submodels are given by Heinrich (1984), LeNeveu (1986) and Mehta (1985).

Figure 2 shows the histogram of frequency versus maximum annual effective dose equivalent or maximum "dose" for about 1000 simulations. (At the time of these assessments, criteria had not been formally defined, and the analysis proceeded only to estimation of maximum dose rates.) For each simulation, the maximum dose is taken to be the largest estimated dose from the start of the simulation to a prespecified time cutoff. The five curves in Figure 2 are for five different time cutoffs. For comparison purposes, a vertical reference line identifies 1% of the average annual dose from natural background radiation in Canada (0.18 mSv).

Figure 3 shows an alternative way of presenting these results, using downward cumulative fractions. In this case the curve shows the fraction of simulations giving doses that exceed specified levels (for 10^7 years only). For example, the solid line shows that 92% of the simulations had a dose exceeding 10^{-10} mSv, and only 7% of the simulations exceeded the vertical reference line. Figure 3 also includes the 99% confidence bound on these results. It indicates that 99 times out of 100 the "true" curve would be below the confidence bound at any dose level.

Figure 4 shows a histogram of frequency versus estimated concentrations of boron in well water (this particular result is from an assessment of reprocessing wastes, with a sodium borosilicate waste form). The vertical reference line indicates a typical acceptable concentration of boron in man's drinking water.

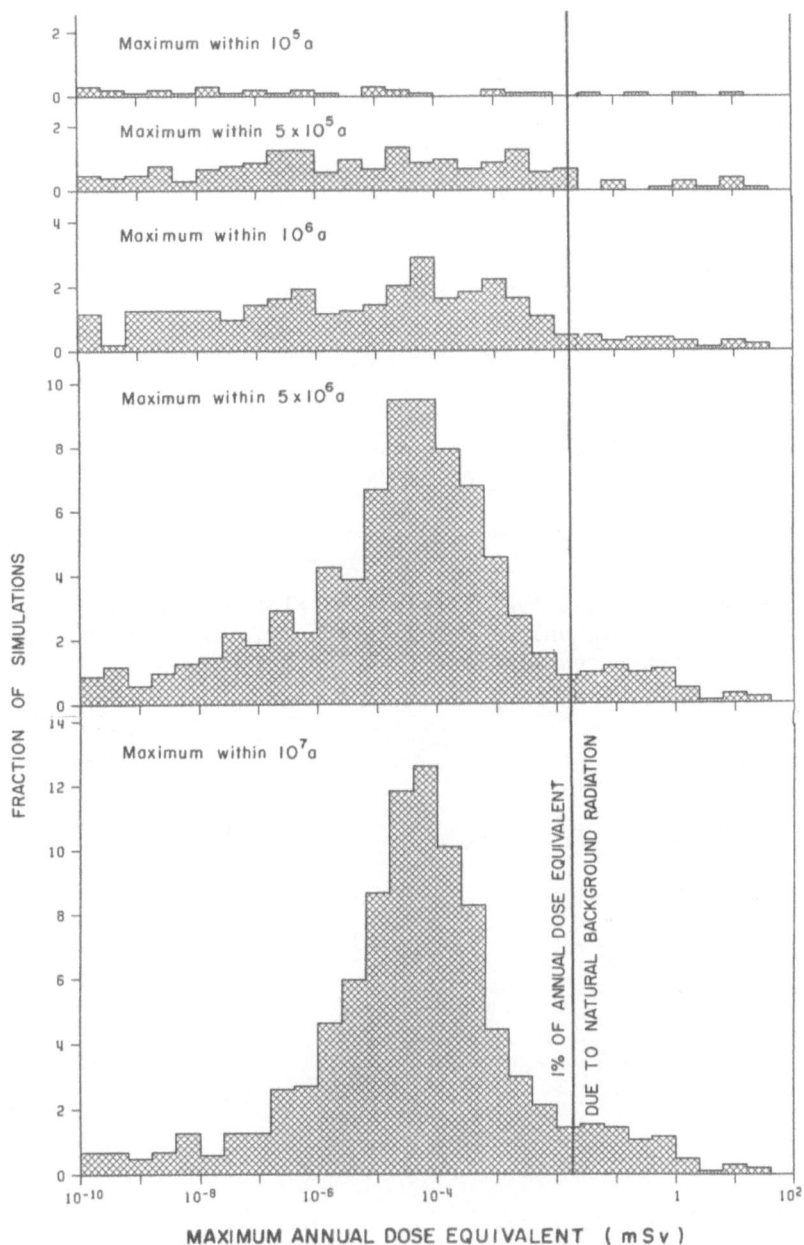

Figure 2. Histogram of the frequency versus maximum annual effective dose equivalent (maximum "dose") for 1037 simulations. The five histograms show maximum dose up to 10^5, 5×10^5, 10^6, 5×10^6 and 10^7 years following closure of the vault.

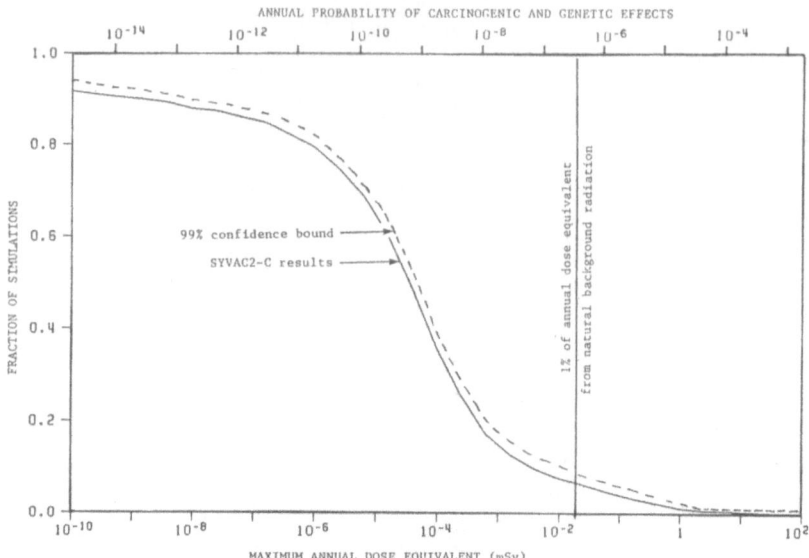

Figure 3. Downward cumulative distribution plot of the results shown in Figure 2, for 10^7 years only. Cumulative curves are more useful than histograms to compare results (see examples below).

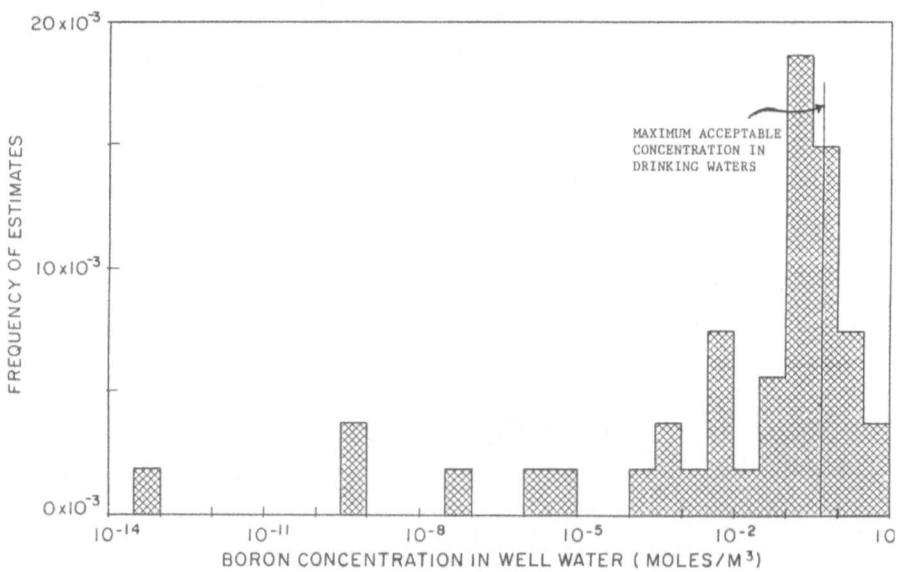

Figure 4. Histogram of the frequency versus maximum concentration of boron in the well compartment. The boron is associated with a fuel reprocessing waste form, sodium borosilicate glass. The plot is discontinuous because the well was selected as a source of drinking water in only about 100 of the 1000 simulations.

Figures 5 to 8 display some results from sensitivity analysis of the results. In Figure 5, we can see that the dose due to ingestion overlaps the total dose curve. This observation demonstrates that dose due to ingestion is much more important than dose due to inhalation or external exposure. Figure 6 shows doses due to drinking water obtained either from a lake or a well. We see that the high-dose simulations always involved the well as the source of water. Finally, Figures 7 and 8 identify the most important radionuclide that contributes to dose. Figure 7 is a downward cumulative plot, and it is clear that Iodine-129 is the major contributor to dose because its curve coincides with the "all nuclides" curve. Figure 8 is a scatterplot that shows the maximum dose for each simulation versus the time at which it occurred and identifies the radionuclide most contributing to that dose. Again only Iodine-129 is the major contributor.

Although these studies are preliminary, some overall conclusions may be drawn that will likely remain valid in the formal assessment. For a significant impact, the radionuclide or chemically toxic element must

- be present in relatively high quantities in the waste,
- be relatively mobile in groundwater in the geosphere,
- have a relatively high rate of release from the waste form, and
- have a relatively high degree of radiological or chemical toxicity.

SYVAC2 results show that the inventory, release rate and toxicity factor mostly affect the magnitudes of the estimated impacts, while the mobility factor mostly affects the frequency of appearance of impacts.

Finally, the four factors listed above are each qualified since their relative importance varies from one simulation to the next. This last observation is significant because it demonstrates the value of the systems variability analysis approach. Other analysis approaches would tend to pick one or another factor as the single most important factor. Systems variability analysis shows that several factors could be "most" important, taking into account parameter uncertainties.

5. ASSESSMENT OF URANIUM MILL TAILINGS

The National Uranium Tailings Program (NUTP) was established by the Canada Centre for Mineral and Energy Technology, and ran from 1981 to 1986. The NUTP had three major objectives: to gather data, to investigate alternative disposal options, and to develop assessment models and code. This last objective was met with the creation of three versions of the Uranium Tailings Assessment Program (UTAP). Each of these computer codes implement the systems variability analysis approach.

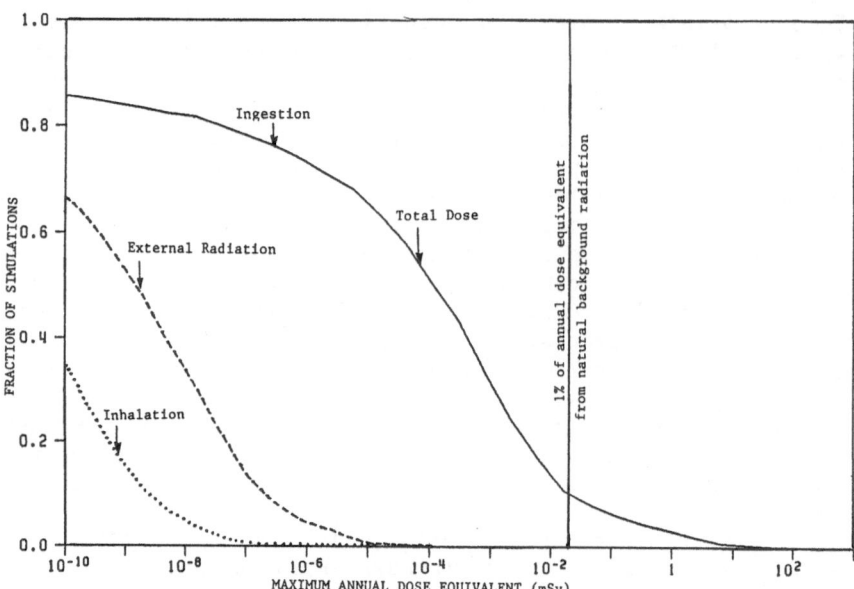

Figure 5. Distributions of maximum dose for times up to 10^7 years, showing curves for total dose and dose due to ingestion, inhalation and external exposure. Ingestion pathways are much more important than the other two.

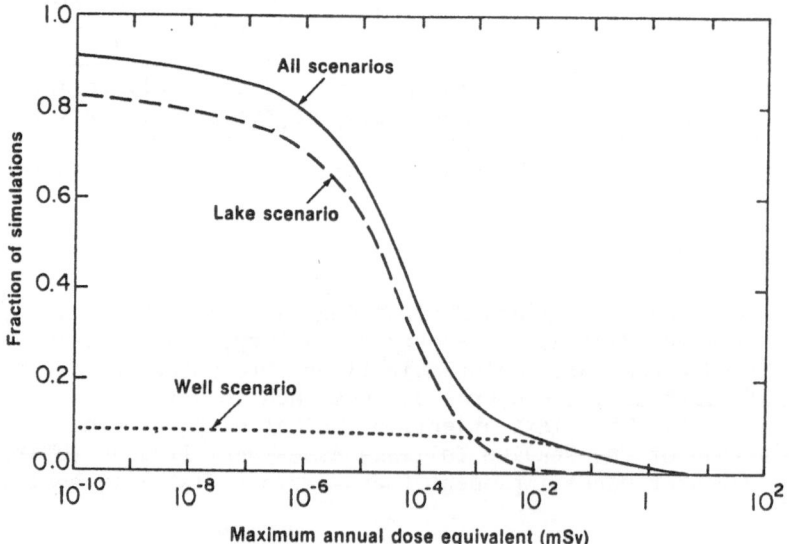

Figure 6. Distributions of maximum dose for times up to 10^7 years, showing curves for total dose ("all scenarios") and for those simulations where the source of drinking water was a lake ("lake scenario") or a well ("well scenario"). The highest observed doses are associated with the use of well water.

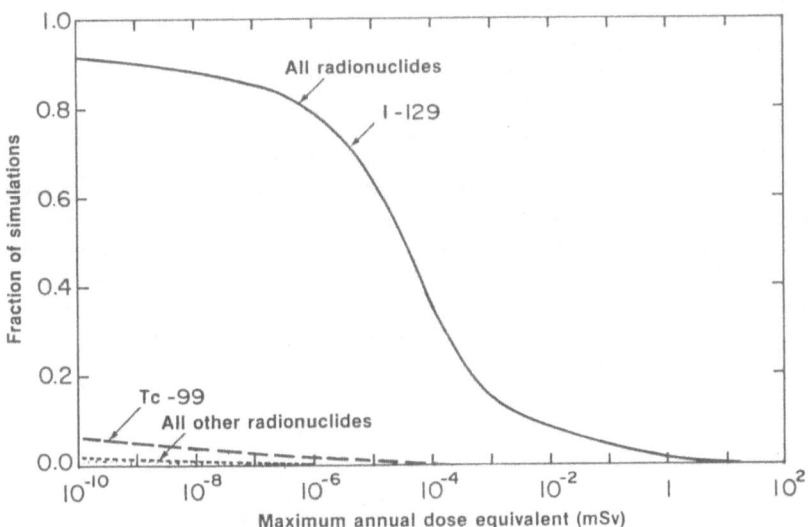

Figure 7. Distributions of maximum dose for times up to 10^7 years, showing total dose ("all radionuclides") and contributions from individual radionuclides. Iodine-129 clearly dominates the total dose curve.

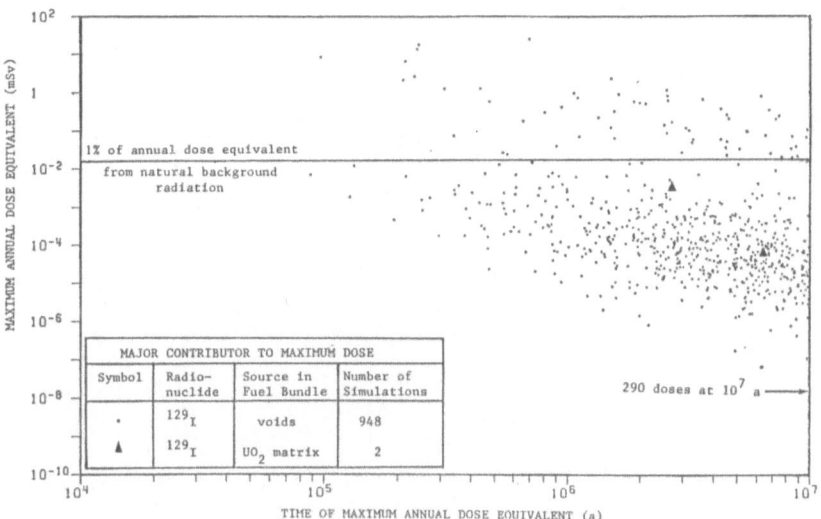

Figure 8. Scatterplot of maximum dose versus time of occurrence for each simulation. The major contributor to the maximum dose is indicated using different symbols. Iodine-129 was the only major contributor in all simulations with maximum doses exceeding 10^{-10} mSv/yr.

The last version, UTAP3 (SENES, 1986a), has been applied in three studies of different uranium tailings sites (SENES, 1986a, 1987; Beak, 1987). The following comments describe the application of UTAP3 to the hypothetical reference site illustrated in Figure 9 (SENES, 1986b). The tailings are located in a natural depression, behind a slightly pervious dam. The site occupies 50 ha and has a total volume of 5 x 10^6 m^3. The tailings consist of residues from the acid leaching of low-grade pyritic uranium ore from which most of the uranium has been removed. Figure 10 shows some details of the processes occurring within the tailings: there is a complex chemical environment with reactions such as oxidation of pyrite to sulphates and the formation and dissolution of sulphate compounds. More details are given by SENES (1985; 1986a; 1986b; 1987) and references therein.

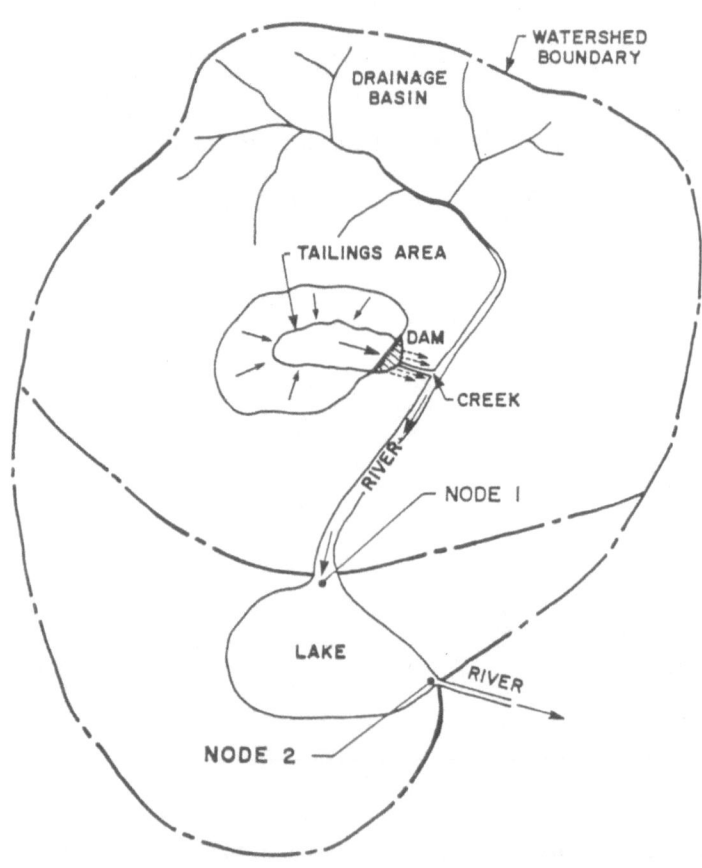

Figure 9. Hypothetical uranium tailings disposal site, showing the tailings site and part of the nearby downstream watershed.

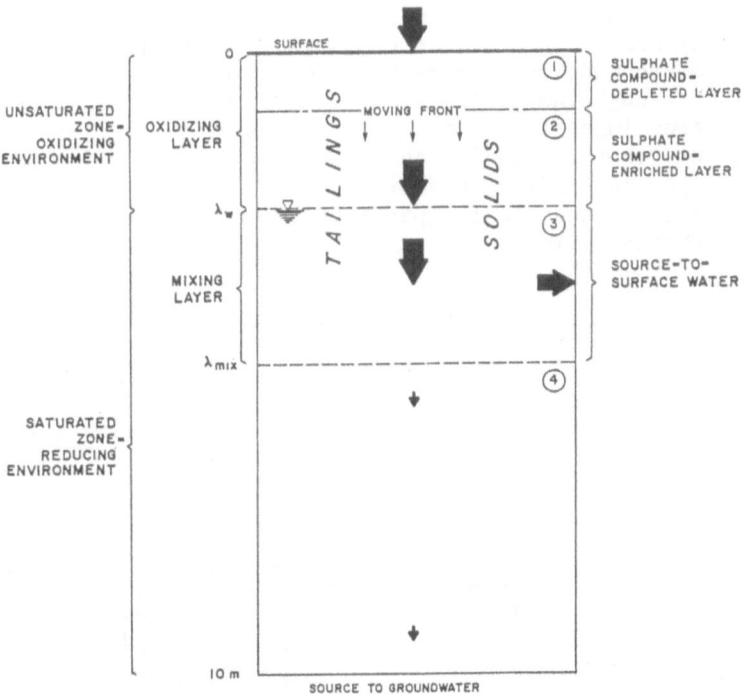

Figure 10. Conceptual model of the uranium tailings source term. Radon gas and radium in dust is assumed to enter the air compartment. Precipitation moving through the tailings pile carries uranium, thorium, radium and lead isotopes directly to surface waters, or indirectly to surface waters through a groundwater flow system. Nonradioactive species such as calcium and sulphate ions are also transported to surface waters, affecting pH.

Figure 11 is an overview of the system model developed for UTAP3 (SENES, 1986a). The code was designed to model the tailings source, and provide estimated releases of contaminants to air, surface water and groundwater compartments. The air compartment describes transport over a circular area with a radius of 50 km from the tailings. The two water compartments model an entire watershed, including groundwater flows passing through the tailings. The major output from UTAP3 is

- dose from ingestion of and external exposure to isotopes of uranium, thorium, radium and lead, and dose from inhalation of radon gas and radium in dust; and
- estimated changes to the pH of bodies of water downstream from the tailings site.

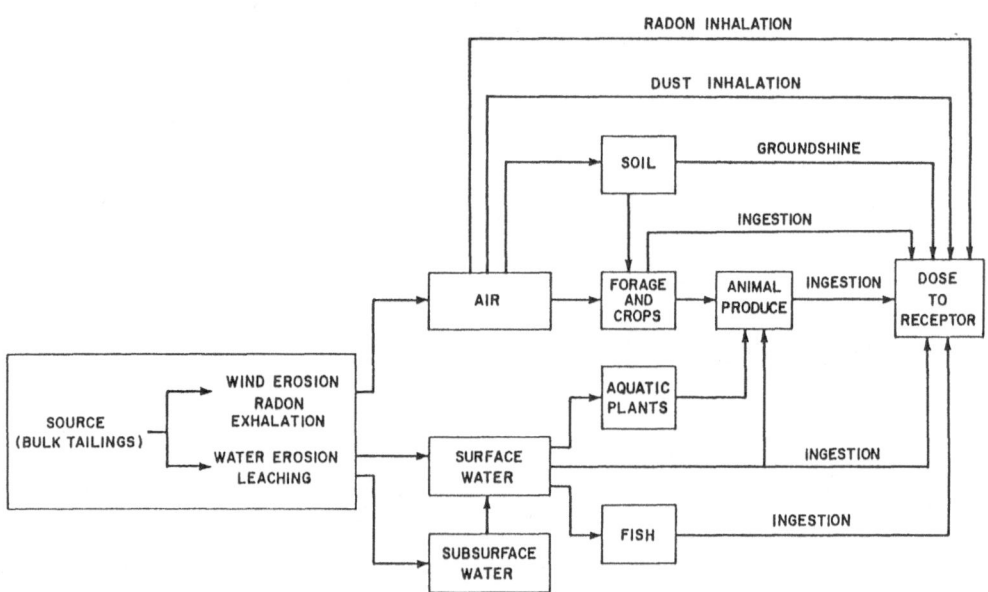

Figure 11. Simplified diagram of the system model contained in UTAP3. Dose to a receptor can occur through inhalation of radon gas and radium in dust, external exposure (groundshine) and ingestion of various foods.

Figure 12 shows a histogram of the frequency versus dose for the "base case" disposal option. (SENES (1986a) also assesses an alternative disposal option characterized by lower pyrite concentrations and added calcite.) The plot is for 200 simulations at time = 250 years, and the estimated doses range from about 6 to 160 µSv/yr. The corresponding histogram for time = 0 years displays a much larger estimated dose range, probably because the tailings source contributes significantly to uncertainty. This uncertainty decreases substantially with time, since many of the complex chemical reactions in the tailings are complete after about 150 to 200 years. Histogram plots at 0 and 250 years show that the distributions of estimated doses are highly skewed.

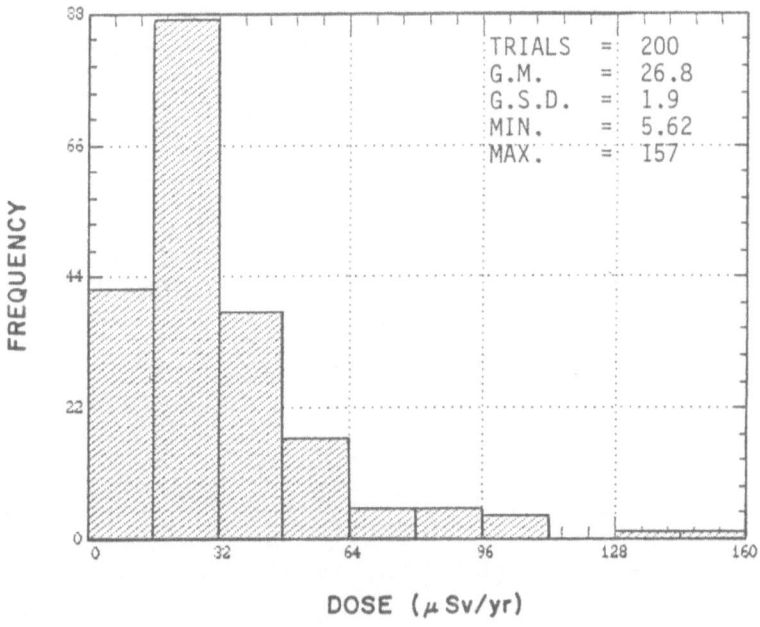

DOSE (μ Sv/yr)

Figure 12. Histograms of the total dose predicted by UTAP3, for 200 simulations at 250 years. The geometric mean, geometric standard deviation, minimum and maximum doses are 26.8, 1.9, and 5.62 and 157 µSv/yr, respectively.

Figure 13 presents similar results for dose at 500 years versus upward cumulative frequency. In this case the curve shows the frequency of simulations giving doses less than specified levels. Figure 13 also includes upper and lower 95% confidence bands, which enclose the "true" curve that would have been obtained if a very large number of simulations had been run.

Figure 14 summarizes an analysis of the doses from different pathways as a function of time. The dose quantities plotted are the geometric mean doses from 200 simulations. The total dose curve is dominated by dose due to ingestion of drinking water, and dose due to ingestion of fish. Both of these pathways are in the surface water compartment. Doses due to inhalation ("radon dose" and "inhalation dose") are relatively unimportant. The figure does not show the curve for doses due to external exposure (groundshine), which is orders of magnitude lower than other doses (SENES, 1986a).

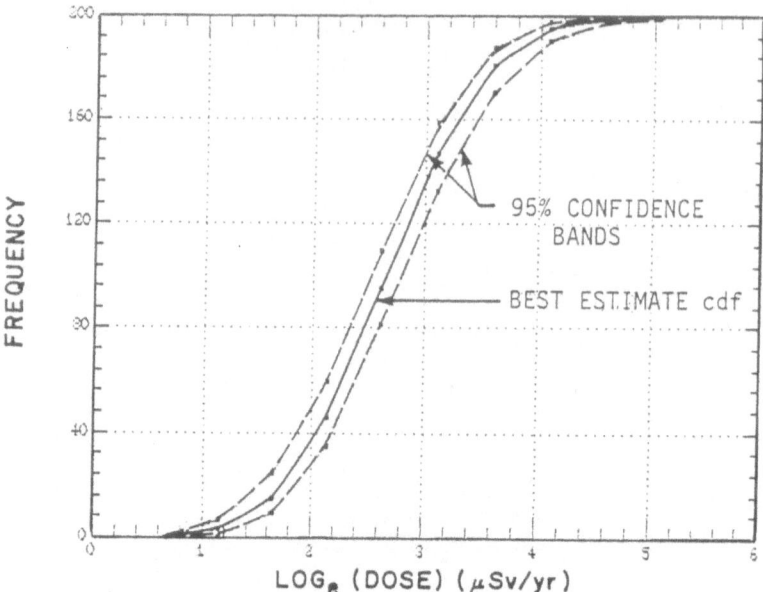

Figure 13. Upward cumulative frequency plot of the total dose at 500 years, for 200 simulations. The upper and lower 95% confidence bands (SENES, 1986a) are also shown.

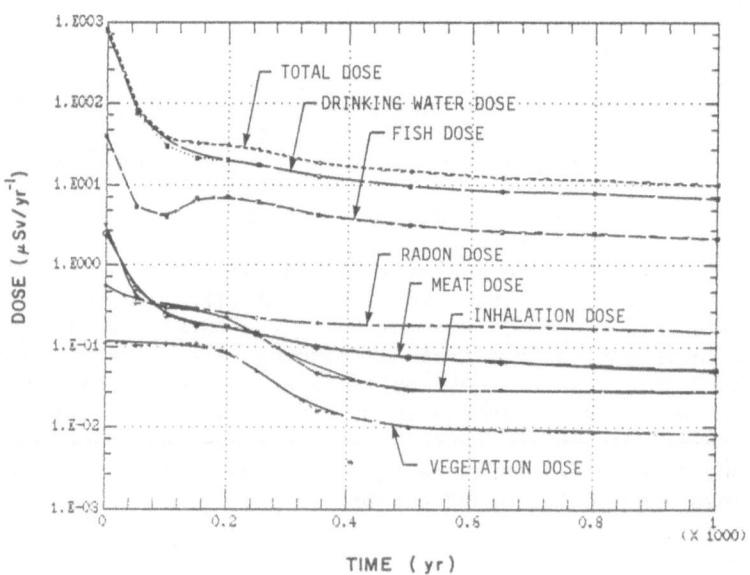

Figure 14. Plots of geometric mean dose versus time, showing total dose and dose due to different pathways. The dose due to ingestion of drinking water dominates the total dose curve.

Another analysis summary is given in Figure 15. This figure identifies those elements contributing to the dose that are due to ingestion of drinking water. The early contributors are mostly isotopes of lead and thorium, changing to radium-226 after about 150 years. Much of the time dependence of these curves can be explained in terms of the chemistry of the tailings pile.

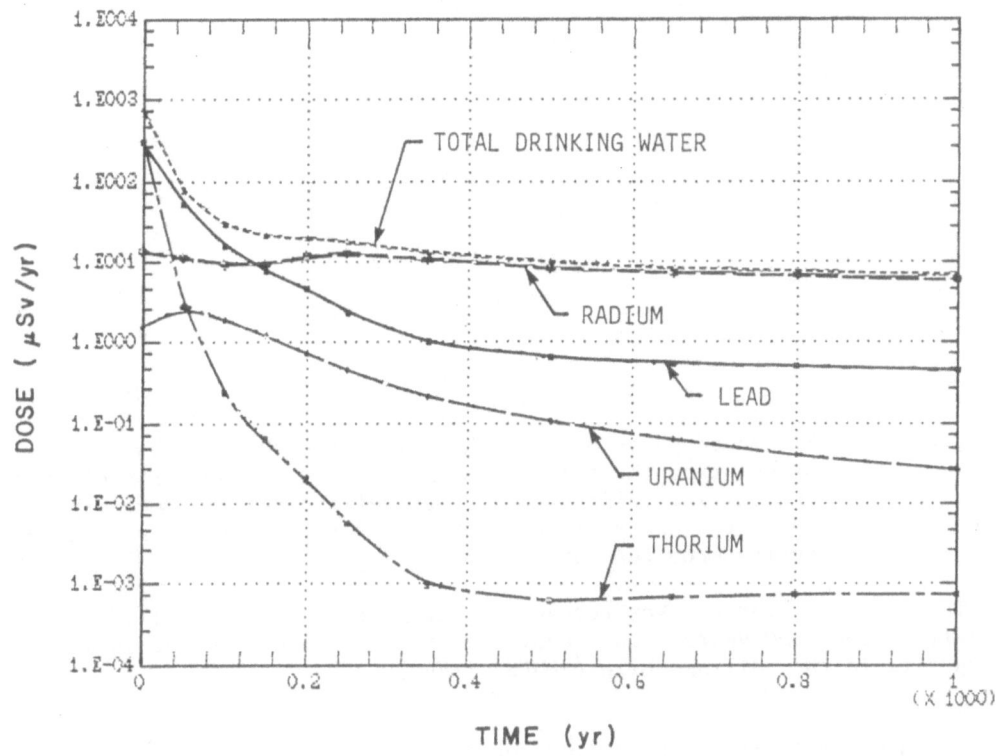

Figure 15. Major contributors to the dose due to ingestion of drinking water. The pattern of results can be explained in terms of the chemical reactions occurring in the tailings source term.

As noted previously, UTAP3 was also developed to estimate changes in pH in the watershed. Figure 16 shows the dependence of pH on time for three lakes immediately downstream from the tailings pile. Lake 3 is least affected because it is farthest from the tailings and any perturbations are buffered by processes such as dilution.

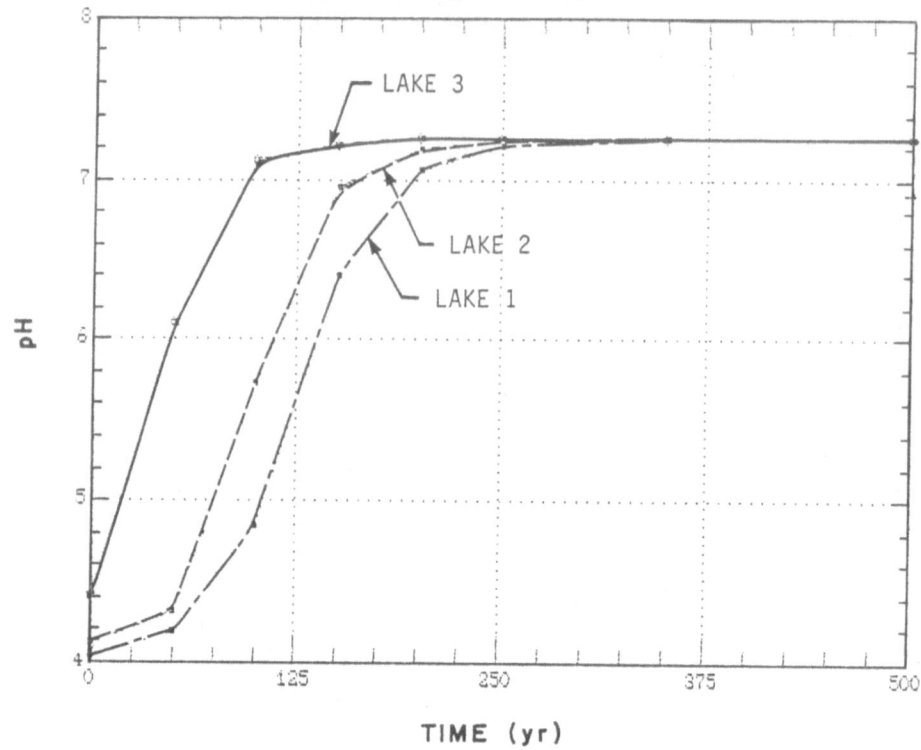

TIME (yr)

Figure 16. Estimated pH as a function of time for three lakes downstream from the tailings pile. Lake 3 is the farthest from
the tailings pile, and is least affected. The recovery of
pH to the unperturbed value of 7.2 parallels the removal of
pyrite from the tailings pile.

The pH of all three lakes recover to their original pH level of 7.2
after about 250 years. The recovery period parallels the depletion of
pyrite in the tailings pile (SENES, 1986a).

6. OTHER POTENTIAL APPLICATIONS OF SVA

It is worthwhile to mention the potential scope of SVA in a wider
context. The SVA approach has been developed to deal with complex
environmental systems and to account for variability and uncertainty.
The latter characteristic is unique to SVA. As such, SVA will find use
in assessments that must deal with complicating factors such as long
time scales and/or unavailability of models and data.
    Some potential applications include assessments of
    - Other concepts for the disposal of nuclear wastes. SVA is
      well-suited to deal with concepts such as shallow land burial
      for low- and intermediate-level wastes.

- Facilities for the disposal of chemical wastes. These types of waste may be more persistent (and are certainly more common) than many radioactive wastes. As a consequence, an assessment of these facilities should deal with long time scales as well as uncertainty and variability.
- Mine and mill tailings. The visible wastes produced by industrial societies are often much smaller than "invisible" wastes such as large mine tailings piles located in remote areas. Changes in future land use may require an assessment of such wastes.

To generalize further, we can find application of the SVA approach in many other types of modelling and simulations studies. The strength of the SVA approach lies with its ability to deal with uncertainty and variability. Consequently, SVA could be useful in studies of the effects of acid rain or the effects of rising levels of atmospheric CO_2 (the greenhouse effect). When used in these studies, SVA would help quantify the uncertainties, and thereby help improve our understanding of the potential consequences of complex processes.

References

Beak Consultants (1987). Application of the National Uranium Tailings Program probabilistic system model (UTAP) to the Rabbit Lake uranium tailings site. Research report prepared for the National Uranium Tailing Program, Canada Centre for Mineral and Energy Technology, Energy, Mines and Resources, Canada; CANMET (DSS) serial number 23317-6-1729/01-SQ.

Goodwin, B.W., Andres T.H., Davis, P.A., LeNeveu, D.M., Melnyk, T.W., Sherman, G.R. and Wuschke, D.M. (1987a). Post-closure environmental assessment for the Canadian nuclear fuel waste management program. Radioact. Waste Manag. Nuc. Fuel Cycle 8, 241.

Goodwin, B.W., Garisto, N.C. and Barnard, J.W. (1987b). A probabilistic systems assessment of the long-term impacts of chemically toxic contaminants from the disposal of nuclear fuel waste. Atomic Energy of Canada Limited Report, AECL-8367.

Heinrich, W.F. (1984). Geosphere submodel for the second interim assessment of the Canadian concept for nuclear fuel waste disposal: post-closure phase. Atomic Energy of Canada Limited Record, TR-286*.

LeNeveu, D.M. (1986). Vault submodel for the second interim assessment of the Canadian concept for nuclear fuel waste disposal: post-closure phase. Atomic Energy of Canada Limited Report, AECL-8383.

Mehta, K.K. (1985). Biosphere submodel for the second interim assessment of the Canadian concept for nuclear fuel waste disposal: post-closure phase. Atomic Energy of Canada Limited Record, TR-298*.

Robinson, P.C. and Hodgkinson, D.P. (1987). Exact solutions for radionuclide transport in the presence of parameter uncertainty. Radioact. Waste Manage. Nucl. Fuel Cycle 8(4).

Saltelli, A., Sartori, E., Andres, T.H., Goodwin, B.W. and Carlyle, S.G. (1987). PSACOIN level 0 intercomparison. NEA/OECD, Paris.

SENES (1985). Probabilistic model development for the assessment of the long-term effects of uranium mill tailings in Canada - Phase II. Volume 1 - Main report. Volume 2 - Workshop notes and topical discussion. Volume 3 - Probabilistic analysis user's manual and documentation for the Uranium Tailings Assessment Program (UTAP). Research report prepared for the National Uranium Tailing Program, Canada Centre for Mineral and Energy Technology, Energy, Mines and Resources, Canada.

SENES (1986a). Probabilistic model development for the assessment of the long-term effects of uranium mill tailings in Canada - Phase III. Volume 1 - Summary report. Volume 2 - System Description and Control Module Documentation. Volume 3 - Component Model Documentation. Volume 4 - Uranium Tailings Assessment Program (UTAP) User's Manual. Research report prepared for the National Uranium Tailing Program, Canada Centre for Mineral and Energy Technology, Energy, Mines and Resources, Canada; contract number 0SQ85-00182.

SENES (1986b). Risk assessment for Canadian uranium mill tailings. Research report prepared for the National Uranium Tailing Program, Canada Centre for Mineral and Energy Technology, Energy, Mines and Resources, Canada; CANMET (DSS) serial number 0SQ85-00225.

SENES (1987). Probabilistic system model development and application to a real site - Phase IV. Research report prepared for the National Uranium Tailing Program, Canada Centre for Mineral and Energy Technology, Energy, Mines and Resources, Canada; CANMET (DSS) serial number 23317-6-1728/01-SQ.

Sherman, G.R., Donahue, D.C., King, S.G. and So, A. (1986). The systems variability analysis code SYVAC2 and submodels for the second interim assessment of the Canadian concept for nuclear fuel waste disposal. Atomic Energy of Canada Limited Technical Record, TR-317*.

Wuschke, D.M., Gillespie, P.A., Mehta, K.K., Heinrich, W.F., LeNeveu, D.M., Guvanasen, V.M., Sherman, G.R., Donahue, D.C., Goodwin, B.W., Andres, T.H. and Lyon, R.B. (1985). Second interim assessment of the Canadian concept for nuclear fuel waste disposal. Volume 4: Post-closure assessment. Atomic Energy of Canada Limited Report, AECL-8373-4.

Bibliography

Note: The following compilation attempts to list all recent reports dealing with applications of SVA. These reports are not referenced directly in the text.

AECB (1987). Regulatory objectives, requirements and guidelines for the disposal of radioactive wastes - long-term aspects. Atomic Energy Control Board (of Canada) Regulatory Document R-104, P.O. Box 1046, Ottawa, ON K1P 5S9.

Apted, M.J., Liebetrau, A.M. and Engel, D.W. (1987). Spent fuel as a waste form: analysis with AREST performance assessment code. In proceedings of the '87 Waste Management Symposium, Tucson, AZ, March 1-5, 1987; Vol. 2, High-Level Wastes, R.G. Post (ed.), University of Arizona, Tucson, AZ.

Apted, M.J., Liebetrau, A.M., Engel, D.W. and Alexander, D.H. (1986). Analysis of spent fuel performance in a geologic repository. In proceedings of the 88th Annual Meeting of the American Ceramic Society, Chicago, IL, 1986 April 27 to May 1. PNL-SA--13928; CONF-860418--9.

Avogadro, A., Murray C.N. and De Plano A. (1980). Transport through deep aquifers of transuranic nuclides leached from vitrified high-level wastes. C.J.M. Northrup, Jr. (ed.), In Scientific Basis for Nuclear Waste Management, Vol. 2, Plenum Press, NY, pp. 665-671.

Avogadro, A., Murray, C.N., De Plano, A. and Bidoglio, G. (1981). Underground migration of long-lived radionuclides leached from a borosilicate gass matrix. In IAEA/CEC/NEA International Symposium on Migration in the Terrestrial Environment of Long-Lived Radionuclides from the Nuclear Fuel Cycle, Knoxville, TN, STI/PUB/597, pp. 527-540.

Beak Consultants (1987). Application of the National Uranium Tailings Program probabilistic system model (UTAP) to the Rabbit Lake uranium tailings site. Research report prepared for the National Uranium Tailing Program, Canada Centre for Mineral and Energy Technology, Energy, Mines and Resources, Canada; CANMET (DSS) serial number 23317-6-1729/01-SQ.

Bergstroem, U., Andersson K. and Roejder, B. (1984). Variability of dose predictions for cesium-137 and radium-226 using the PRISM method. Studsvik Energiteknik AB, Sweden, STUDSVIK NW-84/656.

Bertozzi, G. and D'Alessandro, M. (1981). Radioactive waste disposal into a plastic clay formation: a site specific exercise of consequence analysis. Tech. Note N. 1.07.03.81.66.

Bertozzi, G. and D'Alessandro, M. (1982). A probabilistic approach to the assessment of the long-term risk linked to the disposal of radioactive waste in geological repositories. Radioact. Waste Manage. Nucl. Fuel Cycle 3, 117.

Bertozzi, G., D'Alessandro, M., Girardi, F., Vanossi, M. (1977). Evaluation of the safety assessment of radioactive waste disposal into geological formations: preliminary applications of fault tree analysis to salt formations. In Risk Analysis and Geologic Modelling in Relation to the Disposal of Radioactive Wastes into Geological Formations. Proceedings of ISPRA, Italy, 1977 May 23-27; OCDE/NEA, Paris, pp. 13-33.

Bertozzi, G., D'Alessandro, M., Saltelli, A. and Bonne, A. (1984). A site-specific probabilistic risk evaluation exercise for radioactive waste disposal in a continental clay formation. Radioactive Waste Management, Vol. 4, International Atomic Energy Agency, Vienna, IAEA-CN-43/53, STI/PUB/649, pp. 289-301.

Bertozzi, G., D'Alessandro, M. and Saltelli, A. (1984). Waste disposal into a plastic clay formation: risk analysis for probabilistic scenario. EUR 9641 EN.

Bertozzi, G., D'Alessandro, M. and Saltelli, A. (1985). Waste disposal into a plastic clay formation: risk analysis for probabilistic scenarios. European Applied Research Reports - Nuclear Science and Technology 6 (4), 869-947.

Bonano, E.J. and Cranwell, R.M. (1988). Treatment of uncertainties in the performance assessment of geologic high-level radioactive waste repositories. Journal of Mathematical Geology (to be published).

Bonano, E.J. and Davis, P.A. (1988). Demonstration of a performance assessment methodology for nuclear waste isolation in basalt formations. Sandia Laboratories, Albuquerque, NM, SAND-86-2125C; CONF-880201-6.

Bonano, E.J., Davis, P.A., Cranwell, R.M., Shipers, L.R., Brinster, K.F., Beyeler, W.E., Updegraff, C.D., Shepherd, E.R., Tilton, L.M. and Wahi, K.K. (1987). Demonstration of a performance assessment methodology for high-level waste disposal in basalt formations. NUREG/CR-4759; SAND86-2325 (Draft).

Bonne, A. and Marivoet, J. (1985). Safety analysis of a HLW repository in a clay formation. In proceedings of the ANS meeting on High-Level Nuclear Waste Disposal Technology and Engineering, Pasco, WA, 1985 September 24-26, Battelle Press, Columbus, OH, pp. 629-636.

Cadelli, N., Girardi, F. and Orlowski, S. (1986). PAGIS, Common European methodology for repository performance analysis. Radioactive Waste Management and Disposal. In proceedings of the 2nd European Community Conference, 1986, EUR-10163, pp. 628-638.

Cranwell, R.M. and Bonano, E.J. (1987). Sources/treatment of uncertainties in the performance assessment of geologic radioactive waste repositories. In proceedings of an NEA workshop on Uncertainty Analysis for Performance Assessments of Radioactive Waste Disposal Systems, Seattle, WA, 1987 February 24-26; OECD/NEA, Paris. pp. 53-68.

Cranwell, R.M. and Helton, J.C. (1982a). Application of a risk-assessment methodology to a hypothetical high-level waste repository in bedded salt. SAND-81-2169C; CONF-820303-28, Sandia Laboratories, Albuquerque, NM.

Cranwell, R.M. and Helton, J.C. (1982b). Uncertainty analysis for geologic disposal of radioactive waste. In Proceedings of the Symposium on Uncertainties Associated with the Regulation of the Geologic Disposal of High-Level Radioactive Waste. D.C. Kocher, (ed.), NUREG/CP-0022; CONF-810372, Oak Ridge National Laboratory, Oak Ridge, TN.

Cranwell, R.M., Campbell, J.E., Helton, J.C, Iman, R.L., Longsine, D.E., Ortiz, N.R., Runkle, G.E. and Shortencarier, M.J. (1987). Risk methodology for geologic disposal of radioactive waste: final report. NUREG/CR-2452; SAND81-2573, Sandia Laboratories, Albuquerque, NM.

Crick, J.J., Hill, M.D. and Charles, D. (1987). The role of sensitivity analysis in assessing uncertainty. In proceedings of an NEA workshop on Uncertainty Analysis for Performance Assessments of Radioactive Waste Disposal Systems, Seattle, WA, 1987 February 24-26; OECD/NEA, Paris, pp. 116-130.

D'Alessandro, M. and Bonne, A. (1980). Radioactive waste disposal
into a plastic clay formation: probabilistic assessment of
geological containment. C.J.M. Northrup, Jr. (ed.), Scientific
Basis for Nuclear Waste Management, Vol. 2, Plenum Press, NY,
pp. 711-720.

D'Alessandro, M. and Bonne, A. (1982). Fault-tree analysis for
probabilistic assessment of radioactive-waste segregation: an
application to a plastic clay formation at a specific site. In
Predictive Geology with Emphasis on Nuclear-Waste Disposal,
G. de Marsily and D.F. Merriam (eds.), Pergamon Press, Oxford,
p. 45.

D'Alessandro, M. and Bonne, A. (1983). Radioactive waste disposal
within a plastic clay formation: preliminary investigation on
glacial action as a cause of containment failure and radioactive
waste dispersion. Tech. Note EUR 8709 EN.

D'Alessandro, M., Girardi, F. and Bertozzi, G. (1981). Geologic
segregation of radioactive wastes: probabilistic analysis of
scenari of 'far field' type for containment breakings. In
proceedings of the Workshop on Radionuclide Release Scenarios for
Geologic Repositories; Paris, 1980 8-12 September. OECD/NEA, Paris,
ISBN92-64-02172-8, p. 240.

D'Alessandro, M., Murray, C.N., Bertozzi, G. and Girardi, F. (1980).
Probability analysis of geological processes: a useful tool for the
safety assessment of radioactive waste disposal. Radioactive Waste
Manage. Nucl. Fuel Cycle 1, 25.

Dames and Moore International Department of the Environment, London
(UK) (1988). Evaluation of environmental change and its effects on
the radiological performance of a hypothetical shallow engineered
disposal facility at Elstow, Bedfordshire. Department of the
Environment, London (UK) Report, DOE/RW-87.124; PECD-7/9/; TR-D/M-9.

De Marsily, G., Behrendt, V., Esminger, D.A., Flebus, C., Hutchinson,
B.L., Kane, P., Karpf, A., Klett, R.D., Mobbs, S., Poulin, M.,
Stanners, D.A. and Wuschke, D. (1987). Radiological assessment of
the consequences of disposal of high-level radioactive waste in
subseabed sediments. Sandia Laboratories, Livermore, CA, SAND-
87-2598C; CONF-871101-66.

De Marsily, G., Broyd, T.W., Come, B., Hodgkinson, D., Put, M. and
Saltelli, A. and Simon, R. (1985). Modelling of radionuclide
release and migration. In Radioactive Waste Management and
Disposal, Cambridge University Press, NY, pp. 652-673.

De Marsily, G., Hill, M.D., Mobbs, S.F., Webb, G.A.M., Murray, C.N.,
Talbert, D.M. and van Dorp, F. (1982). Application of System
Analysis to the Disposal of High-Level Waste in Deep Ocean
Sediments. Radio. Waste Manage. Nucl. Fuel Cycle 3(2), 199-213.

Eslinger, P.W. and Sagar, B. (1987). Incorporation of uncertainty
analysis in assessment of the basalt site for high-level radioactive
waste disposal: statistical aspects. In proceedings of an NEA
Workshop on Uncertainty Analysis for Performance Assessments of
Radioactive Waste Disposal Systems, Seattle, WA, 1987 February
24-26; OECD/NEA, Paris, pp. 221-236.

ERL (1985). Handling uncertainty in environmental impact assessment. Report (volume 18) prepared for the Ministry of Public Housing, Physical Planning and Environmental Protection by Environmental Resources Limited, 79 Baker Street, London.

Flebus, C. and Stanners, D.A. (1987). Coupling feasibility study of the Mark-A ocean box model and the probabilistic code LISA. In Study of the Feasibility and Safety of the Disposal of Heat Generating Wastes into Deep Oceanic Geological Formations. Part 1: Radiological Safety Assessment, C.N. Murray (ed.), JRC Report Series N. SP-1.87.31, Commission of the European Communities, p. 42.

Flebus, C. and Stanners, D.A. (1988). A stochastic assessment using LISA. In Study of the Feasibility and Safety of the Disposal of Heat Generating Wastes into Deep Oceanic Geological Formations. Part 1: Radiological Safety Assessment, C.N. Murray (ed.), JRC Report Series, EUR 11754 EN, Commission of the European Communities, p. 159.

Frizelle, C.J.G., Mutton, J.E., Upton, P.S. and Thompson, B.G.J. (1986). Modelling the long term evolution of radioactive waste disposal environments. In International Conference on Radioactive Waste Management, Canadian Nuclear Society, Toronto, pp. 508-516.

Gardner, R.H. (1984). A unified approach to sensitivity and uncertainty analysis. In Applied Simulation and Modelling, M.H. Hamza (ed.), Proceedings of the IASTED International Symposium, San Francisco, CA, 1984 June 4-6. ACTA Press, Anaheim, CA, p. 155.

Gardner, R.H., Trabalka, J.R. and Emanuel, W.R. (1984). Methods of uncertainty analysis for a global carbon dioxide model. Environmental Sciences Division, ORNL (draft).

Girardi F. (1983). The function of waste immobilization as a barrier against radionuclide release in the frame of an overall risk assessment. In the Intern. Symp. on the Conditioning of Radioactive Wastes for Storage and Disposal, IAEA, 1982 June 21-25, Utrecht (The Netherlands) ORA 30439; IAEA-SM-261/44, pp. 57-67.

Girardi, F., Avogadro, A., Bertozzi, G., D'Alessandro, M., Lanza, F., and Murray, C.N. (1980). A risk analysis methodology for deep underground radioactive waste repositories and related experimental research. In Underground Disposal of Radioactive Wastes, Vol. 2, IAEA-SM-243/161, p. 407.

Girardi, F., Bertozzi, G., D'Alessandro, M. (1978). Long-term risk assessment of radioactive waste disposal in geological formations. Trans. Am. Nucl. Soc. 26, 275. Report EUR 5902.

Girardi, F., Orlowski, S. and Myttenaere, C. (1983). PAGIS; An European performance assessment of geological isolation systems. Int. Conf. on Radioactive Waste Management, Seattle, WA, 1983 May 16-20, IAEA-CN-43/123.

Girardi F. and Sabbioni, E. (1982). Long-term health impacts of energy production. Preliminary assessment of environmental pollution by heavy metals and long-lived radionuclides for identification of research priorities. In proceedings of International Symposium on Health Impact of Different Sources of Energy, Nashville, TN, 1981 June. Health Impacts of Different Sources of Energy, IAEA-SM-254/56, pp. 429-430.

Goodwin, B.W. (1985). The SYVAC approach for long-term environmental assessments. In proceedings of the Symposium on Groundwater Flow and Transport Modelling for Performance Assessment of Deep Geological Disposal of Radioactive Waste and Critical Evaluation of the State of the Art, Albuquerque, NM, 1985 May 20-21.

Goodwin, B.W. (1987). Application of SYVAC to the Canadian concept for nuclear fuel waste disposal. In proceedings of an NEA workshop on Uncertainty Analysis for Performance Assessments of Radioactive Waste Disposal Systems, Seattle, WA, 1987 February 24-26; OECD/NEA, Paris, pp. 149-166.

Goodwin, B.W. and Andres, T.H. (1986). Comparison of the probabilistic systems assessment codes, UTAP and SYVAC3, applied to the long-term assessment of uranium mill tailings in Canada. Research report prepared for the National Uranium Tailing Program, Canada Centre for Mineral and Energy Technology, Energy, Mines and Resources, Canada; contract number OSQ85-00293.

Goodwin, B.W., Andres, T.H., Davis, P.A., LeNeveu, D.M., Melnyk, T.W., Sherman, G.R. and Wuschke, D.M. (1987a). Post-closure environmental assessment for the Canadian nuclear fuel waste management program. Radioact. Waste Manag. Nuc. Fuel Cycle 8, 241.

Goodwin, B.W., Garisto, N.C. and Barnard, J.W. (1987b). A probabilistic systems assessment of the long-term impacts of chemically toxic contaminants from the disposal of nuclear fuel waste. Atomic Energy of Canada Limited Report, AECL-8367.

Goodwin, B.W., Andres, T.H., Lemire, R.J. and Ho, P. (1987). Comparison of the National Uranium Tailings Program's probabilistic code UTAP2 with Atomic Energy of Canada Limited's probabilistic code SYVAC3-UT2. Research report prepared for the National Uranium Tailing Program, Canada Centre for Mineral and Energy Technology, Energy, Mines and Resources, Canada; contract number 15SQ.23317-6-1731.

Gralewski, Z.A., Kane, P. and Nicholls, D.B. (1987). Development of a methodology for post closure radiological risk analysis of underground waste repositories. Illustrative assessment of the Harwell site. Department of the Environment, London (UK) Report, DOE/RW-87.028

Gralewski, Z.A., Kane, P. and Nicholls, D.B. (1987). Dry run 2: development of a methodology for post-closure radiological risk analysis of underground waste repositories. Illustrative assessment of the Harwell site. Vols. 1 and 2. Department of the Environment, London (UK) Report, DOE/RW-87.034-035.

Guvanasen, V.M. (1985). Preliminary assessment for the disposal of intermediate-level reactor waste. Atomic Energy of Canada Limited Technical Record, TR-293*.

Guvanasen, V.M. (1987). Seabed disposal of fuel recycle waste: Preliminary assessments of a reference case and four accident scenarios. Atomic Energy of Canada Limited Record, TR-367*.

Hall, P.A. and Thompson, B.G.J. (1985). Case studies, using SYVAC 'A', of the performance of a LAND3/4 repository in a clay/shale site. DOE SYVAC Technical Report, TR-DOE-2.

Heinrich, W.F. (1984). Geosphere submodel for the second interim assessment of the Canadian concept for nuclear fuel waste disposal: post-closure phase. Atomic Energy of Canada Limited Record, TR-286*.

Hofer, E. and Hoffman, F.O. (1987). Selected examples of practical approaches for the assessment of model reliability - parameter uncertainty analysis. In proceedings of an NEA Workshop on Uncertainty Analysis for Performance Assessments of Radioactive Waste Disposal Systems, Seattle, WA, 1987 February 24-26; OECD/NEA, Paris, pp. 132-148.

Hoffman, F.O. and Baes III, C.F. (eds.) (1979). A statistical analysis of selected parameters for predicting food chain transport and internal dose of radionuclides. Oak Ridge National Laboratories Report, ORNL/NUREG/TM-282; NUREG/CR-1004.

Hoffman, F.O. and Gardner, R.H. (1983). Evaluation of uncertainties in radiological assessment models. In Radiological Assessment: a Textbook on Environmental Dose Analysis, J.E. Till and H.R. Meyers (eds.), US Nuclear Regulatory Commission, Washington, DC, NUREG/CR-3332; ORNL-5968.

Hoffman, F.O., Gardner, R.H. and Eckerman, K.F. (1982). Variability in dose estimates associated with the food chain transport and ingestion of selected radionuclides. Oak Ridge National Laboratories Report, ORNL/TM-8099; NUREG/CR-2612.

Hoffman, F.O., Miller, C.W. and Ng, Y.C. (1984). Uncertainties in radioecological assessment models. In proceedings of the Seminar on Environmental Transfer to Man to Radionuclides Released from Nuclear Installations. Commission of the European Communities, 1983 October 17-21, Brussels; Vol. 1, p. 331, IAEA-SR-84/4; CONF-831032-1.

Holmes, R.W. (1987). Uncertainty analysis for the long-term assessment of uranium mill tailings. In proceedings of an NEA Workshop on Uncertainty Analysis for Performance Assessments of Radioactive Waste Disposal Systems, Seattle, WA, 1987 February 24-26; OECD/NEA, Paris, pp. 167-190.

Iman, R.L. (1985). A FORTRAN 77 program and user's guide for the calculation of partial regression coefficients; NUREG/CR-4122.

Iman, R.L. and Conover, W.J. (1980). Small sample sensitivity analysis techniques for computer models, with an application to risk assessment. Communications in Statistics A9(17), 1749.

Iman, R.L. and Shortencarier, M.J. (1984). A FORTRAN 77 program and user's guide for the generation of Latin hypercube and random samples for use with computer models. NUREG/CR-3624; SAND83-2365 RG.

Jarvis, R.G., Adam, R.Y., Bretzlaff, C.I., Laurens, J.M. and Wilkinson, S.R. (1986). The COSMOS-S/D Assessment Code Complex for a SLB Repository at CRNL. Atomic Energy of Canada Limited Report, AECL-9350.

Johnston, P.D. and Thompson, B.G.J. (1987). The aims and uses of uncertainty analysis in the context of development and regulation of underground disposal of radioactive waste. In proceedings of the Workshop on Uncertainty Analysis for Performance Assessments of

Radioactive Waste Disposal Systems, Seattle, WA, 1987
February 24-26; OECD/NEA, Paris, pp. 44-52.

Kane, P. (1987). Radiological assessment of the disposal of high
level radioactive waste on or within the sediments of the deep ocean
bed. Department of the Environment, London (UK)
Report, DOE/RW-87.100.

Kane, P., Karpf, A., Klett, R.D., Mobbs, S., Poulin, M. and
Stanners, D.A. (1988). Subseabed disposal of high level radioactive
waste. Volume 2, Radiological Assessment. OECD/NEA, Paris (to be
published).

Karlberg, O. (1987). A demonstration sensitivity/uncertainty analysis
of a reactor accident consequence calculation. In proceedings of an
NEA Workshop on Uncertainty Analysis for Performance Assessments of
Radioactive Waste Disposal Systems, Seattle, WA, 1987 February
24-26; OECD/NEA, Paris, pp. 191-203.

Kjellbert, N.A., Papp, T. and Thegerstroem, C. (1986). Role, status
and development of performance assessment in the Swedish nuclear
waste management programme. In proceedings of a symposium on
Siting, Design and Construction of Underground Repositories for
Radioactive Waste organized by the IAEA, Hannover, 1986 March 3-7,
IAEA-SM-289/32, pp. 669-681.

LeNeveu, D.M. (1986). Vault submodel for the second interim
assessment of the Canadian concept for nuclear fuel waste disposal:
post-closure phase. Atomic Energy of Canada Limited Report,
AECL-8383.

Liebetrau, A.M. (1986). The adaption and use of research codes for
performance assessment. Discussion paper prepared for presentation
at the fourth meeting of the OECD/NEA PSAC Users Group meeting,
London, England, 1986 November 11-14. AREST, plans.

Liebetrau, A.M., Apted, M.J., Engel, D.W., Altenhofen, M.K., Reid,
C.R., Strachan, D.M. and Erikson, R.L. (1987). AREST: A
probabilistic source-term code for waste package performance
analysis. In proceedings of the '87 Waste Management Symposium,
Tucson, AZ, 1987 March 1-5; Vol. 2, High-Level Wastes (R.G. Post,
ed.), University of Arizona, Tucson, AZ, pp. 535-544.

Liebetrau, A.M., Apted, M.J., Engel, D.W., Altenhofen, M.K.,
Strachan, D.M., Reid, C.R., Windisch, C.F., Erikson, R.L. and
Johnson, K.I. (1987). The analytical repository source-term (AREST)
model: Description and documentation. Pacific Northwest Laboratory
Report, PNL-6346.

Marivoet, J. (1988). Probabilistic calculations for the reference
site at Mol. PAGIS, Phase 2, Clay Option, Comprehensive Final
Report, Vol. 2, Chapter 4 (to be published).

Marivoet, J. (1988). Probabilistic computer codes: LISA.SCK and
LISGAN. PAGIS, Phase 2, Clay Option, Comprehensive Final Report,
Vol. 2, Chapter 4 (to be published).

Marivoet, J., Saltelli, A. and Cadelli, N. (1987). Uncertainty
analysis techniques. EUR 10934 EN, 20 p.

Marivoet, J. and Van Bosstraeten, C. (1988). Probabilistic
performance assessment for radioactive waste disposal: a simplified
biosphere model. Health Physics (to be published).

Mehta, K.K. (1985). Biosphere submodel for the second interim
assessment of the Canadian concept for nuclear fuel waste disposal:
post-closure phase. Atomic Energy of Canada Limited Record,
TR-298*.

Mobbs, S.F. (1988). PAGIS, Phase 2, Subseabed Option (to be published
in Volume 4).

Mobbs, S.F., Charles, D., Delow, C.E and McColl, N.P. (1988). The
assessment of subseabed disposal of vitrified high level waste for
the PAGIS project (to be published as NRPB R218 and EUR 11483).

Murray, C.N. (1988). The disposal of heat generating nuclear waste in
deep ocean geological formations: a feasible option. In Advances in
Underwater Technology, Ocean Science and Offshore Engineering,
Vol. 16, pp. 139-144.

Murray, C.N. (ed.) (1988). JRC Report Series: Study of the Feasibility
and Safety of the Disposal of Heat Generating Waste into Deep
Oceanic Geological Formations. Part 1: Radiological Safety
Assessment; Part 2: Geochemical and Geotechnical Studies; Part 3:
Engineering Investigations for Sub-Seabed Emplacement; Part 4: Deep
Ocean Instrumentation Development; Part 5: Simulation and Testing
Facilities.

Murray, C.N. and Avogadro, A. (1979). Effect of a long-term release
of plutonium and americium into an estuarine-coastal ecosystem. 1.
Development of an assessment methodology. Health Phys. 36, 573-585.

Murray, C.N. and Stanners, D.A. (1982). Development of an assessment
methodology for the disposal of high-level radioactive waste into
deep ocean sediments. Radioact. Waste Manage. 2 (3), 239-293.

Nancarrow, D.J., Jones, C.H. and Martin, A. (1986). A preliminary
assessment of the potential post-closure radiological impact of
repositories for radioactive wastes located at Elstow, Bedfordshire.
Part 1: summary report. Department of the Environment, London (UK)
Technical Report, TR-ANS-20.

NEA (1984). Long-term radiation protection objectives for radioactive
waste disposal. OECD/NEA, Paris.

Nies, A. (1987). Application of uncertainty analysis in assessments
of radioactive waste disposal in the Federal Republic of Germany.
In proceedings of an NEA workshop on Uncertainty Analysis for
Performance Assessments of Radioactive Waste Disposal Systems,
Seattle, WA, 1987 February 24-26; OECD/NEA, Paris, pp. 237-251.

PAGIS, (1988). Performance Assessment of Geological Isolation
Systems: summary report, (to be published).

Robinson, P.C. and Hodgkinson, D.P. (1987). Exact solutions for
radionuclide transport in the presence of parameter uncertainty.
Radioact. Waste Manage. Nucl. Fuel Cycle 8(4).

Saltelli, A. (1985). Improvement and extensions of the LISA code.
Joint Research Centre of Ispra., Tech. Note 1.07.C2.85.56.

Saltelli, A. (1986). The links between the individual process models
and integrated model in the probabilistic risk assessment codes. In
proceedings of the Second NEA Workshop on System Performance
Assessments for Radioactive Waste Disposal, Paris, 1985
October 22-24; OECD/NEA, Paris, pp. 89-99, ISBN92-64-02831-5.

Saltelli, A. (1987). PREP and SPOP utilities. Two FORTRAN programs
for sample preparation. Uncertainty analysis and sensitivity
analysis in Monte Carlo simulation. Programs description and user's
guide. EUR 11034 EN.

Saltelli, A. and Bertozzi, G. (1982). Risk assessment for geological
disposal of radioactive waste: uncertainty propagation and sampling
techniques. Joint Research Centre of Ispra, Tech. Note N.
1.07.03.82.49.

Saltelli, A. and Bertozzi, G. (1987). Data uncertainty in long-term
prediction. Present trends in risk analysis with Monte Carlo
techniques. In proceedings of the Workshop on Mathematical
Modelling for Radioactive Waste Repositories, Madrid, 1986,
December 10-12, ENRESA.

Saltelli, A. and Marivoet, J. (1988). Safety assessment for nuclear
waste disposal. Some observations about actual risk calculations.
Radioact. Waste Manage. Nucl. Fuel Cycle 9 (4), 309-321.

Saltelli, A., Bertozzi, G. and Stanners, D.A. (1984). LISA - A code
for safety assessment in nuclear waste disposals program
description and user guide. Commission of European Communities,
Joint Research Centre Report, EUR 9306 EN.

Saltelli, A., Sartori, E., Andres, T.H., Goodwin, B.W. and Carlyle,
S.G. (1987). PSACOIN level 0 intercomparison. NEA/OECD, Paris.

SENES (1985). Probabilistic model development for the assessment of
the long-term effects of uranium mill tailings in Canada - Phase
II. Volume 1 - Main report. Volume 2 - Workshop notes and topical
discussion. Volume 3 - Probabilistic analysis user's manual and
documentation for the Uranium Tailings Assessment Program (UTAP).
Research report prepared for the National Uranium Tailing Program,
Canada Centre for Mineral and Energy Technology, Energy, Mines and
Resources, Canada.

SENES (1986a). Probabilistic model development for the assessment of
the long-term effects of uranium mill tailings in Canada - Phase
III. Volume 1 - Summary report. Volume 2 - System Description and
Control Module Documentation. Volume 3 - Component Model
Documentation. Volume 4 - Uranium Tailings Assessment Program
(UTAP) User's Manual. Research report prepared for the National
Uranium Tailing Program, Canada Centre for Mineral and Energy
Technology, Energy, Mines and Resources, Canada; contract number
OSQ85-00182.

SENES (1986b). Risk assessment for Canadian uranium mill tailings.
Research report prepared for the National Uranium Tailing Program,
Canada Centre for Mineral and Energy Technology, Energy, Mines and
Resources, Canada; CANMET (DSS) serial number OSQ85-00225.

SENES (1987). Probabilistic system model development and application
to a real site - Phase IV. Research report prepared for the
National Uranium Tailing Program, Canada Centre for Mineral and
Energy Technology, Energy, Mines and Resources, Canada; CANMET
(DSS) serial number 23317-6-1728/01-SQ.

Sherman, G.R., Donahue, D.C., King, S.G. and So, A. (1986). The
systems variability analysis code SYVAC2 and submodels for the
second interim assessment of the Canadian concept for nuclear fuel

waste disposal. Atomic Energy of Canada Limited Technical Record,
TR-317*.

Stanners, D.A., Murray, C.N. and Saltelli, A. (1983).   The modelling
of natural processess: a probabilistic approach in the assessment
of radionuclide transfer in the geosphere and biosphere.   In
Environmental Transfer to Man of Radionuclides Released from
Nuclear Installations. IAEA/CEC Seminar Proceedings, Vol. II,
Brussels, 1983 October 17-21,   Luxembourg, pp. 489-499.

Sundstroem, B. (1986).   Development of uncertainty analyses for the
transport of radionuclides from a repository for high-level
radioactive waste.   Stage one.   Swedish Nuclear Power Inspectorate
Technical Report, SKI-TR-87-1.

Thompson, B.G.J. (1986).   A probabilistic risk assessment of an
underground radioactive waste disposal facility.   Department of the
Environment, London (UK) Technical Report, TR-DOE-8.

Thompson, B.G.J. (1987a).   The development by means of trial
assessments, of a procedure for radiological risk assessment of
underground disposal of low and intermediate level radioactive
wastes.   Department of the Environment, London (UK) Technical
Report, TR-DOE-13.

Thompson, B.G.J. (1987b).   The development of procedures for the risk
assessment of underground disposal of radioactive wastes: Research
funded by the Department of Environment 1982-1987.   Radio. Waste
Manage. Nucl. Fuel Cycle 9, 215.

Thompson, B.G.J., (1987c).   Uncertainty analysis in assessing the risk
associated with the disposal of low and intermediate level
radioactive wastes.   In proceedings of an NEA workshop on
Uncertainty Analysis for Performance Assessments of Radioactive
Waste Disposal Systems, Seattle, WA, 1987 February 24-26; OECD/NEA,
Paris, pp. 206-220.

Thompson, B.G.J., and Broyd, T.W. (1986).   'DRY RUN 1':   An initial
examination of a procedure for the post-closure radiological risk
assessment of an underground disposal facility for radioactive
waste.   Department of the Environment, London (UK) Technical
Report, TR-DOE-4.

Thompson, B.G.J, Duncan, A.G. and Hall, P. (1984).   The development of
post-closure radiological assessment methods by the Department of
the Environment.   Radioactive Waste Management, British Nuclear
Energy Society, London.

Wuschke, D.M., Gillespie, P.A., Mehta, K.K., Heinrich, W.F.,
LeNeveu, D.M., Guvanasen, V.M., Sherman, G.R., Donahue, D.C.,
Goodwin, B.W., Andres, T.H. and Lyon, R.B. (1985).   Second interim
assessment of the Canadian concept for nuclear fuel waste disposal.
Volume 4:   Post-closure assessment.   Atomic Energy of Canada
Limited Report, AECL-8373-4.

Wuschke, D.M., Mehta, K.K., Dormuth, K.W., Andres, T., Sherman, G.R.,
Rosinger, E.L.J., Goodwin, B.W., Reid, J.A.K. and Lyons, R.B.
(1981).   Environmental and safety assessment studies for nuclear
fuel waste management.   Vol. 3: Post-closure assessment.   Atomic
Energy of Canada Limited Record, TR-127-3*.

Wuschke, D.M., Rice A.M. and Gillespie, P.A. (1983). Environmental assessment of sub-seabed disposal of nuclear wastes: a demonstration probabilistic systems analysis. Atomic Energy of Canada Limited Record, TR-206*.

*Unrestricted, unpublished report, available from SDDO, Atomic Energy of Canada Research Company, Chalk River, ON, K0J 1J0.

THE ROLE OF THE CODE INTERCOMPARISON EXERCISES: ACTIVITIES OF THE
PROBABILISTIC SYSTEM ASSESSMENT CODES GROUP

Andrea Saltelli
Commission of the European Communities
Joint Research Centre - Ispra Establishment
Institute of the Environment
21020 Ispra (Varese) - Italy

FOREWORD. This lecture is related to my previous one on uncertainty -
sensitivity analysis techniques, as most of the worked examples given
there are part of the intercomparison exercises described here. The
reason for devoting an entire lecture to the activities of the Probabilistic Safety Assessment Code (PSAC) User's Group is because it constitutes the most appropriate arena for testing methodologies, procedures and codes related to probabilistic risk assessment. The intercomparison activities described here represent only a part of the activities of the group, but are the most challenging and instructive. This
lecture provides a worked example of how such intercomparisons can be
programmed, implemented and finalised, showing how risk assessment
techniques develope under the pressure of implementing an exercise
which has to pass peer review. A chronology of all the Group's activities is also included (Appendix).

INTRODUCTION

The NEA Probabilistic Systems Assessment Code User Group (PSAC) was
established in early 1985, following a proposal by the United Kingdom
Department of the Environment (UK DoE), to help the development in OECD
Member countries of Probabilistic Safety Assessment (PSA) codes for
assessment of the performance of radioactive waste disposal systems.
The activities of PSAC embrace information exchange, peer review, joint
code development, discussion of topical issues and comparison of codes.
The latter, which represents a major PSAC activity, provides confidence
in the ability of PSA codes to adequately represent the radioactive
waste disposal systems being modeled and hence their capability to
provide reliable information for safety assessments.
    Since December 1984, when the terms of reference of this group
were first discussed, six meetings have taken place where several
technical issues related to the use of stochastic assessment codes have
been discussed. Two code intercomparison were planned which constitute
the object of the present lecture: Level 0 (completed), and level E
(still under way). These and other actions undertaken by the Group are

A. Saltelli et al. (eds.), Risk Analysis in Nuclear Waste Management, 129–160.
© 1989 ECSC, EEC, EAEC, Brussels and Luxembourg.

summarised in the Appendix (see also Carlyle, 1987).

This note is mostly completely devoted to the Level 0 exercise taken as a worked example. It represent the first attempt to intercompare probabilistic risk assessment codes, and is based on a very simple model of the repository site.

The test case for the second intercomparison, Level E, is presently being run. In this intercomparison the results from the PSA codes can be compared against an analytical (the E stands for Exact) stochastic solution. Only some data on the case specification are given.

1. LEVEL 0 INTERCOMPARISON

1.1. Level 0 Intercomparison

The PSAC User Group aims to establish a basis for confidence in PSA and related codes. Code intercomparisons provide evidence that different codes operated by different groups produce similar results when applied to the same problem. Such evidence contributes to the verification and quality assurance of the codes involved. In this section the Level 0 Intercomparison is described.

1.2. Objectives

The intercomparison aims:
(i)   To gain experience in the speinincation, coding and comparison of PSA codes.
(ii) To compare "executive" modules and associated code of different PSA codes to verify that they meet their design specifications.

The codes to be compared includes:
a) The "executive" modules of the PSA code. The executive modules perform a set of basic functions, such as the sampling of parameter values from their probability distributions, and the transfer of parameter values to an implemented version of the system model.
b) The "post-processor" codes. Most PSA codes generate a large amount of data, and the post-processor codes are used to extract summary information, such as statistics for the mean dose and variance in mean dose.
c) Sensitivity analysis packages.

1.3. Intercomparison

The essential element of the intercomparison are given here. A comprehensive description of the study can be found in the Level 0 final report (Saltelli et al., 1987).

The problem specification note (Goodwin and Saltelli, 1986) requests predictions for annual effective dose equivalent to the maximum exposed individual of a critical group due to the nuclides Cs135, I129, Pd107, Se79, Sm151, Sn126 and Zr93. For simplicity decay chains are not considered. The dose arises after leaching from the primary containment in a repository, transport through a buffer and the geosphere, and ingestion by drinking from a contaminated well or stream. The source term is represented by a constant leach rate model including decay. Transport through the buffer is described as a pure delay function, i.e. the output from the buffer sub-model equals its input delayed by a time t_B, computed as a function of the buffer characteristic and of the radionuclide dispersion-retention parameters. Radioactive decay is also considered. For the geosphere the Gaussian transfer function corresponding to transport by advection and dispersion is simplified to a rectangular transfer function, whose width simulates the effect of dispersion. The pulse is delayed by the retention and decreased by radioactive decay. The biosphere model considers water ingestion only, where the water is pumped from a well constituting the geo/biosphere interface. Each segment of the radionuclide migration path is defined in terms of a number of parameters, some of which have uncertain values. Probability density functions are given for the uncertain parameters which have to be sampled by the probabilistic assessment code. In total there are 22 variable parameters.

Results include:
1) the effective dose equivalent probability distributions at specific points in time (histograms);
2) the maximum effective dose equivalent probability distributions up to specific points in time (histograms);
3) the mean values of these distributions and their standard deviation;
4) the ranking of the influential nuclides and parameters at the various time points (Sensitivity Analysis).

The test case proves to be non-trivial. The consequence for a given parameter set are non-zero only over a limited time interval. The position of this interval varies over a large time span and hence only a few parameter sets give non-zero results at a particular time point.

The intercomparison history is quite instructive. Before the stage was reached when the comparisons could be finalized four different iterations of preparing the case specification, running the case and analyzing the results were needed. The iterations helped in solving problems and difficulties arising in this exploratory exercise.

For example:
- The case specification was substantially changed after the first iteration as the flow versus time profile generated by the original test case was so spiked as to create time-stepping and interpolation problems.
- Following the second iteration, it was necessary to agree on a common format for data presentation. This was evident from the large differences in scale, format, type of result, etc. produced by different participants. A standard results request form was therefore prepared to ensure a common format for all the presentations.
- During the third iteration (also called the next-to-last iteration in the discussion below) the first comprehensive analysis of the results was carried out and program coding errors and post processing problems were recognized. The standard data format enabled a more

careful investigation of the results to be carried out and a restricted document containing the case results analysis was prepared by the NEA and distributed to the participants. This revealed a number of outliers due to coding errors or misinterpretation of the case specification. In one case natural logarithms were used in place of decimal ones. In another, the code computed mean doses for those runs giving non-zero doses rather than including them in the calculation. Also one code gave the maximum dose between subsequent time steps rather than the dose at the precise time point requested. Truncation of normal and lognormal distributions was another source for discrepancies.
- The fourth iteration (called the final iteration in the discussion below) gave participants about one month to identify and correct any error and an acceptable degree of consistency was obtained.

Twenty-one contributions were received in all. However, two were withdrawn because errors were detected but no corrected results were available at the closing date for acceptance.

Twelve different groups from eight countries and one international organisation participated; some submitted more than one solution in which different sampling methods or number of runs were used.

Each contribution is identified by a fictitious name in Table 1, which also gives the sampling method and the number of runs used. In the figures each participant is indicated by the letter (identifier) given in Table 1.

1.4. Results

Typically a PSA code may produce large amounts of raw data from each simulation. This data is often condensed into summary information using post-processors associated with the PSA codes. Summary information is then used to evaluate the performance and safety of a typical radioactive waste disposal facility. Examples of summary information derived from the simulation data include: the effective annual dose equivalent frequency distribution at specific points in time; the maximum effective annual dose equivalent frequency distributions up to specific points in time; the mean values of these distributions and their standard deviations. These quantities are considered in the comparison of the results.

To demonstrate agreement among participating Groups, several different comparison strategies were investigated. The method explored during the initial phase of the comparison exercise was principal component analysis (PCA). PCA provides a quick and effective method for the identification of globally erroneous results (also called outliers), but does not identify the reason for the errors.

Other methods were also explored, including: classification trees, box and whiskers plots, and convergence analysis. Also investigated were more direct comparison methods, using, for example, histogram plots of dose data, downward cumulative distribution plots of dose data (with a calculated confidence bound), and simple graphical displays of various calculated variables.

In the following some of the methods applied to a comparison of the Level 0 results are described.

Table 1. List of Level 0 intercomparison participants. Each participant is identified by a fictitious name in the text and by a letter in the figures (MC=purely random sampling; LHS=Latin hypercube sampling).

Partecipant	Identifier	Sampling Methods	Number of Runs	Comments
Steve	A	MC	1000	1000 time points
Jan	B	LHS	1000	500 runs for the mean of maximum dose equation
Budhi	C	MC	3000	
Andy1	D	MC	1000	1000 time points
Toshi	E	LHS	1000	
Timo1	F	MC	500	
Timo2	G	LHS	500	
Peter1	H	MC	1000000	Special code optimised for Level 0
Peter2	I	MC	1000	Special code optimised for Level 0
Withdrawn1	J			
Withdrawn2	K			
Ron	L	MC	2000	500 runs for the mean of maximum dose estimation
Bruce1	M	MC	1000	Automatic time step size
Bruce2	N	MC	7865	Automatic time step size
Shelley1	O	LHS	100	
Shelley2	P	LHS	1000	
Niels1	Q	MC	946	
Niels2	R	LHS	858	
Niels3	S	LHS	196	Spurious correlation removed
Andy2	T	LHS	1000	Spurious correlation removed
Alex	U	LHS	2000	

1.4.1. Histograms. Most participants calculated the frequency distribution (histograms) both for the dose at the specified times and the maximum dose up to the specified times.

The dose distributions at one million years are shown here for graphical comparison. This time point has the highest number of runs with non-zero dose (24%). In addition, the histograms show more structure than at the other time points studied and are therefore preferable for a detailed comparison. Three peaks can be distinguished. From the lowest to highest dose, the peaks can be attributed to doses from Se-79, from Pd-107 together with Sn-126, and from I-129 (overlapped with contributions from Zr-93 and Cs-135). All data for this point in time are plotted in Figures 1.1 to 1.15.

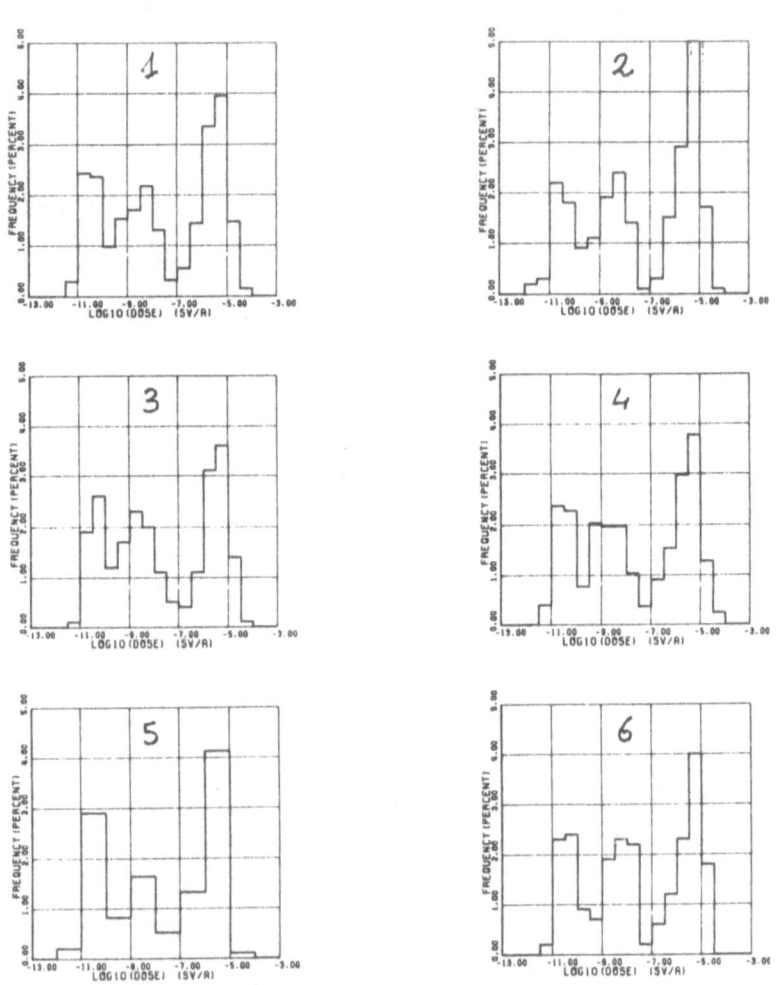

Figure 1. Histogram plots of frequency versus at 10^6 years for the different participants. The bin size of the LOG10(dose) axis is either one or one half logarithmic unit (from Saltelli et al. 1987).
1-Peter1 2-Steve 3-Jan 4-Budhi 5-Andy1 6-Toshi.

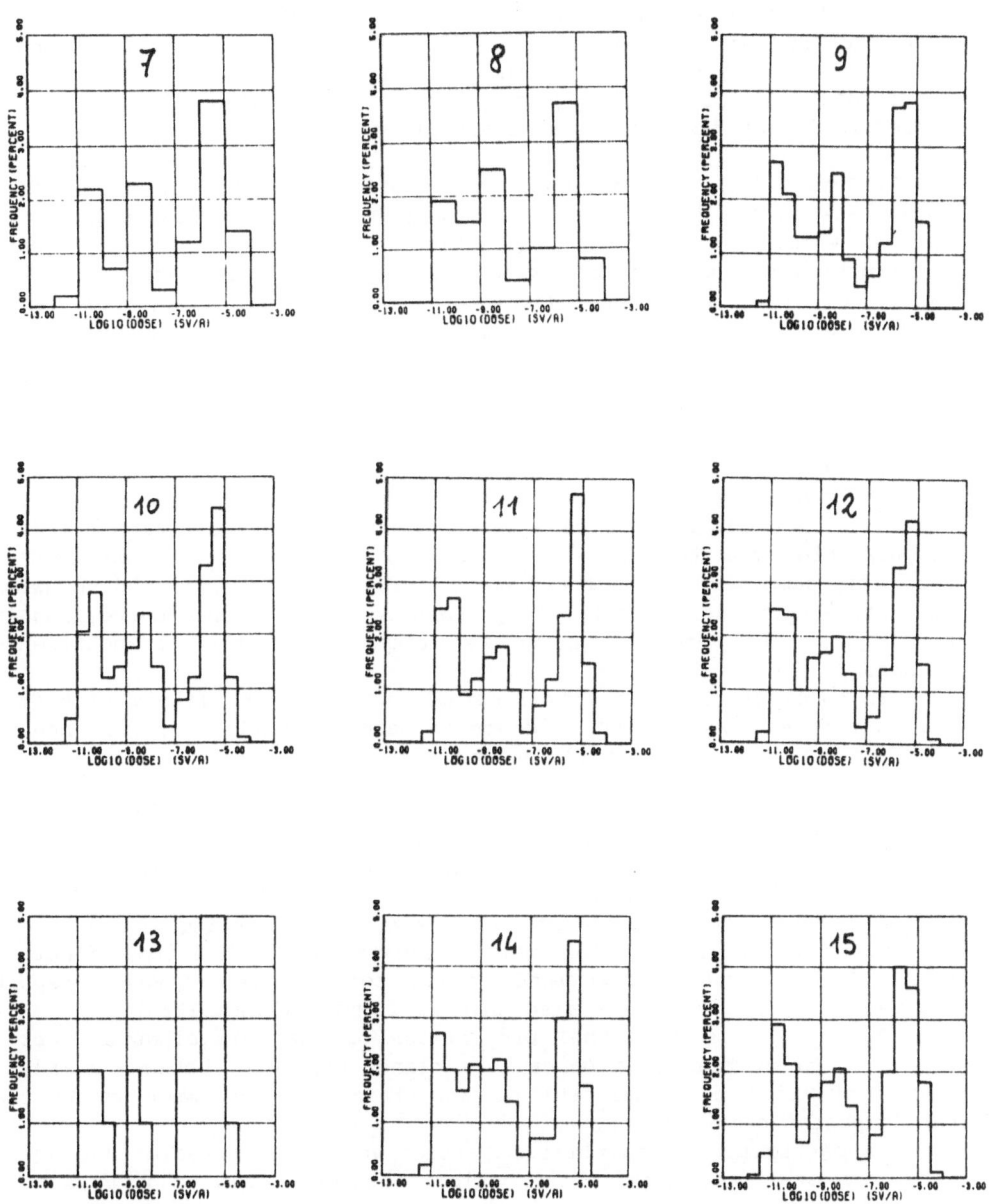

Figure 1 (continued). Histogram plots of frequency versus at 10^6 years for the different participants. The bin size of the LOG10(dose) axis is either one or one half logarithmic unit (from Saltelli et al. 1987). 7-Timo1 8-Timo2 9-Peter 10-Ron 11-Bruce1 12-Bruce2 13-Shelley1 14-Shelley2 15-Alex.

The three peaks show up clearly in all histograms. This is true even for the set of results involving only 100 simulations, although the set involving one million simulations provides more detail. Three participants provided histograms with a double-size bin range and therefore do not display the same detail as the others.

1.4.2. Cumulative Distributions. To investigate whether or not differences between the histogram data from different participants could be due to simple statistical fluctuations, confidence bounds on each distribution can be compared with a reference distribution. This can easily be done using the downward cumulative function, for which Kolmogorov statistics can be calculated to produce confidence bounds on each distribution, accounting for different sample sizes.

From the raw histogram data the downward cumulative distribution function with 95% Kolmogorov confidence bounds were calculated following the algorithm of SPOP (Saltelli, 1987). The graphical representation is shown in Figures 2.1 to 2.15. The thick line in Figure 2.1 is the one from participant Peter1 (one million run) together with the confidence bounds. In this case the confidence bounds are almost indistinguishable from the downward cumulative function itself because of the large sample size. In the other Figures (2.2 to 2.15) the downward cumulative curve is generally very close to the reference curve, and in all Figures the reference curve is always within the 95% confidence bounds.

In conclusion, apart from fluctuations which are likely statistical in nature, there is no evidence of pathologically different behaviour amongst the various code results.

1.4.3. Principal Components Analisis (PCA) was used in a global evaluation of the values provided by the participants for the different variables under investigation in order to identify which sets of results were consistent with each other.

PCA is a powerful mathematical tool for carrying out such an investigation especially for values subject to statistical fluctuations. PCA is briefly described in the comprehensive Level 0 report (Saltelli et al., 1987; see also Chatfield and Collins, 1983).

PCA allows bidimensional projections of the participants, representing each of them as a point in a plane. Proximity between adjacent points provides a visual aid to ascertain the agreement among the participants' predictions.

In particular it identifies participants whose predictions agree with each other and highlights possible outliers. It also gives a measure of improvement achieved in the consistency of the results over more than one iteration of the comparison exercise.

For the PCA the following quantities (variables) were considered:
a) Mean and standard deviation of the effective dose equivalent and the relative fraction of non-zero values at specified time points.
b) Mean and standard deviation of the maximum effective dose equivalent and the relative fraction of non-zero values up to specified time points.

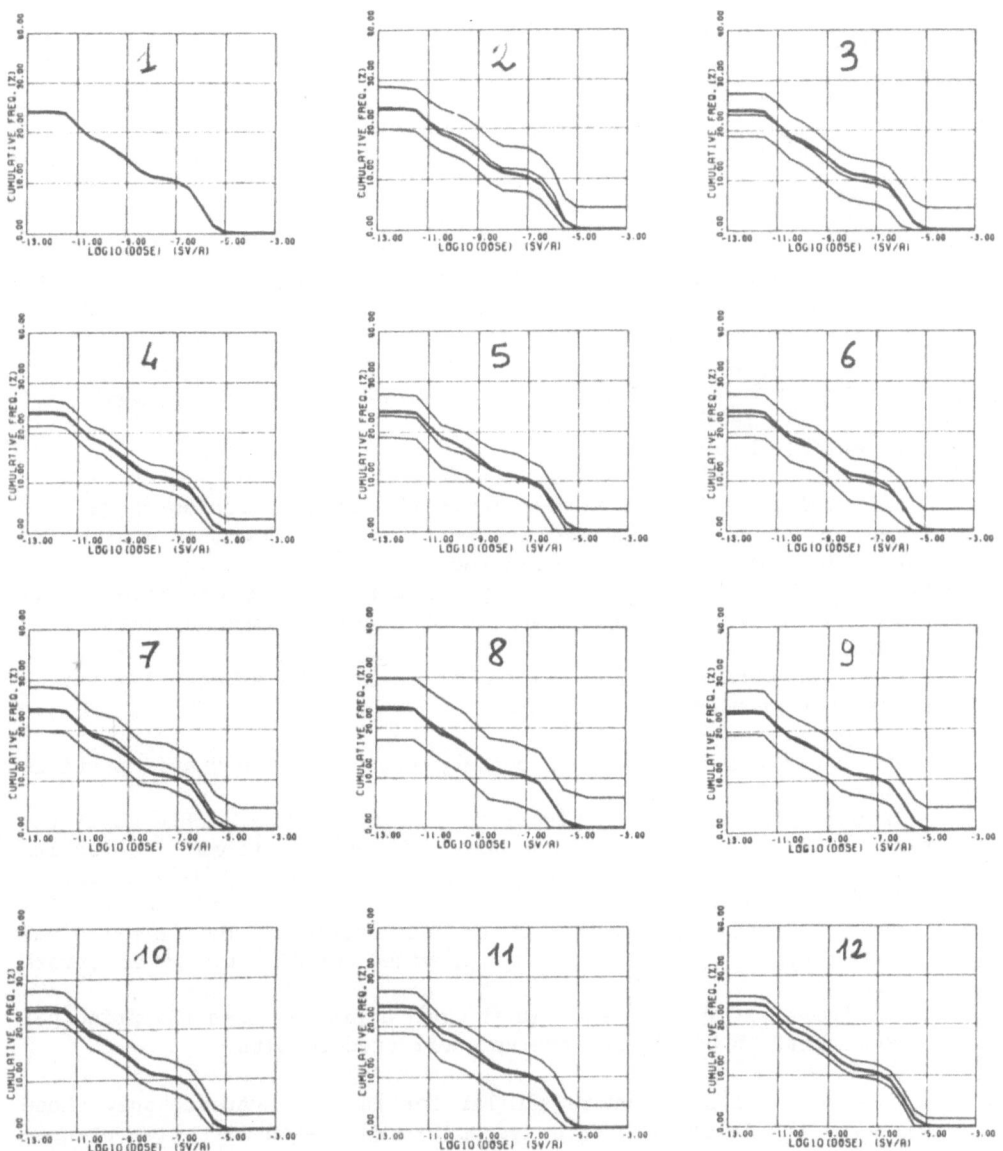

Figure 2. Downward cumulative frequency versus dose at 10^6 years for the different participants. The thick line is the reference curve from the one million run contribution (Participant Peter1 in Table 1). The other three lines are the downward cumulative frequency curve surrounded by its upper and lower 95% confidence bounds (from Saltelli et al. 1987).
1-Peter1 2-Steve 3-Jan 4-Budhi 5-Andy1 6-Toshi 7-Timo1 8-Timo2 9-Peter2 10-Ron 11-Bruce 1 12-Bruce2.

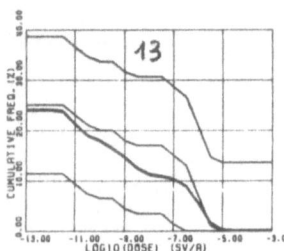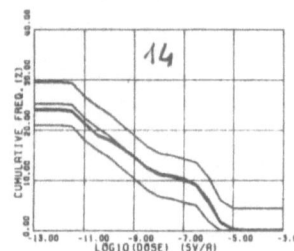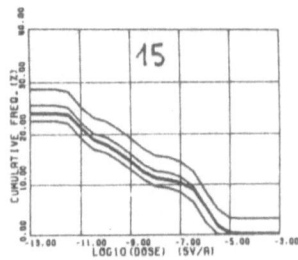

Figure 2 (continued). Downward cumulative frequency versus dose at 10^6 years for the different participants. The thick line is the reference curve from the one million run contribution (Participant Peter1 in Table 2). The other three lines are the downward cumulative frequency curve surrounded by its upper and lower 95% confidence bounds (from Saltelli et al. 1987).
13-Shelley1 14-Shelley2 15-Alex.

In total 36 variables were considered (the 6 variables above at 6 different time points). As not all participants provided results for all variables two cases were considered:
1) The subset of data involving all participants but only nine variables (mean, variance and fraction of non-zero doses at three time points).
2) The set of data involving all variables and the subset of participants that provided all information.

The above distinction is necessary because PCA discards participants with missing values.

In order to effectively visualize the improvements achieved in the consistency of the results during the last two iterations, the results of the analysis were displayed on the PCA planes in two different coordinate systems:
(i) The "old" reference system is the one having as axes the principal components obtained from data provided in the iteration before last.
(ii) The "new" reference system is the one having as axes the principal components obtained from the new corrected results.

Because results are quite similar for the two systems, only those relative to the "old" reference system are discussed here. Two principal components are sufficient to summarize the variability of the results in our study, which accounts - in general - for about 60-70% of the total variance. The projections of the points (representing the participants) onto planes formed by the two principal axes are displayed graphically before and after the corrections.

In order to avoid the participants' names being superimposed on the planes each participant is identified by a single alphabetic letter. Three additional points have also been added. The first one (identified by asterisk) provides values equal to the unweighted average of all the results provided in the last iteration; the other four,

Table 2. Contributors to mean dose, ranked in decreasing order of importance (from Saltelli et al. 1987).

PARTICIPANT

		nuclide ranking at $t=10^5$ years					
Timo2	I						
Shelley1	I	Se					
Jan	I	Se	Sn	Cs	Pd		
Andy1	I	Se	Sn	Cs	Pd		
Peter2	I	Se	Sn	Cs	Pd	Sm	Zr
Alex	I	Se	Sn	Cs	Pd	Zr	Sm
Peter1	I	Se	Sn	Cs	Zr	Pd	Sm
Toshi	I	Se	Sn	Cs	Zr	Pd	Sm
Budhi	I	Se	Sn	Pd	Cs	Zr	
Timo1	I	Se	Sn	Pd	Cs	Zr	Sm
Ron	I	Se	Sn	Zr	Cs	Pd	Sm
Bruce	I	Se	Sn	Zr	Cs	Pd	Sm
Steve	I	Se	Sn	Zr	Cs	Pd	Sm
		nuclide ranking at $t=10^6$ years					
Timo2	I						
Shelley1	I	Cs					
Budhi	I	Se	Pd	Sn	Cs	Zr	
Andy1	I	Zr	Cs	Pd	Sn	Se	
Jan	I	Zr	Cs	Pd	Sn	Se	
Alex	I	Zr	Cs	Pd	Sn	Se	Sm
Ron	I	Zr	Cs	Pd	Sn	Se	Sm
Peter1	I	Zr	Cs	Pd	Sn	Se	Sm
Peter2	I	Zr	Cs	Pd	Sn	Se	Sm
Toshi	I	Zr	Cs	Pd	Sn	Se	Sm
Bruce	I	Zr	Cs	Pd	Sn	Se	Sm
Steve	I	Zr	Cs	Pd	Sn	Se	Sm
Timo1	I	Zr	Cs	Pd	Sn	Se	Sm
		nuclide ranking at $t=10^7$ years					
Timo2	I						
Jan	I	Cs	Zr	Pd			
Andy1	I	Cs	Zr	Pd			
Alex	I	Cs	Zr	Pd	Sn	Se	Sm
Ron	I	Cs	Zr	Pd	Sn	Se	Sm
Bruce	I	Cs	Zr	Pd	Sn	Se	Sm
Steve	I	Cs	Zr	Pd	Sn	Se	Sm
Peter1	I	Cs	Zr	Pd	Sn	Se	Sm
Péter2	I	Cs	Zr	Pd	Sn	Se	Sm
Toshi	I	Cs	Zr	Pd	Sn	Se	Sm
Timo1	I	Cs	Zr	Pd	Sn	Se	Sm
Budhi	Zr	Pd	Sn	Cs	I		

identified by a + (plus) or a - (minus) bear values deviating systematically from the population average by plus or minus one or two standard deviations.

Figures 3 to 6 display graphically a selected set of results. Figures 3 and 5 display the results sent in for the next-to-last iteration and Figures 4 and 6 the ones after the correction (last iteration). The following plots were produced:
- Plot of the first and second principal component in the old reference system coordinates before and after correction (all participants, nine quantities) (Figures 3 and 4);
- Same as above for the analysis involving 12 participants and all the quantities (Figures 5 and 6).

These graphs clearly show that in the final iteration of the exercise the points representing the participants form a unique cluster. The spread around the average is believed to be mostly random in nature in the final iteration and therefore the results show a high degree of consistency among each other. In the graphs including the "old" values, clear systematic deviations are observed indicating which participant had incorrectly simulated in previous iterations the model proposed in the exercise. The graphical comparison of the results

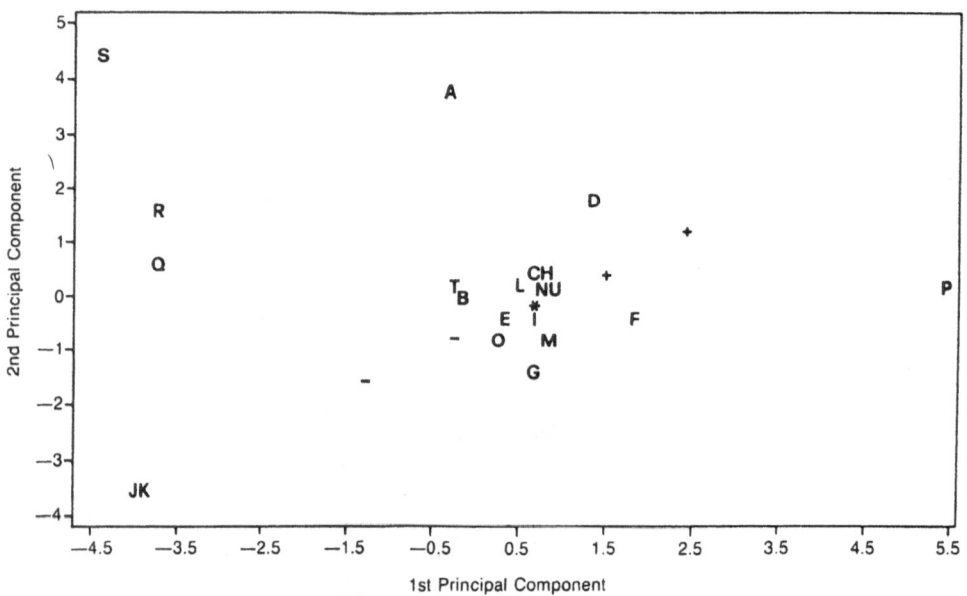

Figure 3. Plot of the first two principal components before errors were corrected (old reference system). All Participants - 9 Quantities; letters indicate participants (from Saltelli et al. 1987).

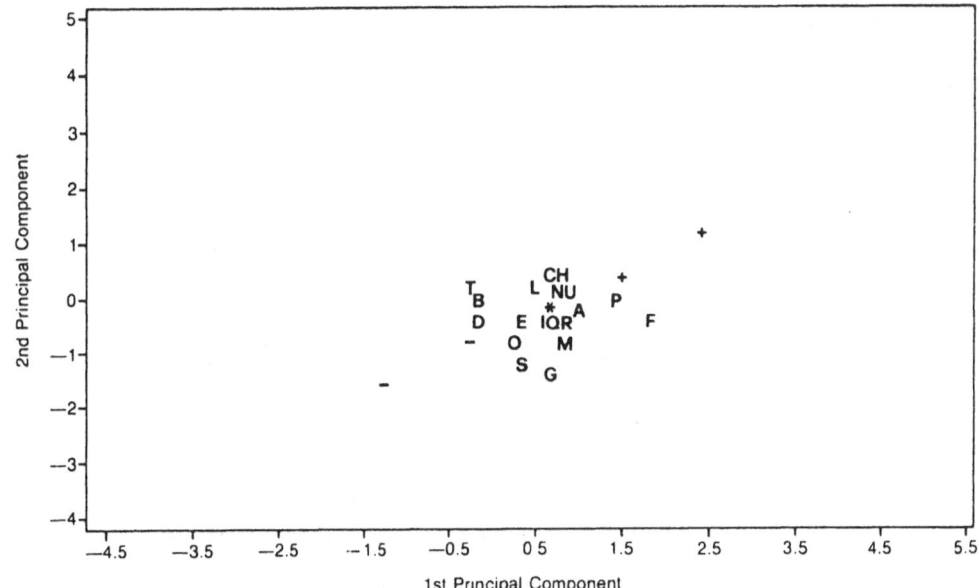

Figure 4. Plot of the first two principal components after corrections of errors (old reference system). All Participants - 9 Quantities; letters indicate participants (from Saltelli et al. 1987).

proved to be effective enough to convince these participants to subject their calculation to a careful re-analysis. In fact most of the "outliers" provided new results after they had detected their errors.

The letter H identifies a one million run contribution (participant Peterl in Table 1), which lies close to the the average (✳), as expected.

In conclusion PCA has proved to be an effective tool for identifying outliers in a comparison exercise and to visualize the consistency of results obtained in the different phases of the exercise.

1.4.4. Box and whiskers Plots. Box-and-whisker plots were developed by John Tukey (Tukey, 1977) as a means of illustrating the location and spread of a set of data points. A box-and-whisker plot shows asymmetries about the median value, and identifies outliers that do not seem to belong to the central cluster of points.

In the Level 0 Intercomparison, box-and-whisker plots were used to compare the simulation results of the participants. They helped to identify outliers from the main group of results at particular simulation times, for particular result variables.

To make a box-and-whisker plot, five statistics must be calculated from a data set: median, upper and lower hinges (i.e. 25 and 75 percentiles of the sample set), and the upper and lower adjacent points

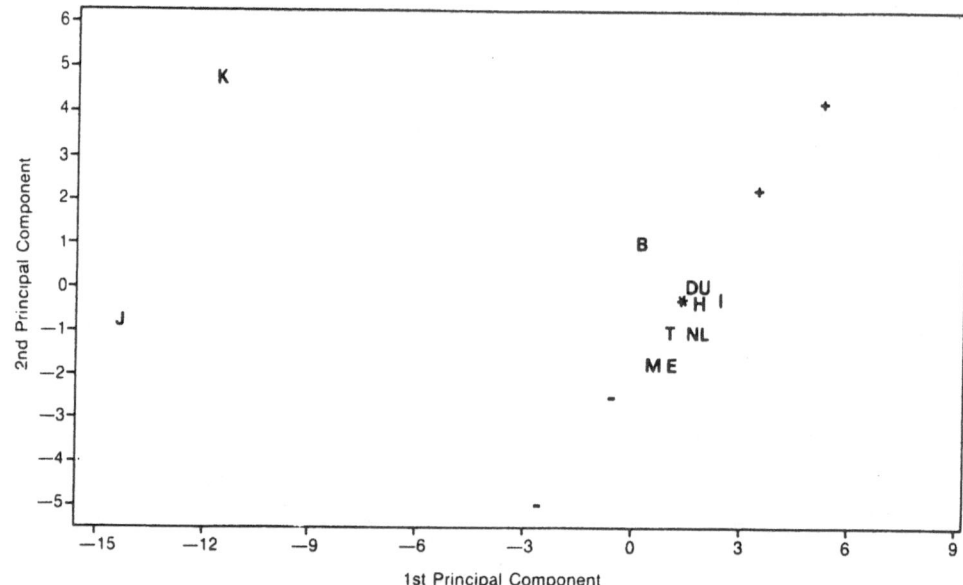

Figure 5. Plot of the first two principal components before errors were corrected (old reference system). 13 Participants - All Quantities; letters indicate participants (from Saltelli et al. 1987).

(defined below). These points are shown in Figure 7, which contains an expanded box-and-whisker construction. The box stretches vertically from the lower hinge to the upper hinge. The median is marked by a line across the box. Fifty per cent of the sample points lie in the box, with 25% above the median, and 25% below. The height of the box is called the "H-spread". The data point that is above the box and furthest from it, while remaining within 1.5 H-spreads of the box, is termed the upper adjacent point, and it is marked by a cross bar. The lower adjacent point is defined and marked in a similar manner. Points further out than the adjacent points are marked and individually labelled. If the data were sampled randomly from a normal distribution, there would be only about a 0.0035 probability of selecting an outlying point on each end (or about 0.007 for both ends).

In analysis of Level 0 results, box-and-whisker plots were constructed for mean dose, standard deviation of dose, and probability of non-zero dose. An example is shown in Figure 8, relative to the mean dose at the six times at which data was available for the results from the last iteration. The data was transformed by taking the logarithm base 10 before calculating statistics, so that the data, which varied by several orders of magnitude from one time to another, could be more easily plotted. Outliers in the plot were labelled with the letters given in Table 1.

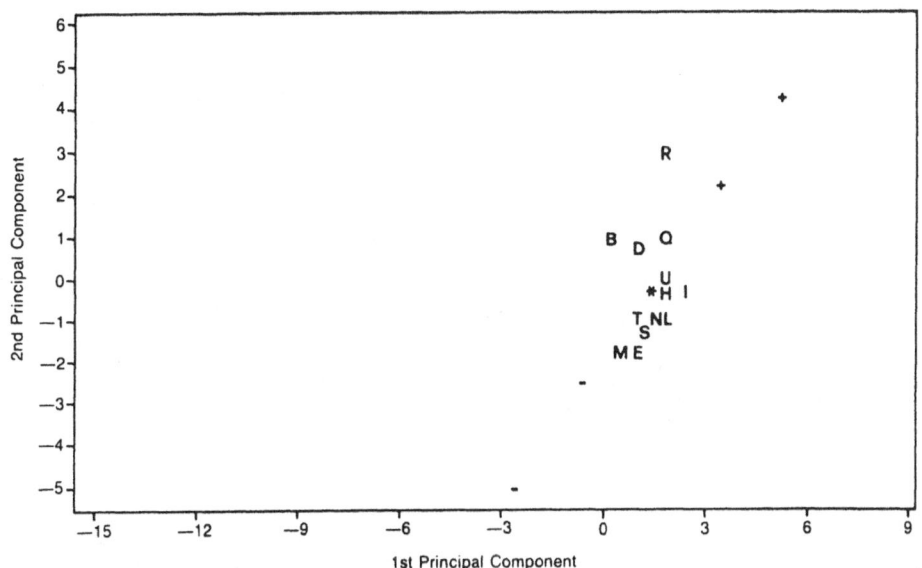

Figure 6. Plot of the first two principal components after corrections of errors (old reference system). 13 Participants - All Quantities; letters indicate participants (from Saltelli et al. 1987).

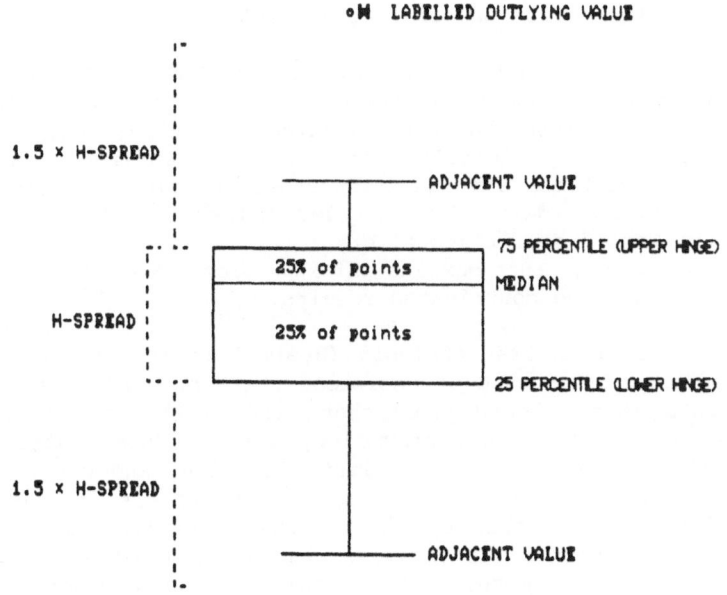

Figure 7. Description of a box and whiskers plot (from Saltelli et al. 1987).

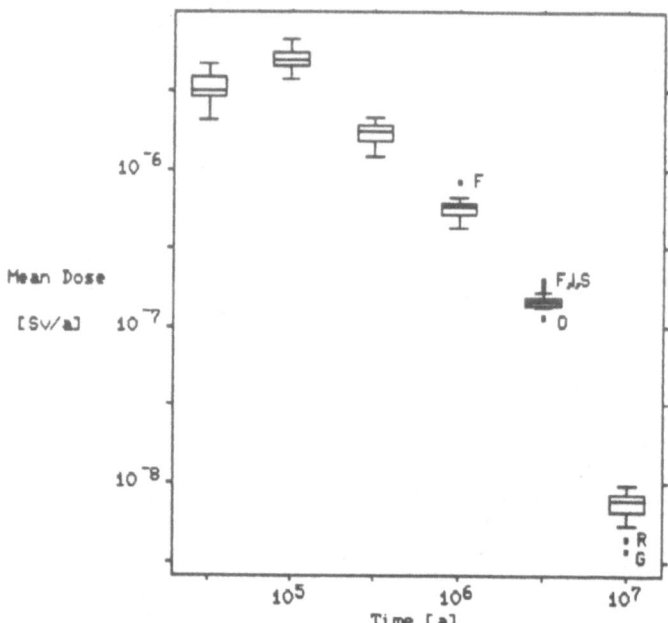

Figure 8. Box and whiskers plot showing the variation of mean dose at six points in time (from Saltelli et al. 1987).

A few outliers can be seen in Figure 8. However there is no evidence that these are due to anything but random variation. Specifically, there are no points that are far from the central box, and no codes that are consistently identified as generating outliers. In contrast, plots from the next-to-last iteration showed outliers that were shifted from the boxes by orders of magnitude, and showed that some specific contributors were frequently outliers.

In conclusion, the box-and-whisker plots demonstrate excellent agreement amongst the contributed results.

1.4.5. Classification Tree analysis (Breiman et al., 1984) was carried out by constructing a tree of decisions that would split the available Level 0 data into subcategories that were similar in their results. This type of analysis was performed by Terry Andres (Saltelli et al. 1987) on the data of the next-to-last iteration, and was mainly meant to the identification of possible systematic errors in the data. The outliers identified by this type of treatment were the same pointed out by the principal components analysis, whose results were thus confirmed. This technique is a multi-step procedure, slightly more complex to implement than the PCA and the Box and whisker plots described previously; for this reason its description is not reported here. The interested reader is referred to the comprehensive Level 0 report for a description of the method and the analysis of the results.

1.4.6. Convergence of Results as a Function of Sample Size. A description of this analysis, intended to study the dependence of sample mean dose on sample size, was given at a previous lecture in the present seminar and is not repeated here. It ensured that the residual error in the data, after correction of the systematic errors, was of random nature and was compatible with the sample size employed by the various participants.

1.5. Level 0 Sensitivity Analysis

At the fourth PSAC Meeting held in London in December 1986 it was agreed to expand the intercomparison exercise to include a section on Sensitivity Analysis (SA), although it was not part of the original specifications. Sensitivity analysis is normally not carried out with the PSA code itself, but with what is called a post-processor. This part of the analysis therefore compares the different post-processors in use to perform SA.

The participants were asked to specify the technique used in the sensitivity analysis study and, for the method used:
- To rank in decreasing order of importance the radionuclides contributing to the mean dose at 10^5, 10^6 and 10^6 years.
- To identify and rank in decreasing order of importance the significant parameters affecting mean dose at 10^5, 10^6 and 10^6 years.

While the ranking of contributing radionuclides was a well defined problem the ranking of parameters proved to be particularly difficult.

Nuclide Ranking. The results for ranking nuclides in decreasing importance of contribution to the mean dose at the selected reference times are given in Table 2.

There is, in general, very good agreement amongst the various codes. At time=10^5 years, for example, differences start at the fourth most important radionuclide. At time=10^6 years, most participants give identical ranking for all nuclides. Exceptions are the rankings from Shelley1 and Budhi. The result from Shelley1 is not surprising because of the small number of runs. At time=10^7 years, only I, Cs, Zr and Pd contribute to the dose, the dose from Sn, Se and Sm being neglegible. Here again an excellent agreement in ranking the contributing nuclides is observed, except for the Budhi results.

Parameter Ranking An extensive study of sensitivity analysis was conducted on the Level 0 test case (Saltelli and Marivoet, 1988) as part of a technique intercomparison. One of the findings of this study was that the test case was particularly difficult, as far as SA is concerned. This is due to:
- the non-linearity of the model;
- the fact that the distribution functions of the input parameters range over orders of magnitude;
- the existence of many ties (same values) in the output vectors, (i.e. many output values are zero).

The same situation was observed with the Level E model discussed in the sensitivity analysis example of the previous lecture. We shall not report here all the results of the Level 0 sensitivity analysis, which can be found on the comprehensive report (Saltelli et al. 1987), and shall illustrate instead some of the problem encountered.

The large percentage of zeros in the output has a crucial impact on the SA results. This can be clearly seen in Figures 9 and 10, where the percentage of non-zero runs is plotted together with the model coefficient of determination R^2. The R^2 coefficient is a measure of the performance of a linear regression model based on the Standardized Regression Coefficient. It can be shown that R^2 equals the fraction of the total variance of the output data accounted for by the regression model (Iman et al, 1985; see also the previous lecture).

In Figures 9 and 10, R^2 has been computed on the ranks of the input-output vectors, using a random sample of 1000 for a simulation of the test case and analyzing the results with the SPOP code (Saltelli, 1987). In Figure 9 the output quantity is the dose at a certain time point, whereas in Figure 10 the maximum dose between zero and the time point is considered. It can be seen in Figure 9, where the percentage of non-zero output is very low (below 25%), that the model coefficient of determination is also low (never higher than 0.24). In Figure 10, where the maximum doses are considered and more non-zero output is available, R^2 increases proportionally to the percentage of non-zero doses.

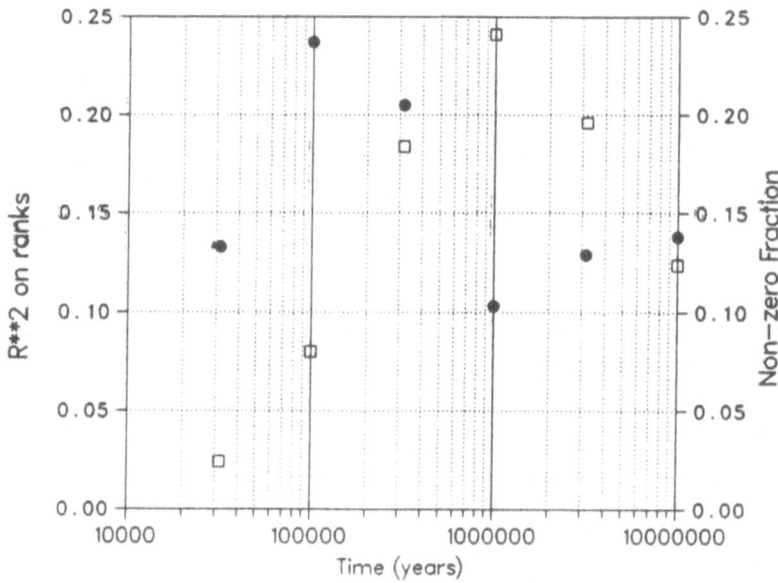

Figure 9. Plot of the fraction of non-zero doses (open boxes) and of the model coefficient of determination (solid circles) versus time (from Saltelli et al. 1987).

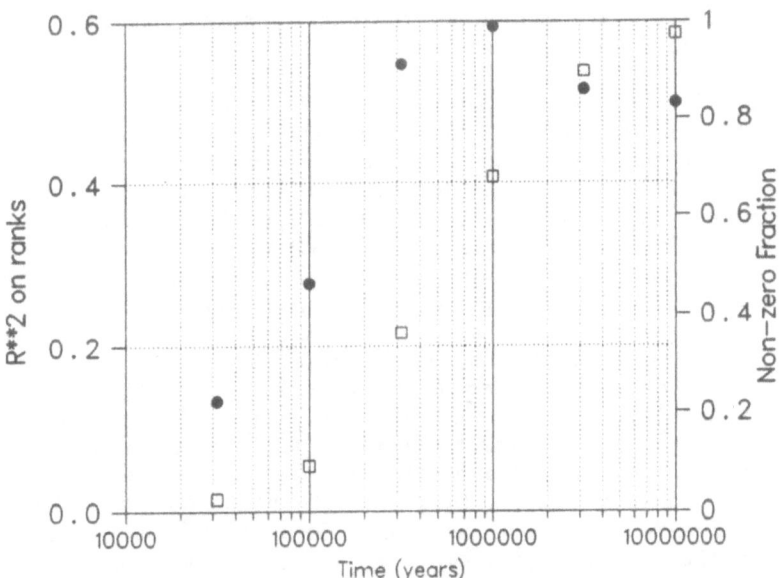

Figure 10. Plot of the fraction of non-zero maximum doses (open boxes) and of the model coefficient of determination (solid circles) up to six specific times (from Saltelli et al. 1987).

As a result, SA of the output doses at the established time points is quite difficult; when the regression model based on the ranked variables fits the data poorly, it is risky to rank the data based on the regression coefficients. Instead it is easier to rank a parameter with respect to the maximum dose up to the established time points.

Most differences in the participants' responses were due to the different selections of output quantity used in their analyses: dose, rank of dose, logarithm of dose and dose raised to a fractional exponent were tried by the various participants. In fact the ranking depends strongly on which function of the output variable is taken in the analysis. Using dose, the rank of the dose or its logarithm results in different rankings. Quoting from one of the participants:

"The varying orders of importance, depending on the variant analyzed, can sometimes be explained easily. For example, the water extraction rate is very important to maximum dose, and of middle importance to (maximum dose)$^{0.1}$ and log10(maximum dose). The differences in indices for water extraction rate stem from the fact that dose is inversely proportional to the water extraction rate. Since this parameter varies by almost an order of magnitude, its effect on maximum dose can be quite important on a linear scale. On a log scale, it can cause a linear shift of roughly one order of magnitude, but this shift is to be compared with the twenty orders of magnitude variation present in the maximum dose sample. In this context, the importance of water extraction rate is much reduced."

If the results are to be used to build a regression model, the ranks or a small power of the dose should be taken, as the degree of linearity between dose and the input parameters is very low. (A power of dose is better than the logarithm of dose for this analysis because powers can deal with zero dose values, and logarithms cannot.) Because non-linear dependences are more difficult to analyze, a regression model for "dose" will always be very poor compared to the same regression scheme for "rank of dose" or "power of dose".

As far as the SA techniques are concerned most of the participants have taken an approach based on correlation-regression coefficients, working on some normalized transformation of the output dose (mostly logarithms or ranks). The Smirnov test has also been used and its performance was quite similar to those of the regression based methods.

Just to give an idea of the level of agreement (or disagreement!) among the various participants part of the SA output is given in Table 3, relative to the output variable "dose at 10^7 years". For a description of the SA techniques employed and of the meaning of the variables see the Level 0 comprehensive report. The influential variables are given in Table 3 in order of decreasing importance, and it can be seen that non-negligeable differences are present.

Table 3. Variable ranking based on dose at 10^7 years as the analysis variable. Input variables are listed in order of decreasing importance (from Saltelli et al. 1987).

PARTICIPANT	VARIABLE RANKING
Toshi	V_G, $ALPHA_G$, K_D^G (Pd), K_D^G (Cs)
Alex	V_G, K_D^G (I), $ALPHA_G$, W
Andy	$ALPHA_G$, V_G, K_D^G (Cs), K_D^G (Zr)
Shelley1	$ALPHA_G$, V_G, K_D^B (Zr), K_D^G (Zr)
Timo	$ALPHA_G$, K_D^B (Zr)
Bruce	X_G, V_G, K_D^G (I), W
Budhi	X_G, K_D^G (Pd), V_G, W
Steve (I129 only)	R^O, K_D^G (I), K_D^B (I), X_B
Andy (I129 only)	K_D^G (I), V_G, X_B, W

In conclusion, given the non-trivial nature of the exercise, the participants tried to produce, in a single iteration, results which could be compared. The experience gained in this first SA intercomparison shows that much can still be done in the field of intercomparing post-processors, and that such activities, far from being an academic exercise, are a useful and perhaps essential tool in code verification.

1.6. Conclusions of the Level 0 Exercise

In conclusion the results of the Level 0 exercise showed that the participating codes are consistent in their results, and that excellent agreement was obtained for estimates of the mean dose for 19 sets of results, involving 12 different PSA codes, different sampling strategies and sample sizes ranging from 100 to 10^6 simulations. There were also several secondary objectives of the Level 0 Intercomparison, including:

a) Testing of the statistical post-processors associated with the PSA codes. Since the results actually compared were the outputs from some post-processor codes, the intercomparison implicitly includes a test of those post-processors. The consistency of the results indicates that they are functioning correctly and as expected.

b) Testing of sensitivity analysis techniques. This objective met with limited success. Most of the participants consistently identified and ranked the same set of radionuclides that contributed most to the mean dose. Most participants also consistently identified the most significant parameters affecting mean dose, but there was less satisfactory agreement in the ranking of these parameters. This exercise and previous experience by members of the PSAC User Group has shown that sensitivity analysis of PSA code results is a difficult task, and further work is required.

2. LEVEL E INTERCOMPARISON

The Level E test case study is based on models of radionuclide release and migration which are simple, but somewhat more realistic than those used in Level 0. These models allow the exact calculation of the mean dose rates (hence E, for exact, in the title; PSAC/DOC 87(13)). The exact results arise from analytical expressions for integrals over parameter space of the model dose rate, expressed as a Laplace transform rather than as a function of time (Robinson and Hodgkinson, 1987). A high precision Laplace transform inversion technique allows the results to be calculated to a precision approaching that of the computer used. In the Level E test model the repository is represented by a simple leach rate controlled release of the radionuclide into the groundwater of a host rock. The radionuclides are transported through two geosphere layers with different hydrogeological properties. Thence, contamination is assumed to enter a stream from which the critical group obtains drinking water. Four nuclides are considered: I129 and the decay chain Np237-U233-Th229. The model has 33 parameters, 12 of

which are taken as random variables. In order to assist the participants in setting up the case a set of results is provided with three deterministic runs, ie. three runs where all the input parameters have pre-fixed constant value. The two main objectives of the Level E test case, which make it superior to the previous Level 0, are:
(i)   to compare the PSA codes against a known solution;
(ii) to provide a case study with sufficient similariry to practical assessment requirements that production sub-model codes can be used rather than special modules written for the purpose of the exercise

and, possibly
(iii) to provide a common basis for assessing the relative efficiencies of different parameter sampling schemes.

It should be mentioned in fact that in the Level 0 exercise the influence of the sampling scheme on the convergence of the results could not be estimated. The same classification tre analysis could not reach a conclusive statement on the differences between the random and the hypercube sampling.

The results from the Level E exercise cannot be dicussed here, as they are presently beeing produced and a first comparison is expected at the Braunschweig meeting next June (see part 1 of this lecture). A discussion of the sensitivity analysis for the Level E test case has been given at a previous lecture at the present seminar.

REFERENCES

Important note

All the references quoted as SYVAC/DOC and PSAC/DOC are internal documents of the PSAC Users' Group and can only be obtained through direct request to the OECD/NEA offices at the following address:

Daniel Galson
OECD/NEA, Division of Radiation
Protection and Waste Management,
38 Boulevard Suchet 75016 Paris

Breiman, L., Friedman, J.H., Olshen, R.A. and Stone, C.J. Classifica-
    tion and Regression Trees; Wadsworth International Group, Belmont,
    California, (1984).
Carlyle, S.G. The Activities, Objectives and Recent Achievements of the
    NEA Probabilistic Systems Assessment Codes (PSAC) User Group; Pro-
    ceedings of Waste Management '87, Ed. R.G. Post and M.E. Wacks,
    Tucson (1987).
Chatfield, C. and Collins, A.J. Introduction to Multivariate Analysis;
    Editors: Chapman and Hall, London (1983).
Dalrymple, G.J. A method for encoding probability density functions
    from subjective judgement, CAP Scientific Limited report 3097/WP/15

prepared for the UK DoE (1987).

Frizelle, C.J.G., Dames & More, London. Environmental change and post-closure risk assessment of radioactive waste disposal. Report No. CJGF-DAS-054 for the UK DoE London, 1986.

Goodwin B.W. and Saltelli, A. Probabilistic System Assessment Codes. Level 0 Intercomparison: Problem Specification Note. Atomic Energy of Canada Ltd. report (1986), (also as appendix in Saltelli et al., 1987).

Gralewski, A. Lump parameter method for use in UK SYVAC-A/C Electrowatt Engeneering Services (EES) report produced for the UK DoE, rep. No. 2529 (1986).

Iman, R.L., Shortencarrier M.J. and Johnson, D.J. A FORTRAN 77 program and user's guide for the calculation of partial correlation and standardised regression coefficient. SANDIA Natl. Laboratories report NUREG/CR 4122, SAND85-0044 (1985).

Marivoet J. and Bosstraeten, C. Van. A simplified biosphere model applicable in computer codes for probabilistic performance assessment for radioactive waste disposal (in preparation).

Prust, J.O. and Dalrymple, G.J. Development of SYVAC sampling techniques, UK DoE report No. DoE/RW/85.072 (1985).

PSAC/DOC 86(5). Nies, A. The presentation of results from PSA codes. Summary of the discussion held at the second PSAC meeting. NEA, Paris, 1986.

PSAC/DOC 86(7): NEA PSAC User Group. Summary record of the third group meeting, Ispra, 20th-23rd April 1986. NEA, Paris, 1986.

PSAC/DOC 87(1): NEA PSAC User Group. Summary record of the fourth group meeting, London, 11th-14th November 1986. NEA, Paris, 1987.

PSAC/DOC 87(2): NEA PSAC User Group. The reduction of research codes to PSAC sub-models. Compilation of the papers presented at the fourth PSAC meeting. NEA, Paris, 1987.

PSAC/DOC 87(3), revised as PSAC/DOC 87(12): NEA PSAC User Group. Future programme of activities (1987-1989). NEA, Paris, 1987.

PSAC/DOC 87(7): NEA PSAC User Group. Summary record of the fifth group meeting, Stockholm 2nd-5th June 1987. NEA, Paris, 1987.

PSAC/DOC 87(11): Goodwin, B. NEA PSAC User Group; Future topical meetings for the PSAC user group. NEA, Paris 1987.

PSAC/DOC 87(13): Oldfield, S., Sinclair, J., Andres, T. and Robinson, P. NEA PSAC User Group: PSACOIN Level E intercomparison case specification. NEA, Paris 1987.

PSAC/DOC 88(1): NEA PSAC User Group. Summary record of the sixth group meeting, Paris 7th-9th December 1987. NEA, Paris, 1988.

Rae, I.C. SCICON quality assurance procedures used in the development of UK SYVAC codes, SCICON Ltd. prepared for the UK DoE (1985).

Robinson, P.C. and Hodgkinson, D.P. Exact solutions for radionuclide transport in the presence of parameter uncertainty. Radioactive Waste Management and the Nuclear Fuel Cycle 8, 1987.

Sagar, B. and Eslinger, P.W. Use of subjective information in probabilistic system analysis. Rockwell International, 1987 (in PSAC/DOC 87(10)).

Saltelli, A., Bertozzi G. and Stanners, D.A. LISA - A Code for Safety Assessment in Nuclear Waste Disposals; Program Description and User

Guide, JRC Ispra Report EUR 9306 EN, Luxembourg (1985).

Saltelli, A. PREP and SPOP Utilities. Two FORTRAN Programs for Sample Preparation, Uncertainty Analysis and Sensitivity Analysis in Monte Carlo Simulation. Programs Description and User's Guide. Nuclear Science and Technology report EUR 11034 EN, Luxemburg 1987.

Saltelli, A., Sartori, E., Andres, T.H., Goodwin B.W. and Carlyle S.G. PSACOIN Level 0 Intercomparison. An international Code Intercomparison Exercise on a Hypothetical Safety Assessment Case Study for Radioactive Waste Disposal Systems. OECD - NEA publication, Paris (1987).

Saltelli, A. and Marivoet, J. Nonparametric statistics in sensitivity analysis for model output. A comparison of selected techniques, 1988, paper in progress.

Sartori, E. Revised version of the standard output format for PSA simulations. NEA Data Bank, Saclay 1987.

Sherman, G.R., Donahue D.C. and King, S.G. SYVAC2 - A System Variability Analysis Code for Assessment of Nuclear Fuel Waste Disposal Concepts, Atomic Energy of Canada Ltd. report TR-317 (1986).

Stephens, M.E., Goodwin, B.W. and Andres, T.H. Guidelines for selecting probability density functions for SYVAC3-CC3 input parameters Atomic Energy of Canada Ltd. report (1987), (also in OECD NEA report PSAC/DOC(87)10, Paris 1987).

SYVAC/DOC(85)1: NEA User's Group for System Variability Analysis Codes. Summary record of the first meeting, Paris 7th May 1985. NEA, Paris, 1985.

SYVAC/DOC(85)4: NEA User's Group for System Variability Analysis Codes. Summary record of the second meeting, Winnipeg 29th-30th September 1985. NEA, Paris, 1985.

Tukey, J.W. Exploratory Data Analysis; Addison-Wesley Publishing Company, Don Mills, Ontario, (1977).

APPENDIX

SHORT HISTORY OF THE GROUP ACTIVITIES

outline

a.1 London, December 4-5, 1984
   - Establishment of the group
   - Standard Executive (or driver)
   - Terms of reference

a.2 Paris, May 7-9, 1985: first PSAC meeting
   - List of modules
   - topical meeting: Input data aquisition, availability and handling
   - topical meeting: Methods and procedures for sensitivity analysis

a.3 Winnipeg, Manitoba (CAN) September 29-30 1985: second PSAC meeting
   - decision to prepare a Level 0 intercomparison

- topical meeting: Presentation of results
- topical meeting: Statistical sampling procedures

a.4 Ispra, April 21-23 1986: third PSAC meeting
- topical meeting: Software quality assurance and validation
- topical meeting: Time-dependent processes
- Level 0 1st iteration
- Idea of a common intermediate file
- First presentation Harwell results on analytical solutions

a.5 London, November 11-14 1986: fourth PSAC meeting
- topical meeting: Reduction of research codes to PSAC sub-models
- Lumped parameter for biosphere
- Discussion of possible future intercomparisons: Level 1 and Level E
- Level 0 2nd iteration; preparation of a questionnaire

a.6 Stockholm, June 2-5 1987: fifth PSAC meeting
- topical meeting: Probability density function justification
- Level 0 3rd iteration
- Plan for the future intercomparisons

a.7 Paris, December 7-9: sixth PSAC meeting
- Level 0 4th iteration; distribution of the final report
- List of possible future topical meetings

a.8 Braunschweig, June 20-23 1988: seventh PSAC meeting
- Level E 1st iteration
- Level 1a,1b,ESA preliminary calculations
- topical meeting: Model Uncertainty
- topical meeting: Sensitivity Analysis

In this section a short and necessarily incomplete summary of the group's activities is given. Technical reports prepared either by the secretariat or by group members shall also be referenced. The summary is organised in a chronological order, following the various PSAC meetings.

a.1. London, December 4-5, 1984. Establishment of the Group; Standard Executive (or driver); Terms of reference. The idea of establishing a group of users of "SYVAC like" codes originates from the UK DoE (September 1984), following development of the Canadian code SYVAC for safety assessment of nuclear waste disposal (Sherman et al., 1986). A first organisational meeting was held in London (December 1984) to establish the terms of reference of the group. Participants were: NEA (including the group secretariat), representatives from the Swedish KBS, from the Canadian AECL, from the British DoE and AERE/Harwell and from the JRC (Table a.1). The main objective of the group were:
" a) exchanging codes, information and experience thus avoiding unnecessary duplication of efforts;
  b) providing mutual peer review;
  c) discussing specific technical issues such as code quality

Table a.1 List of PSAC group participant organisations.

PARTICIPANT	COUNTRY	ORGANISATION OR PROJECT	ABBREVIATION
S. Carlyle Former PSAC Secretary	UK	Department of the Environment Her Majesty's Inspectorate of Pollution, London	DoE/HMIP
B. Goodwin, Chairman PSAC T. Andres and T. Melnyk	CAN	Atomic Energy of Canada Ltd. Whiteshell Establishment, Pinawa, Manitoba	AECL
R. Holmes	CAN	National Uranium Tailing Program, Ottawa, Ontario	NUTP
T. Iijima	J	Japan Atomic Energy Research Institute, Tokai Establishment	JAERI
N. Kjellbert	S	Swedish Nuclear Fuel and Waste Management Co. (SKB) and Nuclear Power Ispectorate, respectively Stockholm	SKB, SKI
B. Cronhjort	S	Swedish National Board for spent nuclear fuel	SKN
A. Libetrau	USA	Battelle Pacific Northwest Laboratory, Richland, Washington State	PNL
J. Marivoet	B	Studiecentrum voor Kernenenergie/ Centre d'Etude Nucleaire, Mol	SCK/CEN
S. Mobbs	UK	National Radiaction Protection Board, Chilton, Oxfordshire	NRPB
A. Nies	FRG	Gesellshaft fur Strahlen und Umweltforschung mbH Bruanschweig	GSF
B. Sagar	USA	Basalt Waste Isolation project Rockwell Hanford, Richland, Washington State	BWIP
A. Saltelli	CEC	Joint Research Centre of Ispra, I	JRC
E. Sartori	NEA	Nuclear Energy Agency of the OECD, Data bank of Saclay	NEA/DB
J. Sinclair	UK	Atomic Energy Research Establishment, Harwell	AERE
B. Sundstrom	S	Swedish Nuclear Power Inspectorate, Stockholm	SKB
C. Thegerstrom Present PSAC Secretary	NEA	Nuclear Energy Agency of the OECD, Division of Radiation Protection and Waste Managm.	NEA
B. Thompson	UK	Department of the Environment HMIP, London	DoE
T. Vieno	SF	Technical Research Centre of Finland, Helsinki	VTT
S. Wilkinson	CAN	Chalk River National Laboratory, Ontario	CRNL
P. Zuidema	CH	National cooperative for the Storage of Radiactive waste, Baden	NAGRA

NOTE FOR TABLE a.1:
This list reflects the December 1987 situation. Now (January 1989)
A. Nies has replaced B. Goodwin as group chairman; Dan Galson has
replaced C. Thegerstrom as group secretary; R. Holmes and A. Libetran
have left the group. The JAERI representatives are now H. Matsuzuru and
O. Tagawa. The list of new group members includes H. Umeki, from the
Power Reactor and Nuclear Fuel Development Corporation, Tokyo (J);
representatives from the spanish ENRESA and CIEMAT, namely:
P. Carboneras-Martinez, C. Torres Vidal and P. Prado; H. Grogan from
the Paul Scherrer Institute of Wurenlingen (CH); J.M. Laurens from
Electrowatt Engeneering Services (UK) Ltd.; R. Codell from the US
Nuclear Regulatory Commission; E.J. Bonano from SANDIA National Labora-
tories. Other group members have changed affiliation while remaining
within PSAC namely: B. Sagar, now at the Battelle Pacific Northwest
Laboratory (PNL) and P. Robinson, formerly at AERE, Harwell and cur-
rently at INTERA/ECL (UK). Given the large number of people attending
PSAC meetings the above list may not be exaustive.

assurance, verification and validation and applying the results of
these discussions to SYVAC-like codes, in order to increase confidence in the approach;
d) identifying aspects of code development and planning intercomparison of potential mutual benefit."

The meeting also discussed a universal driver, or "executive" for
the PSA codes. The driver is that part of a PSA code which performs
sampling and coordinates the modules describing the system barriers
calling them in correct sequence. This project was later abandoned
because differences among the various codes were too great to allow
such a project to be realised within an acceptable time span. The only
PSA codes already implemented at that time were the Canadian and the
British versions of SYVAC (Sherman et al., 1986) and our LISA (Saltelli
et al., 1985)

a.2. Paris, May 7, 1985: First PSAC meeting. List of modules. Topical
meetings: 1) Input data aquisition, availability and handling 2) Methods
and procedures for sensitivity analysis. This meeting was attended by
representatives of the KBS project (S), the UK Department of the Environment, the Belgian CEN/SCK, the Canadian AECL, NAGRA (CH) and the JRC
(Table a.1). The idea of a universal driver was replaced by the suggestion of establishing, at the NEA data bank of Saclay, a library of PSA
modules and subroutines to be exchanged among the various group members
(for sampling, random numbers generation etc.). A first topical discussion was held about the problem of data aquisition and handling. A
series of recommendations were drawn on the procedures to assign distributions to input parameters and on estimating probabilities. These
can be summarised as follows (SYVAC/DOC 85(1):

"Whenever at least one datum is available for a certain
parameter the use of a truncated Log-normal distribution has
been recommended, with the available datum taken as the mode

of the distribution. If more data are available then their
mean is taken as the mean of the distribution. There should
be a fixed number of standard deviations (for instance 3)
between the mean and the truncation value. This should define
the value of the standard deviation if the truncation value
is taken as the highest possible value for that parameter.
Uniform and Log-uniform distributions are to be used where:
a) the parameter is by its nature undefined (eg. time of
   occurrence of a fault);
b) there is a complete lack of data (only physical bounds
   available)."

The use of similarities among different radioelements to attribute
values to parameters has been discussed, as well as the need to establish correlations among the input parameters. A second topical discussion was devoted to Sensitivity Analysis. It was aknowledged that SA
becomes increasingly difficult when increasing the number of parameters. (Actually the difficulty increases exponentially with number of
parameters). A number of procedure were proposed to reduce the number
of parameters with distributions, such as (SYVAC/DOC 85(1)):

"- Drop radionuclides from the analysis or fix their parameters when
   they make no contribution to dose;
 - Fix sets of parameters related to barriers that do not contribute to
   dose;
 - Carry out sub-models reduction based on sensitivity analysis
 - Minimise distribution range by improving the data ..."

a.3. Winnipeg, Manitoba (CAN) September 29-30 1985: Second PSAC meeting. Decision to prepare a Level 0 intercomparison; Topical meetings:
1) Presentation of results and 2) Statistical sampling procedures. At
this meeting new participants joined the PSAC club, from the Canadian
CRNL, the USA BWIP and BPNL, the German GSF, the French IPSN and a
number of private firms working under contract for the British DoE
(Table a.1). The Group agreed there to start a step-by-step intercomparison exercise, starting with a Level zero where simplified submodels
were to be used. A topical discussion was held on the issue of "Presentation of results from PSA codes", whose conclusions were collected in
a PSAC report (PSAC/DOC 86(5)). The use of histograms, cumulative distributions, confidence bounds and risk curves was analysed. The second
topical discussion was devoted to the statistical sampling procedures.
Theory and applications of several sampling procedures such as random,
importance, Latin hypercube and stratified were highlighted, pointing
out merits and drawbacks of each. A paper of importance sampling prepared for the UK DoE was given by Dalrymple (Dalrymple, 1985). Fractional factorial design techniques were also discussed. The following
conclusions were proposed (SYVAC/DOC 85(4):
1) all the continuous sampling methods can be used to estimate both
   mean dose and probability of dose;
2) discrete sampling methods should not be used for estimating mean
   dose and probabilities;

3) simple random sampling is the most robust method for estimating the probabilities, and should be used if possible;
4) Discrete fractional factorial sampling is recommended for sensitivity analysis

There was not a complete agreement on the conclusions (as often happened in the group meetings!) because it was considered that the choice of the sampling strategy depends on the type of model used.

a.4. Ispra, April 21-23 1986: third PSAC meeting. Topical meetings: 1) Software quality assurance and validation and 2) Time-dependent processes. Level 0 1st iteration. Idea of a common intermediate file. First presentation Harwell results on analytical solutions. Some new participants joined the PSAC club, from the Canadian NUTP, the Japanise JAERI, the UK NRPB and AERE (Harwell laboratory), and more private firms working under contract for the British DoE. At the time of this meeting a number of participants had already run the Level 0 test case and a preliminary screening of the results was made. The NEA activity of modules collecton was encouraged, and all the group members were prompted to provide their modules for random number generation and sampling. The data bank was also charged to investigate the possibility to establish a common format for the output file, i.e. the file containing all the results from a PSA code simulation, to be used for subsequent statistical analysis and graphics. At this meeting a method was presented to compute the mean from a MonteCarlo simulations analytically, using simple barrier models and solving the differential equations system with Laplace transformation; according to this method, presently used for the Level E intercomparison, the exact analytical solution for the mean in the Laplace space in numerically inverted using the Talbot algorithm (Robinson and Hodgkinson, 1987). A number of presentations were given covering software quality assurance (SQA) procedures (Rae, 1985). The conclusion was reached that "SQA procedures are vital for the development of a basis for public confidence. To achieve this the models, data and codes must all be quality assured using systematic procedures that include comprehensive documentation. Nevertheless it was equally clear that the need for rigorous SQA should not hinder the early stages of development of codes where methodologies are under initial research and development." (PSAC/DOC 86(7)). The issue of the incorporation of time dependent processes in PSA codes was also tackled, following a review paper prepared by C.J. Frizelle (Frizelle, 1986), under contract to UK DoE.

a.5. London, November 11-14 1986: fourth PSAC meeting. Topical meeting: Reduction of research codes to PSAC sub-models. Use of lumped parameter for biosphere. Level 0 2nd iteration. Discussion of possible future intercomparisons: Level 1 and Level E. A new member joined the group, from the VTT centre, Finland. Many participants had run a revised version of the Level 0 test case; due to the scatter in the results type and formats it was decided to prepare a questionnaire to in order to ensure a common set of results for all the participants. A sub-group was also charged with the preparation of a preliminary report on the

exercise, after receiving the compiled questionnaires. Two other intercomparisons were also proposed; one, Level E, based on the analytical
solutions method discussed at the previous meeting. The other, Level 1,
covering more realistic disposal systems (PSAC/DOC 87(1)). Several
written contribution were presented at the topical discussion on the
reduction of research model to PSAC sub-models, covering near field,
container failure, geosphere and far field modelling, biosphere and
time dependent precesses. For an application of lumped parameter
approach to biosphere see Marivoet and Van Bosstraeten, 1988.

A compilation paper containing all the written contributions was
prepared by the NEA secretariat (PSAC/DOC 87(2)). One of the conclusions of this topical scussion wes that " Systematic methods for model
simplification are available but these require a large amount of expert
judgement. ... The use of simplified models as part of large system
assessment models should be carried out on a well documented basis, ie.
all the steps leading to the simplified model should be traceable. ...
Site specific research models should wherever possible be used as the
basis for simplification when a major safety assessment is to be
performed."

a.6. Stockholm, June 2-5 1987: fifth PSAC meeting. Topical meeting:
Probability density function justification; Level 0 3rd iteration. Plan
for the future intercomparisons. Based on the data provided on the
questionnaire a new set of Level 0 results was analysed by a restricted
task group before the meeting. A few residual systematic errors were
detected, and all the participants were allowed one more iteration of
the exercise (the last) to provide correct results. As in the preceding
two meetings the discussion of the Level 0 results was thorough (and
animate!). The task group was charged with the compilation of the final
report, after receiving the few correct results still missing. It was
also decided that the next interaction would have been the Level E,
rather than the Level 1 (PSAC/DOC 87(7)). The NEA secretariat presented
a plan for the future activitues of the PSAC User Group covering the
period 1987-89 (PSAC/DOC 87(3). Within the plane it was proposed that
"the objectives of PSAC would remain broadly the same but that, following the success of the Level 0 exercise, the main emphasis of future
work would concentrate on code intercomparison. Hence the future
programme contained four main elements: (i) information exchange; (ii)
topical meetings; (iii) code exchange via the Data Bank and (iv) code
intercomparison exercises (PSACOIN)." The plan for the intercomparison
comprises four exercises:
- Level 0   - Test of executive modules and postprocessors (concluded)
- Level E   - Comparison with exact analytical solutions (ongoing)
- Level 1(a)   - Assessment of an hypothetical deep disposal site
- Level 1(b)   - Assessment of an hypothetical shallow disposal site
- Level SA   - Intercomparsison of sensitivity analysis methods and
                 strategies

Task groups were established to implement the text cases for the various intercomparisons. These task groups are responsible for
"- Developing the case study specification;

- Conducting trial runs of the case study;
- Compiling the results provided by individual groups;
- Preparing a report of the findings for discussion by PSAC"

(PSAC/DOC 87(3)). At this meeting the NEA data bank presented a proposal for a standard output formats for PSA codes (Sartori, 1987). The standard was adopted by the JRC for the post-processor code SPOP (Saltelli, 1987). The topical discussion was centered on the problem of assigning probability density functions to input parameters; several papers were presented, including the one prepared by Stephens discussed previously at this seminar (Stephens et al., 1987) and a paper by G. Dalrymple (Dalrymple, 1987). A paper was also presented on the use of Bayesian logic in handling subjective information in probabilistic system analysis (Sagar and Eslinger 1987). A compilation paper containing all the written contributions was prepared by the NEA secretariat (PSAC/DOC 87(10)).

a.7. Paris, December 7-9: sixth PSAC meeting; Level 0 4th iteration and distribution of the final report; List of possible future topical meetings. At this meeting the two year plan for code intercomparison exercise was adopted (PSAC/DOC 87(12)). It implies finalisation by the end of 1988 of the Level E exercise and by 1989 of Level 1a. The final report on the Level 0 intercomparison was distributed (Saltelli et al., 1987). Following the discussion on this report a number of conclusion were drawn, such as the need for a standardised output, the usefulness of having detailed results from one or more deterministic calculations, and the recommendations to use as much as possible computer network links to speed up the comunications between the test case task group and the other participants. The test case specifications for the Level E were de finitively set and agreed upon (PSAC/DOC 87(13)). The exercise is now underway and a first presentation of results is expected for the next meeting in Braunshweig. Progresses were also made in the specification for Level 1a and 1b (PSAC/DOC 88(1)). For Level SA (Sensitivity Analysis) it was decided to make an exploratory calculation using as test case the same Level E model. These calcultions have been described in the previous lecture on SA techniques at the present seminar. Bruce Goodwin prepared and distributed a document describing possible future (and past) topical meetings, to assist in establishing a priority list of topics for discussion (PSAC/DOC 87(11)). The list includes: The Role of PSA Codes. Methods for Selection of Parameter Probability Density Functions. Time Dependence in PSAC. Definition of a Well Model. Sensitivity Analysis. Sampling Strategies. Simulation Cutoff Times. Acceptance Criteria/Objectives. Peer Review of Models/ Data/Code/Assessment. Methods for Interpreting and Presenting Results. Human Intrusion. Validation of a PSA Codes (Natural Analogues). Methods for Handling Model Uncertainties. Context of PSA Code in an Overall Assessment. Quality Assurance and Verification of Models and Code. Reduction of Detailed R&D Model to PSAC Sub-models. Identification of the Critical Group.

a.8. Braunschweig, June 20-23 1988: seventh PSAC meeting. Level E 1st
iteration. Level 1a,1b,ESA preliminary calculations. Topical meetings:
1) Model Uncertainty and 2) Sensitivity Analysis. This is the next
scheduled meeting of the group. It will be mainly devoted to the Level
E exercise, including a first presentation of results and an exploratory sensitivity analysis study. A topical discussion is scheduled on
model uncertainty, how does it relate to data uncertainty, use of lumped parameters and relative justification.

RISK CALCULATION IN LONG TERM PREDICTION

S F MOBBS
National radiological protection board
Chilton
Didcot
Oxfordshire   OX11 0RQ

ABSTRACT. Radiological Protection objectives for solid radioactive
waste disposal have been recommended by ICRP and they contain the
requirements that the risks to an individual should not, at any time,
exceed a specified level and that the ALARA principle should be applied
when deciding between different waste management strategies. Whereas
the concept of individual risk is relatively simple there are a number
of different possible interpretations, eg, lifetime or annual risk,
risk of death or other health effects and these are discussed briefly.
Next, two different methods of calculating individual risk based on the
scenario approach are described. The first method uses best estimate
values for each scenario, the second method uses parameter
distributions and extracts the expectation value of dose. The way in
which the two methods address the different sources of uncertainty are
highlighted and results obtained for the PAGIS project (subseabed
option) are used to illustrate the fact that the two methods do not
necessarily give the same restlts. The results of individual risk
calculations are compared with a limit or target value and at present
this is done by using the expectation value of the dose. Other
properties of the distribution of doses eg, the total range may also be
important and possible approaches are discussed. Finally the concept
of societal risk or total health detriment is introduced and the
possibility of weighing different contributions at different dose
levels is discussed.

The views expressed in this paper are those of the author; they
have not been formally endorsed by the National Radiological Protection
Board and should not be interpreted as Board advice.

1.   INTRODUCTION

     Over the past five years a number of publications have been issued
giving recommendations on the radiological protection objectives for

A. Saltelli et al. (eds.), Risk Analysis in Nuclear Waste Management, 161–174.
© 1989 NRPB.

disposal of solid radioactive wastes[1,2,3]. The concensus arising
from these reports contains two requirements. The first is that in the
post disposal, post institutional management period the risk to an
individual should not, at any time, exceed a specified level, where
risk is defined as the probability that a dose will be received
multiplied by the probability that the dose will give rise to
deleterious health effects. The second requirement is that all
exposures should be kept as low as reasonably achievable, economic and
social factors being taken into account (the ALARA principle).

Having formulated these general criteria, it is necessary to
define precisely what is meant by "risk". This is taken to be a
combination of the risk of release of radionuclides and the risk
associated with subsequent radiation doses. Therefore it is
implicitly assumed that an assessment should include all events and
processes which could lead to a release of radionuclides or influence
the rate at which release occurs.

In the UK, following guidance from the NRPB[4], the concept of an
individual risk limit has been adopted by the authorising departments
in their principles for authorising land disposal facilities for low
and intermediate level wastes[5]. In fact the authorising departments
specify a risk target for each facility of 10^{-6} a^{-1}.

2. DEFINITION OF INDIVIDUAL RISK

In general terms, the formula for calculating the risk to an
individual, in any one year, from waste disposal is:
$$R = H.r(H).p$$
where R   = individual risk (probability of a serious health
              effect)
       H   = dose
       r(H) = probability per unit dose, at dose level H, that a
               serious health effect will result
       p   = probability that dose H will be received.

There are a number of different interpretations of this formula
and the main points of interest are discussed in turn below.

2.1 Health effects

The two main categories of health effects identified by ICRP[6]
are 'stochastic' and 'non-stochastic' effects. Stochastic effects
include the induction of cancer and serious hereditary disease and are
those for which the probability of an effect occurring is a function of
dose and there is no threshold below which effects do not occur.
Non-stochastic effects are those for which the severity of the effect
varies with dose and may have a threshold.

For stochastic effects, ICRP[6] has estimated the risk of fatal
cancer for whole body exposure and for a number of individual tissues.

These risk coefficients are currently under review following new
information about the Hiroshima and Nagasaki survivors[7] and an
increase of a factor of two or three is anticipated[8]. The ICRP risk
figures are based on an average over adult males and females and hence
are representative of the risk to an average member of a population.
Indeed, since the risk of cancer induction depends upon their genetic
constitution, it is not possible to define a single set of risk
coefficients that is appropriate to all members of an exposed
population. One approach which might be appropriate for waste disposal
assessments would be to define a range of values of risk per unit dose
with a best estimate value that corresponds to the ICRP average value.

Another stochastic effect that could be considered is the risk of
contracting non-fatal cancers. The relative numbers of fatal and
non-fatal cancers depend upon factors such as the type of malignancy,
the stage at which it is diagnosed and the type of treatment available.
Obviously, future improvements in cancer treatment could reduce the
number of cancers which are fatal. At present the fractions of
naturally occurring cancers which are non-fatal are less than 40% for
all cancers except cancer of the skin or thyroid[9]. Hence only in the
cases of highly preferential irradiation of the thyroid or skin would
non-fatal cancer rates exceed fatal cancer rates. The latter is
unlikely to be of significance for post closure assessments of waste
disposal.

Other stochastic health effects of interest are genetic effects
and ICRP has published estimates of the risk of serious hereditary
disease in the first two generations and in all generations[6].
Observations on individuals irradiated in utero at Hiroshima and
Nagasaki have also shown a dose related incidence of severe mental
retardation[10].

ICRP has published estimates of threshold doses for non-stochastic
effects[11]. It is expected that for waste disposal studies potential
doses will rarely be in the range where non-stochastic effects have to
be taken into account. However, this may not be the case for human
intrusion scenarios or "worst case" combinations of parameters.

All these health effects can be aggregated into a single index of
harm. The ICRP index assigns equal weight to fatal cancer and serious
hereditary disease in the first two generations, and assigns a zero
weight to other effects. This gives a risk per unit dose of
0.0165 Sv^{-1}, though as mentioned earlier this is under review.
However, the UK authorising departments' guidelines address the
probability that an individual will contract fatal cancer and this
cannot be applied when non-stochastic effects have to be taken into
account. An alternative approach would be to consider the expectation
value of years of life lost, with an allowance for equivalent years
lost as a result of non-fatal cancer[12]. However this approach has
not yet been fully developed. It is therefore recommended at present
that the ICRP definition based on fatal cancer and serious hereditary

disease in the first two generations be adopted. From the preceding discussion, it will be obvious that the ICRP risk factors have an associated uncertainty.

2.2 Dose at what time?

The risk limits and targets given in references 1-5 apply to a member of a critical group (either real or hypothetical) who is assumed to be present at the time and place where maximum risks arise. Since the probability that a dose arises may vary over a lifetime, not to mention the risk per unit dose, it is important to know for what time, within the lifetime of an individual, risks should be calculated. The simplest approach is to calculate the annual individual dose by summing the effective dose equivalent from external irradiation in a year and the committed effective dose equivalent incurred from intakes of radionuclides in that year. Then the risk is obtained by multiplying by the probability of a health effect per unit dose, assuming that the individual lives long enough for the health effect to occur. The resulting risks for a given dose are therefore those over the lifetime of an individual. This approach overestimates the risk from radionuclides with long biological half-lives since the long latent period between exposure and expression of clinically recognisable cancer mean that the risk is not necessarily expressed during the individual's lifetime. Dunster[13] estimates that for external irradiation the likely expressed risk is 60% of the intrinsic risk, whereas for intake of plutonium the expressed risk is less than the intrinsic risk by a factor of 2 to 5. When doses are in the non-stochastic region this approach cannot be taken and it is suggested that for doses above 1 Sv an approach similar to that adopted in accident consequence calculations (see, for example, reference 14) be employed.

2.3 Critical group

Given that the dose is calculated as indicated above, it is proposed that the risk is calculated from the dose in a specified year, the probability that the dose will be received in that year and the probability per unit dose of a health effect. The suggestion[3] that the probability that the annual dose is received within the lifetime of the individual be used instead of the annual probability, will in general result in an overestimate of the actual risk in any one year. Therefore this approach is not recommended for comparison with risk limits. However, if risks in specific years are to be considered then it is appropriate to consider individuals of different ages. To be consistent with assessments for routine discharges the types of individual to be considered would be, in general, a foetus, a one year old child, a ten year old child and a twenty year old adult, living when the concentrations in the environment are highest for any one scenario. However this introduces a level of complexity that is hard to justify in view of the other uncertainties in the calculation. An alternative approach would be to calculate an average annual dose over the lifetime of an individual. This would be straight- forward since, in general, the level of contamination is likely to remain constant over a typical lifetime and hence an age-averaged metabolic model could be used. (In fact this is likely to be very similar to that for an adult). The exception to this occurs in the case of intrusion into the repository, where direct exposure of the intruder gives the highest risks. In this case the intruder can be assumed to be an adult and the maximum dose is incurred in the year of intrusion. The use of an average annual dose over a lifetime would then underestimate the maximum annual risk by a considerable factor. Therefore it is recommended that individual risks be calculated using the dose to a 20 year old adult for all scenarios and risks be calculated using doses to other age groups when it is expected, from dosimetric and metabolic data, that they will be more at risk.

Since the doses and risks may be incurred in the far future, the question arises as to what behavioural patterns should be assumed for the critical group. Obviously it is not possible to predict how lifestyles or attitudes to risk will develop in the future and hence it is usually assumed that the behaviour of the critical group should be based on current behaviour patterns. This approach essentially affords future populations the same degree of protection that we would accept today. It should be remembered, however, that over the long timescales of interest climatic or other changes may occur which change the characteristics of the environment. Therefore the agricultural practices and pathways to man considered in the biosphere model should cover the range of pathways appropriate to these future climatic states.

3.   METHODS OF CALCULATING RISKS TO INDIVIDUALS

Before discussing methods of calculating risks to individuals it is useful to recall the definition of risk as "the probability that a dose will be received multiplied by the probability that the dose will give rise to serious health effects". The problem lies in the "probability that a dose will be received" since this probability has three major components. These are (i) the probability that a particular sequence of events (or scenario) will occur; (ii) the probability that the parameters used in calculating the rates of transfer in the geosphere and biosphere have particular values for each scenario: and (iii) the probability that the parameters relating concentrations in the environment to doses to man will have particular values. These three components mean that it is not easy to present results of an assessment in a way that can be easily understood by those not familiar with the details. As a result, it is suggested that the three components should be kept separate as much as possible, since then the greatest amount of information is available. To date, no assessments have considered the third component and hence have ignored a major source of uncertainty.

3.1 Types of probability

There are two basic types of probability, related to the two types of uncertainty. These are commonly referred to as Type I and Type II uncertainty[15] and correspond to stochastic variability and lack of knowledge, respectively. This distinction will be maintained in this paper by using the term "probability" in conjunction with Type I uncertainty, the classical "frequentistic" interpretation, and calling any model that uses these "probabilistic". On the other hand, probability used in conjunction with Type II uncertainty, the "subjectivistic" interpretation, will be called "subjective probability". As an example of these two types let us consider the case of an earthquake which alters the groundwater flow pattern around the repository. The frequency of occurrence of the earthquake can be determined from measurement, and more data will lead to better estimates of the frequency. Experts will be able to provide a distribution function for the frequency, giving subjective probabilities that it has each value. This is an example of Type II uncertainty. However, even if the exact frequency is well known, the time at which the earthquake occurs cannot be known and has to be considered as a stochastic process. It is not random – it is just that we do not know enough and probably cannot know enough. This is an example of Type I uncertainty. The probability of occurrence at each particular time can be given but no amount of extra data will enable the exact time to be identified. This type of uncertainty is important if the magnitude of the dose depends upon the time of occurrence. It is not clear, however, whether the ICRP definition of risk includes both the frequentistic and subjective components of probability.

Since the two types of probability have such different interpretations it is useful to keep them separate in an assessment. This means that the effect of random variability on the magnitude of the resulting doses can be compared with the effect of lack of knowledge on the same doses. This in turn enables decision makers to discuss research priorities in a meaningful way. Thus, for clarity, the term "risk analysis" could be used only in conjunction with Type I probabilities, and the term "uncertainty analysis" in conjunction with Type II subjective probabilities. This is consistent with the approach taken in assessments of the consequence of reactor accidents.

Four situations can occur: (i) both Type I and Type II uncertainties are negligible – a deterministic model is adequate; (ii) Type I is not negligible but Type II is – a probabilistic model is needed; (iii) Type I is negligible but Type II is not – use a deterministic model together with subjective probabilities; (iv) neither Type I or Type II are negligible – use a probabilistic model and fold in subjective probabilities. The two methods of calculation presently in use reflect to some extent the way in which these two types of probability are addressed.

3.2 Best estimate approach

The potential doses from each scenario are calculated using
relatively complex models and sets of "best estimate" parameter values.
This approach is obviously best suited to the first two cases listed
above. However, as discussed later on, it is simple to understand and
hence provides a useful contribution in all cases. Type I
probabilities such as time of occurrence of a fault are included by
running the same "scenario" for different times of occurrence, these
could be termed "sub-scenarios". The scenario is thus defined as eg,
"perturbation due to occurrence of a fault", not "perturbation due to
occurrence of a fault at 10^4 y". Risks are then obtained by
multiplying the dose from each scenario by the best estimate of the
probability of occurrence of that scenario and the probability of a
health effect per unit dose. (Obviously, for each scenario the dose
vs. time curve is obtained from the dose vs. time curves and
probabilities of the "sub-scenarios"). Hence a best estimate risk
versus time curve is produced for each scenario. If the scenarios are
mutually exclusive, and the same individual could receive doses in each
instance, then the total risk is the sum over all scenarios. The best
estimate risk vs. time is therefore the expectation value of the best
estimate dose vs. time curves and represents the incorporation of Type
I uncertainties. A separate analysis of Type II uncertainties is then
required to give subjective confidence limits on the results, usually
expressed as percentiles of the distribution.

3.3 Parameter distribution approach

This approach, has mainly been used for case (iii) listed in
Section 3.1 but it could be extended to case (iv) and hence include all
the different components of probability. In this method probability
distributions of parameters are used, rather than single best estimate
values. In its simplest form, the probability distributions are
defined so that they represent either Type I or Type II probabilities
or a combination. A series of repeated dose calculations are then
made, with each calculation using a different set of parameter values
sampled from their distributions. The risk is then defined as the
product of a expectation value of the dose distribution and the
probability of a health effect per unit dose. Thus, a single risk
versus time curve is produced. Here the term 'risk' covers
'uncertainty' as well and therefore it can be argued that the
expectation value is not a good measure of the distribution of results.
In another version of this method the concept of 'scenarios' is
retained and the probability distributions reflect both Type I and Type
II probabilities for each scenario. A distribution of doses is
produced by repeated calculations, but this time there is a
distribution for each scenario. The risk is then obtained by summing
the risks from each scenario, where the risk from each scenario is the
product of the probability of occurrence, the expectation value of
dose, and the risk per unit dose. Again 'risk' and 'uncertainty' are
combined and the same comment applies as above.

A third and more complicated version of this method exists which keeps the treatment of Type I and Type II probabilities separate. Again the concept of scenarios is retained but this time the parameter distributions reflect either Type I or Type II probabilities. A number of repeated calculations are then made using sampled parameters, but this time a set of parameter values from a Type I probability distribution is run for each combination of parameters from Type II probability distributions. This enables the effect of Type I and II uncertainties to be separated by looking at the distribution of doses from Type I probability distributions for each combination of Type II probabilities. Thus, for each combination, the risk in the "frequentistic" sense can then be obtained from the expectation value of dose, the risk per unit dose and, of course, the probability of the scenario. A series of risk curves are therefore created, one for each combination of parameter from the Type II distributions, and these are used to describe the uncertainty in the risk from that scenario.

With all versions of this second method of risk calculation, it is necessary to use simpler models because so many dose calculations are needed.

3.4 Brief comparison of methods

The advantage of the best estimate method of risk calculations is that it is conceptually simple. It is also closer in form to the methods used in calculating the risks associated with accidents at nuclear facilities. However, it is eventually more time consuming, because a separate uncertainty analysis is necessary to quantify the uncertainties in the predicted risks. The parameter distribution method is more efficient but problems can arise in interpreting the results because the expectation values of the calculated dose distributions are often very sensitive to the extremes of the probability distributions of input parameters. If Type I and Type II probabilities are not kept separate, then the expectation value used to calculate the 'risk' will depend on extreme parameter values, which are more difficult to establish than best estimate values. If Type I and Type II probabilities are kept separate then the set of risk curves can be used to determine confidence levels.

In general, the estimates of risk using the best estimate method will not be the same as those calculated using the parameter distribution approach. In the end these risks will be compared with a target and used in decisions on the acceptability of the disposal option in question. Risks obtained from "best estimates" depend upon judgements about the most probable values of parameters whereas risks from "parameter distributions" depend upon judgements about extreme values of parameters. The final choice of definition of risk will depend upon the weights to be assigned to these expert judgements.

It is worth noting here another possible confusion arising from the two methods of calculation. In the parameter distribution approach

the "uncertainty" in the final expectation value of dose is often
expressed in the form of confidence levels. This is the uncertainty
due to the finite number of samples that were taken; with a very large
sample, eg, 10^6 runs, this would be negligible. Thus this measure of
uncertainty is not the uncertainty from the parameter distributions.

4.   EXAMPLE RESULTS

An example of both the best estimate and parameter distribution
methods of calculating risk is given here. The calculations were
performed as part of the CEC PAGIS project[16] and are reported in
detail elsewhere[17]. The system under study was the disposal of
vitrified high level waste in subseabed sediments.

4.1 Best estimate approach

In this approach, a number of events were identified which could
affect the waste or its environment. The effects of these events were
then analysed and a series of scenarios, representative of these
effects, were defined. Hence, each scenario may have more than one
event that could cause it. A total of eight scenarios were identified
and best estimates of their probabilities of occurrence were
calculated. Then the dose versus time curve was calculated for each
scenario, multiplied by the best estimate of the probability of
occurrence of the scenario and by the risk per unit dose and summed to
give a total risk curve (see Figure 1). For each time after disposal
the distribution of doses, one from each scenario, can be used to
determine parameters such as the expectation value of the dose or the
95th percentile of dose. The peak risk obtained from this method was
$4 \ 10^{-10} \ a^{-1}$, occurring $7 \ 10^3$ years after disposal. It is dominated by
the risk from scenario 2 and the probability of occurrence of this
scenario is dominated by the probability of human activity at the
disposal site. The peak risk from the normal evolution scenario,
assuming a probability of occurrence of unity, was $10^{-11} \ a^{-1}$ and
occurred $2 \ 10^5$ years after disposal. A separate uncertainty analysis
was performed by repeated calculations using samples from parameter
distributions. The uncertainty analysis was performed separately for
the normal evolution scenario and for scenario 2. If the normal
evolution scenario is assigned a probability of one, the uncertainty
analysis gives a range of peak individual risks (characterised by the
10th and 90th percentiles) of $7 \ 10^{-14} \ a^{-1}$ to $3 \ 10^{-12} \ a^{-1}$. Therefore,
the best estimate was at the extreme of the uncertainty analysis range
for the normal evolution scenario (it corresponds to the 99th
percentile). For scenario 2 the best estimate lay closer to the
median.

4.2 Parameter distribution approach

Again, a scenario approach was taken but only two scenarios were
investigated in detail, the normal evolution scenario and scenario 2.
In both cases parameter distributions were input and used to obtain a

distribution of doses for each scenario. For the normal evolution scenario, the expectation value of the dose peaked at 10^{-10} Sv a^{-1} at 10^5 years after disposal. Assuming a probability of occurrence of unity, this expectation value can be used to calculate a peak risk of 2 10^{-12} a^{-1} from the normal evolution scenario. This is about one fifth of the corresponding best estimate figure for the normal evolution scenario and occurs slightly earlier. For scenario 2, which considers a probabilistic event, the expectation value was again used to calculate the risk and the result (for ^{99}Tc only) was about an order of magnitude greater than the best estimate result.

The Scenario 2 results were reanalysed in terms of the frequentistic and subjective components of probability since the variation due to time of occurrence had been kept separate from that due to the uncertainty in parameter values. Thus, for each set of parameter values, the different times of occurrence produced a distribution of doses at each time and the expectation value was used to compute a risk versus time curve. Since each set of parameters produced a risk versus time curve, a family of such curves was generated giving the range of results. It was immediately apparent that Type I uncertainty dominated and the median risk curve was close to the best estimate risk curve. In fact, when presented in this way these results are identical to the uncertainty analysis results of 4.1. The mean of the family of risk curves is, of course, the result presented earlier as the 'risk' from both type I and II variations.

4.3 Comparison of results

As seen above, the two methods give different results and the way in which they differ depends upon the scenario. The simple parameter distribution method gave lower estimates of risk from the normal evolution scenario (by a factor of about 5) but higher estimates of risk from the other scenario, scenario 2 (by a factor of about 10). This behaviour is partly due to the way in which the best estimate values relate to the parameter distributions and will be dependent upon the waste disposal option studied. It should be noted that the best estimate approach used here was a true probabilistic approach and gives the best estimate risk value. However a separate uncertainty analysis was required to provide an estimate of the range of risks and this was done using a parameter distribution approach identical to that in the second method. In contrast the parameter distribution approach itself can be used to provide either one unique measure of risk incorporating both objective and subjective elements, or, if designed so that Type I and Type II can be kept separate, it can be used to provide the range of risks from a probabilistic event. The latter interpretation provides more information to the analyst and hence is preferred. In the case described here it was found that the risks were associated with the high dose scenario and that the uncertainty in the risks due to other parameter variations was less once the event had occurred. Although this was immediately apparent from the best estimate results and from the parameter distribution results when Type I and II

probabilities were separated, it was not apparent otherwise. This
added amount of information can be important when comparing the results
with a risk target or limit. The best estimate method, together with
uncertainty analysis, gives a value to be compared with the risk limit
or target and an estimate of the range of risk values expected. The
simple parameter distribution function method just gives a single risk
estimate although, if required, it could also be designed to produce a
range of risk estimates. If two disposal options are being assessed,
it may happen that they have similar best estimate risk values but
different uncertainty ranges. The decision maker can then make
judgements on whether, for example, an option with a slightly higher
best estimate risk but a small uncertainty is preferable to one with a
lower best estimate risk and large uncertainty. With a single risk
figure, this distinction is not clear cut. It seems reasonable to
expect acceptance criteria to take account of the treatment of
uncertainties.

In fact, the distinction between the results of the two methods of
calculation disappears if a different measure of the distribution of
results is taken, for example the 95th percentile (roughly equivalent
to the 95% confidence level). This explicitly includes Type II
uncertainty and hence both methods would be presenting the same result,
the 95th percentile of risk.

5.   SOCIETAL RISK

The societal risk, or risk to a population, can be defined in a
way analogous to that of individual risk. In other words, the product
of the probability of each type of deleterious effect and the severity
of the effects. At low dose rates and dose the detriment to health is
taken to be the expected number of fatal cancers and serious hereditary
disease in the irradiated population and collective effective dose
equivalent is taken as a measure of this detriment. Societal risk is
calculated for use mainly in comparisons of the impact of alternative
sites or disposal options. In these studies it may be pertinent to
ask whether the temporal or spatial distribution of risk should be used
to differentiate between options and whether individual risk levels
should be given different weights, for example should there be a level
below which individual risks are considered not to contribute to
societal risk? These questions are being actively addressed by a
number of groups at the moment. Until a consensus is reached,
calculations of societal risk should therefore contain information
about the temporal, spatial and individual risk components.

6.   CONCLUSION

The definition of the "risk to an individual" has many different
interpretations and some of these have been addressed here. One of the
major sources of confusion arises from the interpretation of the
"probability that a dose is received". Following the nuclear plant
accident methodology, probability can be divided into "frequentistic"

and "subjective components. The former are then used to compute the risk whereas the latter compute the uncertainty. This has led to the "best estimate" method. Alternatively both aspects of probability are used to compute risk and a parameter distribution approach is taken. It has been argued here that more is learnt about the behaviour of the system if the frequentistic and subjectivistic components of probability are kept separate. However if the resources for this are not available, it would be preferable to use the parameter distribution approach than to do best estimate calculations without uncertainty analysis.

7.   REFERENCES

1.   ICRP, Radiation protection principles for the disposal of solid radioactive waste. Oxford, Pergamon Press. ICRP Publication 46. Ann. ICRP 15, No 4 (1985).

2.   Hill, M D and Webb, G A M. Radiological protection criteria for waste management. Nucl. Energy 24 (2) (1985).

3.   NEA, Long term radiation protection objectives in radioactive waste disposal (report of the joint CRPPH/RWMC expert group). Paris, NEA (OECD), 1984.

4.   NRPB Radiological protection objectives for the disposal of solid radioactive wastes. Chilton, NRPB-GS1 (1983) (London, HMSO).

5.   Department of the Environment, Scottish Office, Welsh Office, Department of Environment for Northern Ireland, Ministry of Agriculture Fisheries & Food. Disposal facilities on land for burial intermediate level radioactive wastes: Principles for the protection of the human environment, London (HMSO) 1984.

6.   ICRP, Recommendations of the International Commission on Radiological Protection. Oxford, Pergamon Press, ICRP Publication 26. Ann. ICRP, 1, No. 3 (1977).

7.   Preston, D L and Pierce, D A. The effect of changes in dosimetry on cancer mortality risk. Estimates in the atomic bomb survivors. Hiroshima, Radiation Effects Research Foundation, RERF TRO-89 (1987).

8.   NRPB Interim guidance on the implications of recent revisions of risk estimates and the ICRP 1987 COMO statement. Chilton, NRPB-GS9 (1987) (London, HMSO).

9.   Thorne, M C. Principles of the ICRP system of dose limitation. Br. J. Radiol. 60, 32-38 (1987).

10. ICRP. Developmental Effects of Irradiation on the Brain of the Embyro and Fetus. ICRP Publication 49. Annals of the ICRP, 16 (4), 1986.

11. ICRP. Nonstochastic Effects of Ionising Radiation. ICRP Publication 41. Annals of the ICRP, 14(3), 1984.

12. ICRP. Quantitative Bases for Developing a Unified Index of Harm. ICRP Publication 45. Annals of the ICRP, 15 (3), 1985.

13. Dunster, J H, J. Soc. R. P. 6, (1), 15-22, 1986.

14. Hemming, C R, HEALTH-MARC: the health effects module in the methodology for assessing the radiological consequences of accidental releases. Chilton, NRPB-M78 (1982).

15. Hoefer, E and Hoffman, F O. Selected examples of practical approaches for the assessment of model reliability - parameter uncertainty analysis. IN. Uncertainty analysis for performance assessments of radioactive waste disposal systems. Proceedings of a Workshop, Seattle, February 1987. NEA (OECD), Paris, 1987.

16. CEC. PAGIS. Performance Assessment of Geological Isolation Systems for Radioactive waste: Summary. Commission of the European Communities, Nuclear Science and Technology, EUR 11775, Luxembourg 1988.

17. Mobbs, S F, Charles, D, Delow, C E and McColl, N P. The assessment of subseabed disposal of vitrified high level waste for the PAGIS project. NRPB R218/EUR 11483, 1988.

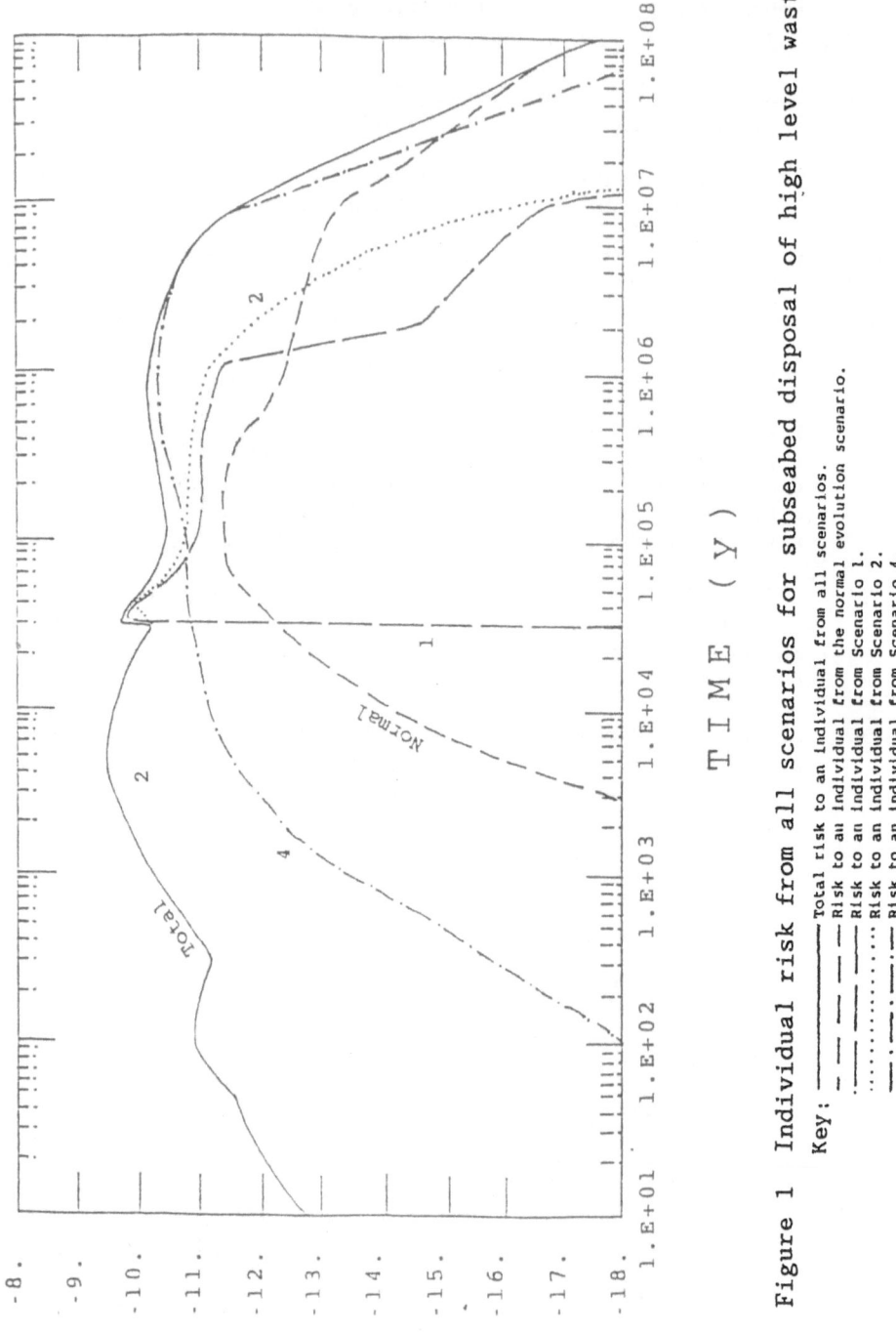

Figure 1 Individual risk from all scenarios for subseabed disposal of high level waste.

Key: ———— Total risk to an individual from all scenarios.
     — — — Risk to an individual from the normal evolution scenario.
     —·—·— Risk to an individual from Scenario 1.
     ········· Risk to an individual from Scenario 2.
     —··—··— Risk to an individual from Scenario 4.

Section 4 - The word to the geologists: what if...?

Predictive geology is aimed at predicting future earth behaviour based on the knowledge of geological processes. In Dr. Fourniguet's presentation the relevant geological phenomena capable of affecting repository performance are screened and their consequences combined with the use of a computer code (CASTOR). The relevance of climatic changes (glaciation) is highlighted. In Dr. D'Alessandro's paper the information relative to a possible future glaciation is parametrised, and the doses arising from an actual disruption of the repository due to glacial bulldozering are computed with the help of a Monte Carlo code.

THE USE OF PREDICTIVE GEOLOGY IN THE ANALYSIS OF REPOSITORY EVOLUTION. DESCRIPTION OF THE RELEVANT NATURAL PHENOMENA

J. FOURNIGUET
Bureau de Recherches Géologiques et Minières
Département Géologie
B.P. 6009
45060 Orléans Cédex 2
France

ABSTRACT. Predictive geology is the extrapolation into the future of the combined geological knowledge of past processes having acted upon the earth, in order to predict the earth's future behaviour. This method is designed and applied to nuclear waste management in order to foresee the kind of wents which could produce some positive or negative effects on a repository. As a first step, it has been necessary to make as exhaustive an inventory as possible of all relevant geological processes that could affect the repository. Those processes appear to belong to two main groups : external geodynamics (climatology, erosion, sea-level changes, ...) and internal geodynamics (tectonics, volcanism, seismicity, ...). Rates and laws have been gathered and introduced in a model specially designed to combine their various effects ; scenarios are derived from the CASTOR model, which show the geological behaviour of a site in the future. It appears that the driving forces that are likely to act upon a repository are of external origin, mainly due to the influence of climatic changes ; vertical movements and strong earthquakes occurring sight on the site could also have negative effects.

1. "PREDICTIVE GEOLOGY"

Predictive geology is the extrapolation into the future of the combined geological knowledge of past processes having acted upon the earth, in order to predict the earth's future behaviour. Obviously, this concept covers a vast domain, but in general each geological subdiscipline will select the predictive tools it requires.

Quantitative forecasting in geology is not a very common practice, even though many scientists try to make projections into the future of the phenomena they study. Nuclear-waste management, with its enormous radioactive half-life constraints that approach a "geological" time span, is nowadays perhaps the most important reason for using such an approach. Geologists are not used to making accurate quantitative predictions for the phenomena they deal with; in general they are better qualified to study the past – and this usually in

A. Saltelli et al. (eds.), Risk Analysis in Nuclear Waste Management, 177–199.

considerable detail. The near geological past, such as the Quaternary period with its nearly intact record, has been fairly well studied and will be the key for our geological predictions.

Forecasting for short periods is common in fields such as the weather and the economy, but for nuclear waste the time span to be taken into account is far longer. Since such waste, whether of high, medium or low activity, has a long half-life, the time involved can be measured in tens or hundreds of thousands, or even millions, of years (Fig. 1).

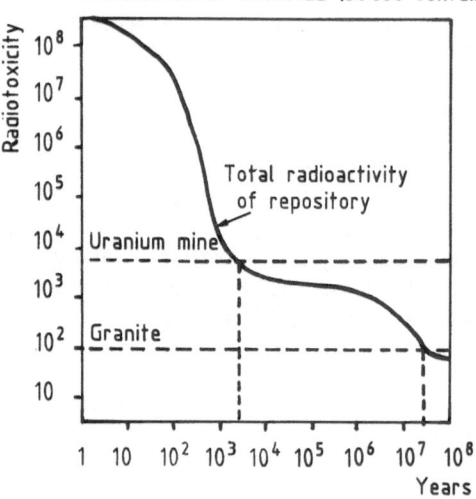

Risk of contamination over periods on a geological time scale.
Need for long-term confinement; disposal of highly-radioactive waste with a long half-life in deep geological formations.

Figure 1. Evolution with time since disposal, of the radioactivity of reprocessing waste (from Masure, 1982)

Such time spans are perfectly compatible with geological periods and the duration of geological phenomena:
- the Quaternary era is officially 1.8 million years long;
- a glaciation lasts roughly 50,000 years;
- plate motions are commonly 3 to 6 cm/year, i.e. 3 to 6 km per 100,000 years;
- vertical movements of the crust are commonly in the range of 1 to 2 mm/year, i.e. 100 to 200 m in 100,000 years or 1 to 2 km over 1 million years;
- certain erosional processes can remove several hundreds of meters over the span of 100,000 years.

Whatever the action of Man, with or without any disposal of nuclear waste, the region where a repository is planned will undergo a natural evolution that may affect the site itself. It is thus necessary to **understand** the processes that are at the root of such evolution, in order to predict their results in the future.

2. HOW PREDICTIVE GEOLOGY WORKS

2.1. The inventory phase

As a first step, it is necessary to make as exhaustive an inventory as possible of all relevant geological processes that could affect a repository site, hereafter called "parameters". During this phase, it must be considered whether **direct** or **indirect**, and **major** or **minor** actions should be taken. Priorities should be allocated at a later stage. It becomes rapidly evident that such parameters are numerous and varied, and that they are highly interactive (Fig. 2). It is, however, clear that they belong to two main groups, pertaining to **exogenic, or external, geodynamics** and **endogenic, or internal, geodynamics** .

Since the prediction must be quantitative, this **qualitative** inventory must be followed or accompanied by a **quantitative** one, i.e. the investigation of rates, speeds, laws, curves, etc., to explain the evolution and mechanism of each parameter.

The second step is to understand all possible interactions between the parameters. None of them acts strictly independently, since the earth is a closed system. The interactions occur between external and internal parameters, as well as between the two categories. Moreover, some feed-back, synergy or cancelling-out effects may appear as well.

2.2. The modelling phase

Because of the number of parameters, their interactivity, and the need for quantified answers, the next step is modelling. In the geopredictive programme developed by the BRGM, a computer simulator known as CASTOR(1) has been designed for this purpose. Without going into too much detail, this simulator can take into account each parameter, with its variations, to combine them and to produce scenarios of their future evolution. This paper deals only with the inventory of the relevant natural parameters.

3. PHENOMENA PERTAINING TO EXTERNAL GEODYNAMICS

An analysis of external geodynamics shows that almost all processes on the earth's surface are related to at least one of three main factors, which are **weathering, erosion** and **sedimentation**. The whole is under the direct influence of climatic variation.

---

(1) Construction Automatique de Scénarios d'évolution d'un site de Stockage de Radionucléides

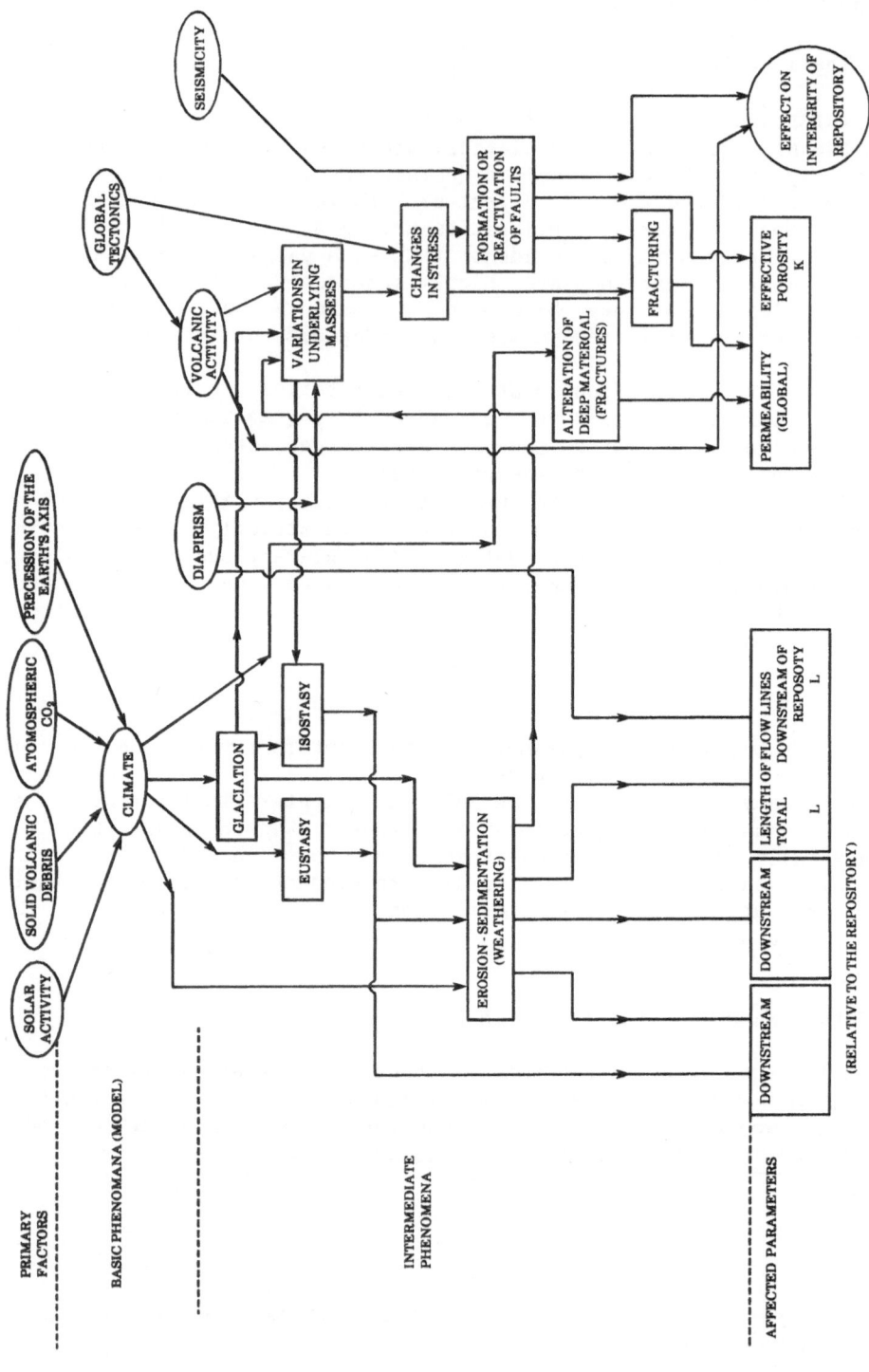

Figure 2. Initial inventory of geodynamic factors and their possible evolution

The three main parameters, governed by climate, act on the surface. A storage-site area, as a part of the surface, is defined by:
- **lithology** or rock-type, which may be granite, schist, carbonates, clay, salt, etc.;
- **structure**, as the rocks will form part of an orogenic belt, a basin, a graben, an old craton, etc.;
- **morphology**, corresponding to specific slopes, altitudes, relief, etc.;
- **geographical location** on the globe, which is obviously significant for the present climate.

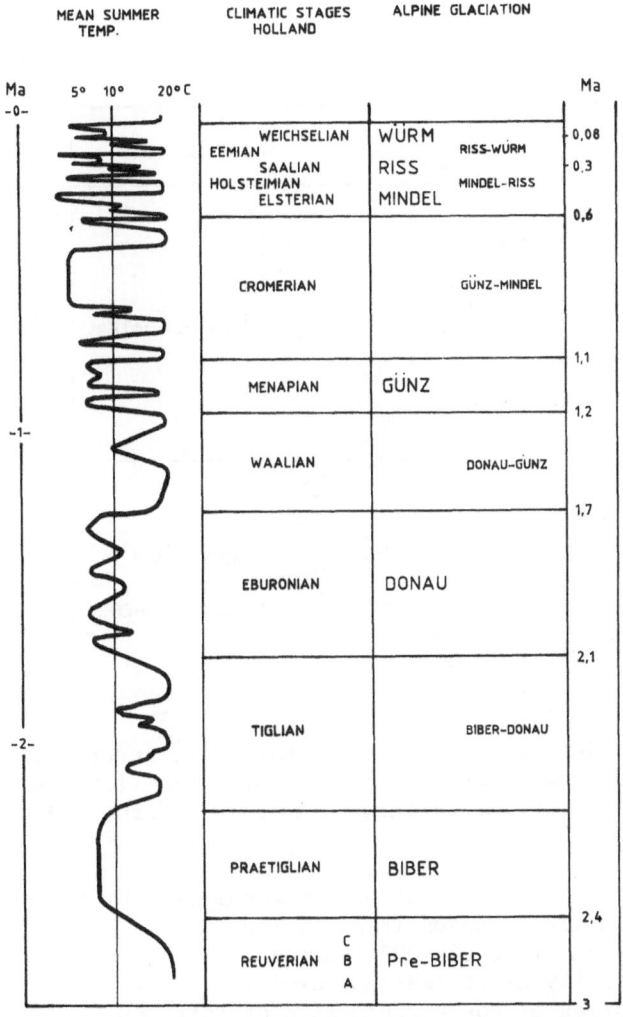

Figure 3. Climatic curve and Pliocene-Quaternary glacial and interglacial periods. The presence of two absolute time scales (Ma) illustrates the uncertainty on the subject. The right-hand scale has been used in the text.

It must be kept in mind that, during the history of the earth, the climate has changed considerably at any given point. Even during the, geologically relatively short, period of the Quaternary and late Pliocene, the climate of the earth underwent numerous severe changes (Fig. 3). It is this period that interests us from the viewpoint of predictive geology, in order to be able to extrapolate with reasonable confidence the geological behaviour of a waste repository.

When considering the last part of the Quaternary era (late Pleistocene) , we can see that the climatic variations were in fact very complex, and that a "cold" period such as the Würm was in reality compounded of very cold stadia, with less cold to temperate inter-stadia (Fig. 4).

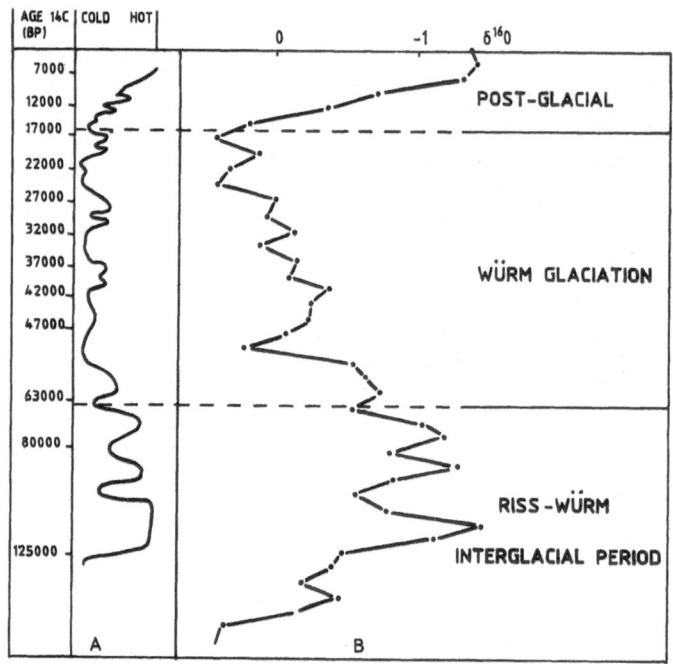

Figure 4. Paleo-climatic curves for the Riss–Würm interglacial period and the Würm glaciation, based on pollen analysis (left) and varia-tions in O18/O16 in marine-foraminifera tests (from: Duplessy et al., 1976)

Such variations can be caused by **astronomical** and by **tectonic** pro-cesses. The astronomical causes (Fig. 5) are:
- the precession cycle (26,000 years);
- the eccentricity of the earth's orbit;
- the inclination of the earth's axis of rotation.

All are well-known and computed, and in combination produce a global insolation curve that can be correlated with climatic curves deduced from geological data (Fig. 6).

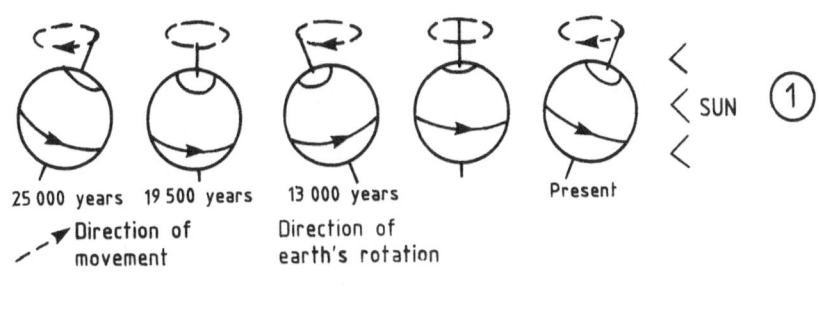

25 000 years   19 500 years   13 000 years   Present

Direction of movement   Direction of earth's rotation

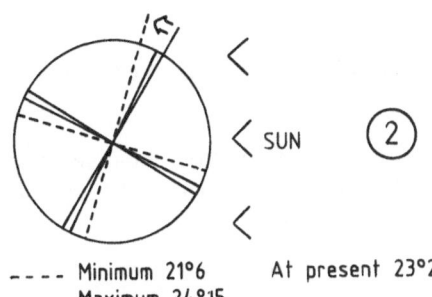

---- Minimum 21°6   At present 23°27
——— Maximum 24°15

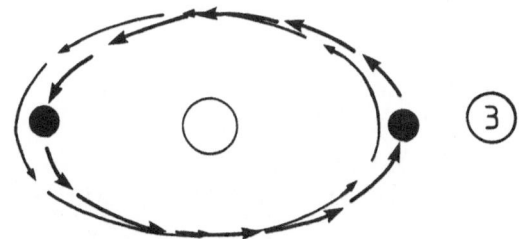

1. The rotation of the earth's axis, causing it to describe a concern in space, is called precession. It completes one revolution in 26,000 years.

2. The inclination of the earth's axis is also affected by a slow oscillation, and varies between 21°6' and 24°15' over a regular period of 40,000 years. The angle at present is 23°27' and is decreasing, thus decreasing the difference between summer and winter.

3. The earth's orbit changes from an almost perfect circle to a pronounced ellipse. During the elliptical phase, the earth is closer to the sun during a certain season. A complete orbital cycle from the circular to the elliptical and back to the circular, lasts between 90,000 and 100,000 years.

Figure 5. The three cycles affecting the earth's orbit according to Milankovitch.
Relation between annual insolation at 45°N and ground-ice cover during the last 160,000 years (the dotted line shows the recalculation of Milankovitch's curve.

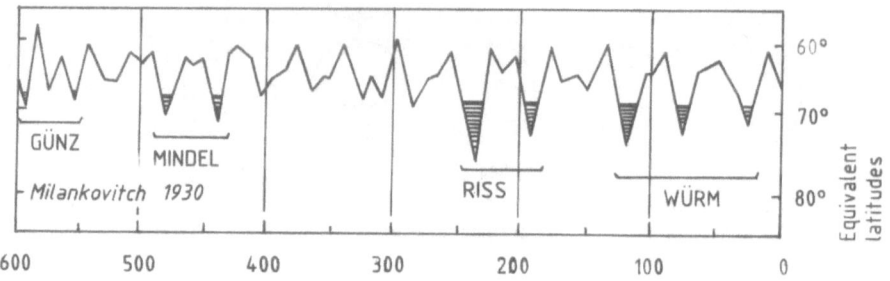

Figure 6. Variations in solar radiation at 65°N over the past 600,000 years

It is thus possible to predict future climatic change, which shows that **there will be another glaciation** within the next 100,000 years, probably about 50,000 years from now (Fig. 7).

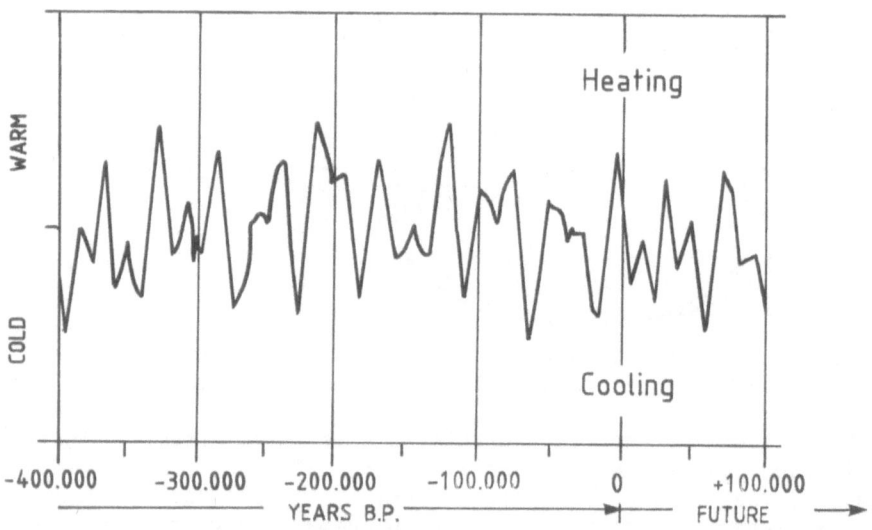

Figure 7. Curve of long-term variations over the last 400,000 years, and forecast for the next 100,000 years (from: Berger et al., 1980)

Global tectonics also have an influence: the present location of crustal plates stops circulation of water between the Pacific and Atlantic Oceans, except in the far south. As the Atlantic Ocean receives most of the cold water of the earth, the global temperature balance will not change, and future ice-caps will be roughly in the same locations as before (Fig. 8).

Figure 8. Present distribution of the continents; the Antarctic cools the oceans and water from the ice-covered Arctic ocean pours into the Atlantic

The main points to keep in mind are:
1. We are at present in an interglacial period that has lasted 13,000 years; comparison with the last interglacial and/or interstadial periods shows that this temperate climate **cannot** continue for very long.
2. The next cold period will probably be as severe as that of the Würm, with a similar extension of ice-caps (Fig. 9).
    The reason why climatic variations are so important in our case, is because they induce two phenomena of major influence on the evolution of the surface.
1. The development of ice-caps, whose existence has three main consequences:
a) isostatic subsidence during the glaciation and isostatic rebound after it; such movements may be of the order of several hundreds of meters and can cover very large areas (Fig. 10). In fact, the increase and decrease of pressure, due to formation and melting of ice, can provoke a certain amount of re-mobilization of diapirs;
b) erosion under the ice-cap and at its edges, leading to U-shaped valleys several tens of km in length and hundreds of meters wide and deep;
c) periglacial conditions at the edge of the ice-cap, and extending up to 1,000 km from it; very strong weathering occurs due to the low temperatures, and soil and rock freeze to several hundred meters depth.

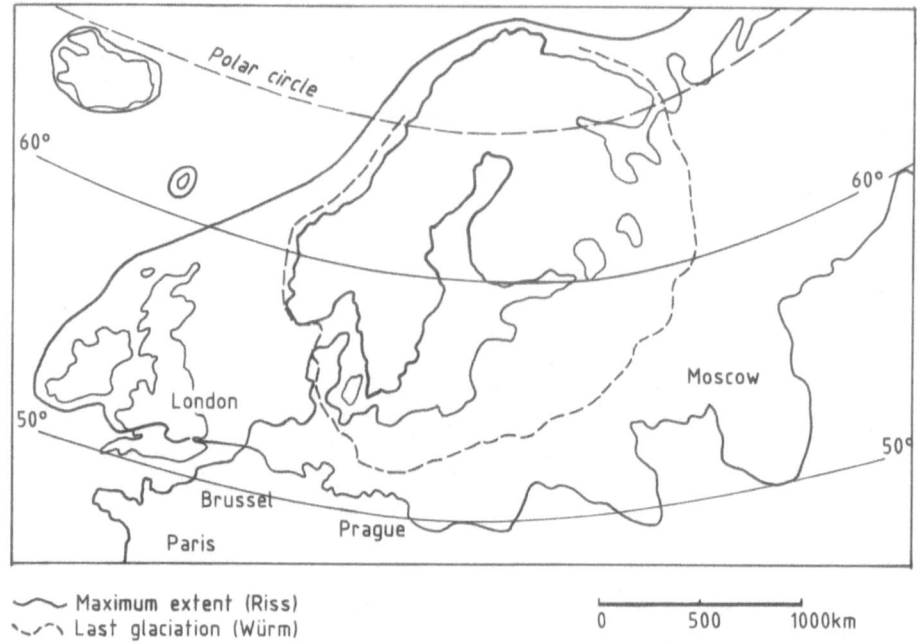

Maximum extent (Riss)
Last glaciation (Würm)

0   500   1000km

Figure 9.   Extent of ice caps in northern Europe

2. Sea-level changes will take place, as the amount of sea-water locked up in ice during glaciations is so great that the sea falls far below its present level; for instance, it is commonly accepted that the sea-shore was 120 to 140 m lower during the Würm glacial period than its present level (Fig. 11). This lowering of the sea-level has 3 main consequences:
- . lengthening of rivers in periglacial conditions;
- . increase in erosion due to the lower base-level;
- . isostatic adjustment of the continental shelf due to the lighter load of the water on it.

Future climatic variations, however, will produce only cold periods in Western Europe. No tropical variation can be predicted on the basis of the near past, and thus no deep lateritic weathering will occur, which can reach 100 or 200 m under such conditions.

To summarize, and considering only the external processes, it is worth keeping in mind, as far as nuclear waste management is concerned, that:
- knowledge of the near past is the key to the near future, as regards the nature and extent of the phenomena;
- events that occurred during the last 100,000 years, will almost certainly recur during a period of similar length in the future, and approximately in the same areas;

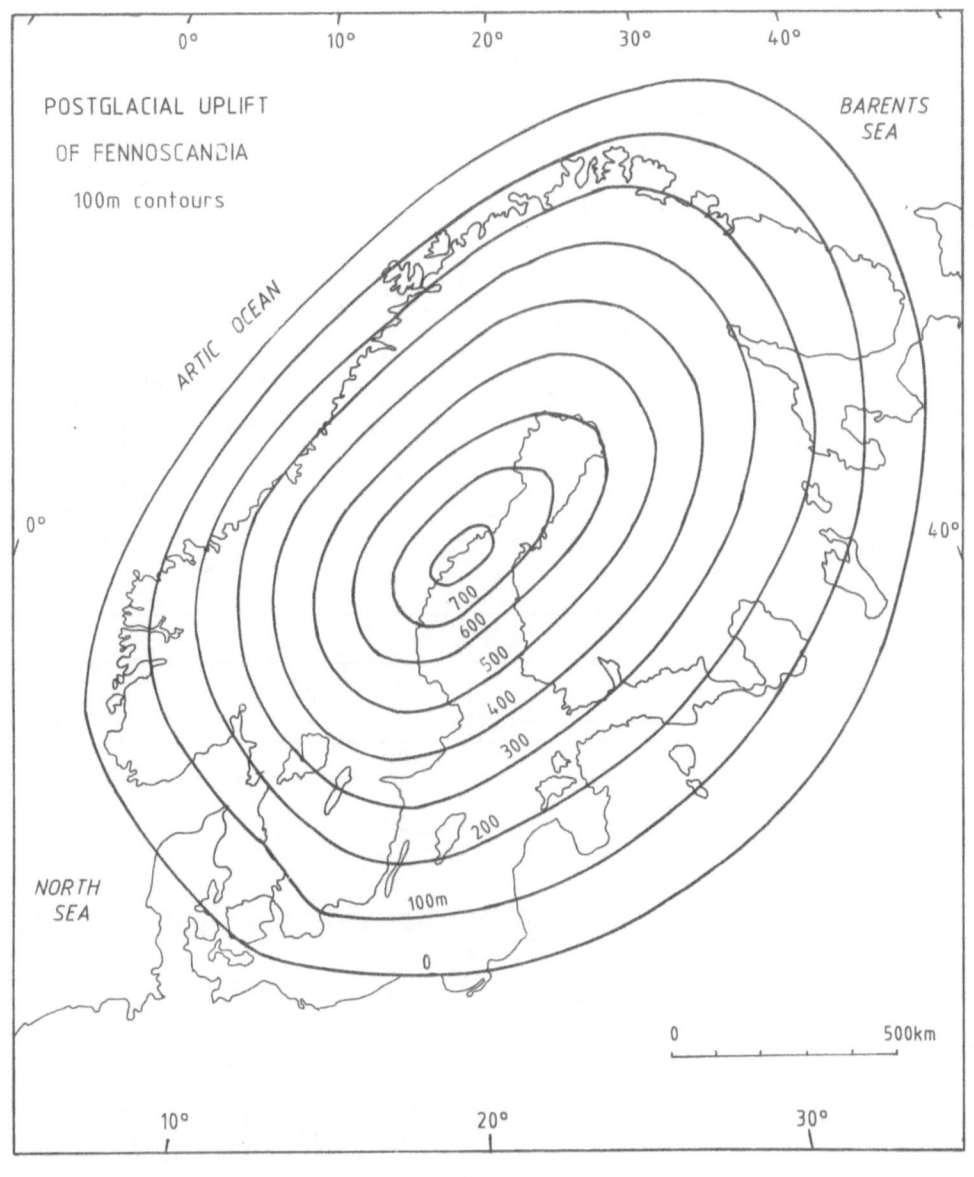

Figure 10. Post-glacial isostatic rise of Fennoscandia (from: Mörner, 1977)

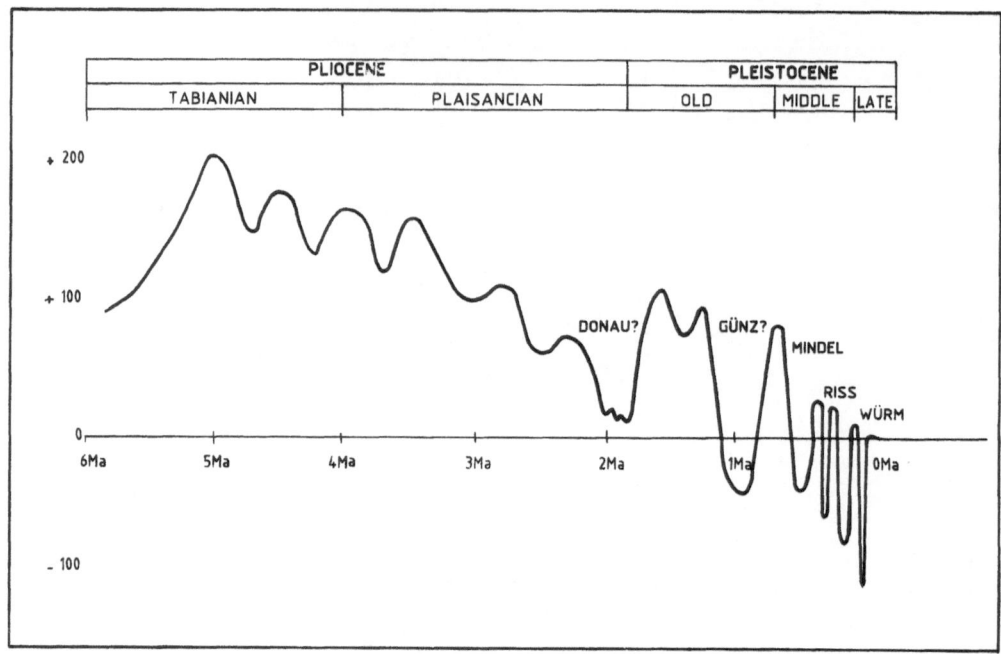

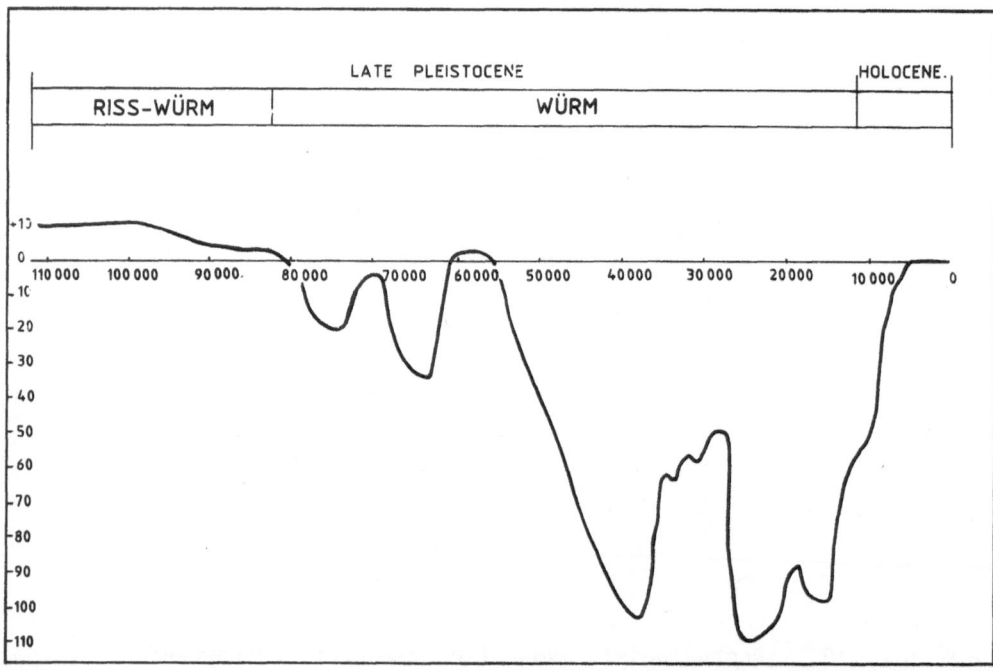

(Modified after: de Lumley, 1972)

Figure 11. Estimated variations in the level of the Mediterranean during the Pliocene-Quaternary (above) and during the last ice age (below) (modified after: de Lumley, 1972)

- the location of a waste repository in relation to shorelines, mountains, and the former extent of ice-caps and periglacial zones, is essential;
- the main problem is erosion, whose rates may vary substantially (Fig. 12):
  . the areas likely to undergo deep erosion, whatever the processes, need to be identified;
  . the types and rates of probable erosion need to be quantified on the disposal site, as a function of the rock successions that might be exposed to erosion, and of the climatic conditions.

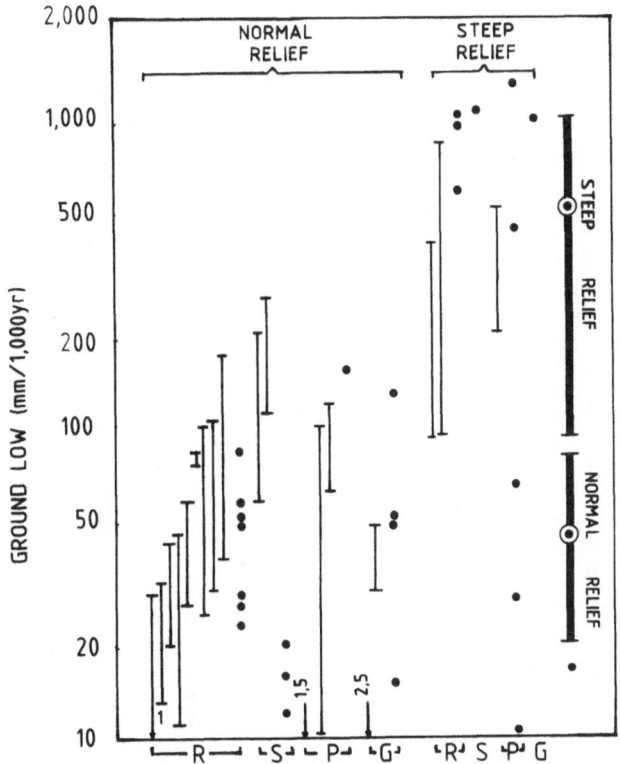

Figure 12. Reported rates of lowering of the land surface. To the right are shown the median values and inter-quartile ranges for normal and steep relief. Types of evidence: R, river load; S, sedimentation in reservoirs; P, surface process measurements; G, geological evidence.

4. PHENOMENA PERTAINING TO INTERNAL GEODYNAMICS

Internal geodynamics are essentially driven by global tectonic processes: plate motions can explain most of the past, present and future deformation and movements. The earth's surface consists of lithospheric plates moving according to several mechanisms:
- convergent movements, producing subduction;
- spreading movements;
- strike-slip along transform or transcurrent faults.

On the continents, and especially in Europe, the first of these processes is dominant. Fig. 13 shows generalized plate motion in Europe.

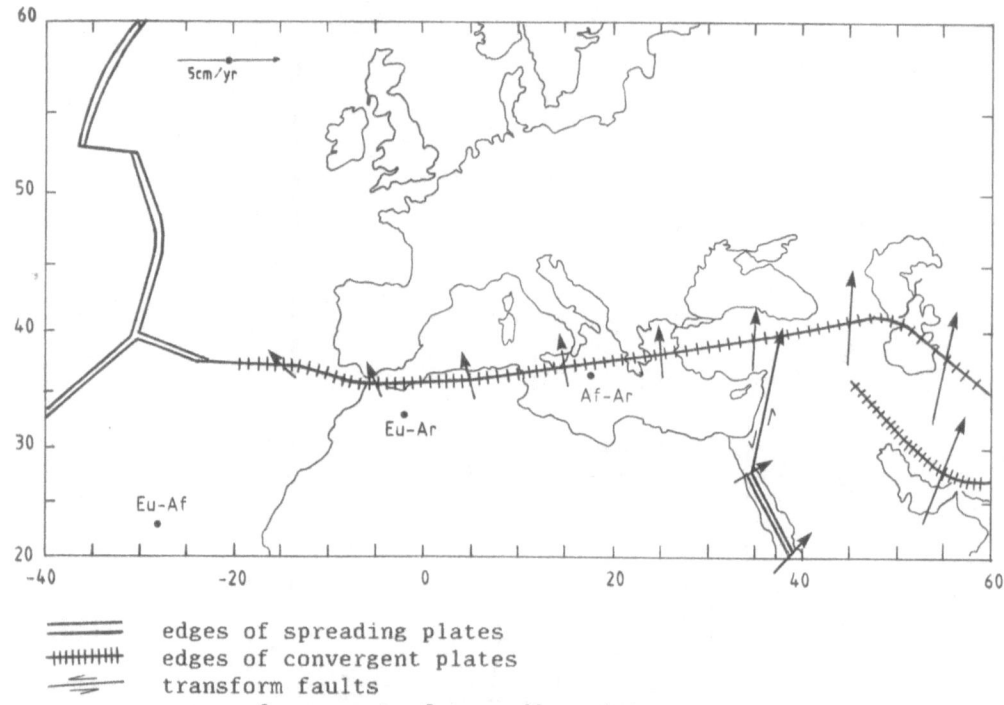

══════	edges of spreading plates
┼┼┼┼┼┼┼┼	edges of convergent plates
══╧══	transform faults
──────►	sense of movement of spreading plate

Figure 13. Simplified geodynamic diagram of the Atlantic-to-Mediterranean Basin area (from: McKenzie, 1972)

Fig. 14 shows the relationships between plates during convergence. Plate motions are the driving forces for:
- seismicity,
- volcanic activity,
- vertical and horizontal movements;
- geothermal gradient and flow,
- tectonic activity.

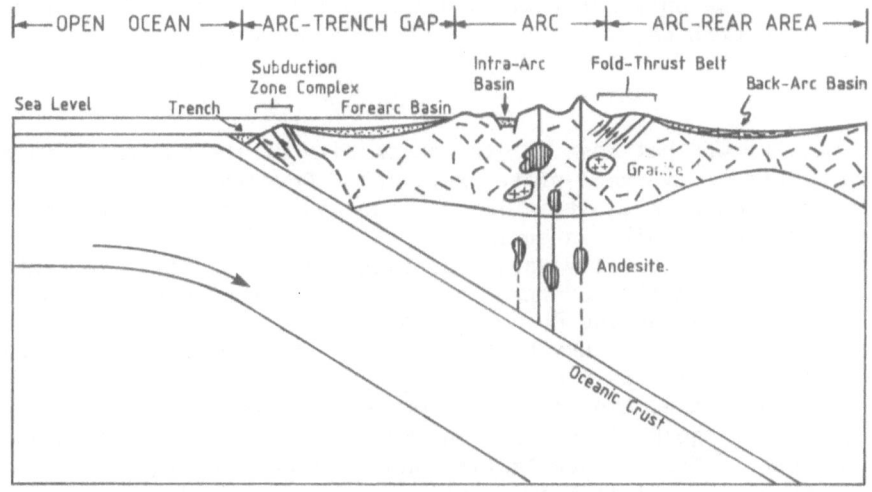

Figure 14a. Morpho-structural units of a subduction zone (from: Condie, 1982)

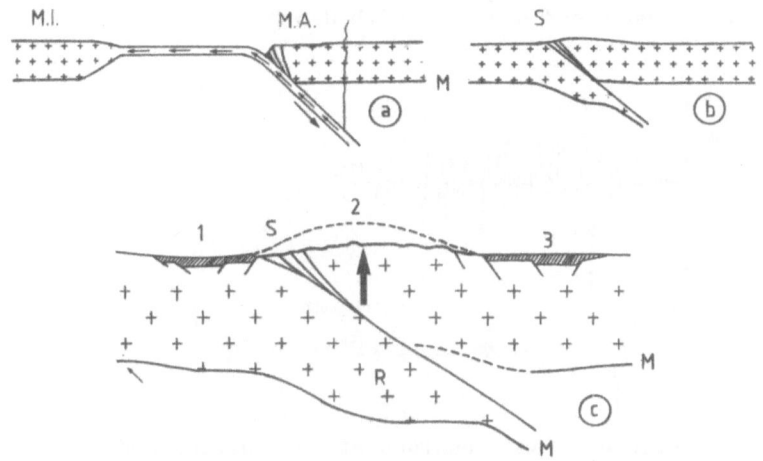

a, subduction preceding collision; b and c, collision
1, external molasse basin; 2, mountain belt; 3, internal molasse basin; M, Moho; R, root; M.I., passive margin; M.A., active margin; S, suture

Figure 14b. Schematic representation of continent-continent collision

Diapirism is the only internal process that is not clearly connected with plate tectonics.

Seismicity

It is well known that most of the seismicity of the earth is concentrated along convergent-plate margins. Nevertheless, intra-plate seismicity does exist; it is less common and randomly scattered, but it may reach quite high intensities.

In Europe, seismicity varies substantially from north to south:
- it is weak in the European plate, except in some places such as the Pyrénées and the Rhine Graben;
- it is strong at the border of the African plate in Spain, southern Italy and Greece.

In either case, two situations may be considered:

1. An earthquake is centred on the site; as this would be due to motion along a fault, substantial damage would be the result. In fact, such a proposal is highly unlikely since we know that **new faults** are very rare, and that all preliminary surveys of sites would have eliminated faulted sites.

2. The earthquake occurs at some distance from the repository, which would thus only undergo the transmitted effects. The farther the site is from the epicenter, the weaker the effects. Moreover, the deeper the focus, the less severe the damage. Fig. 15 shows how the acceleration and speed decrease with depth.

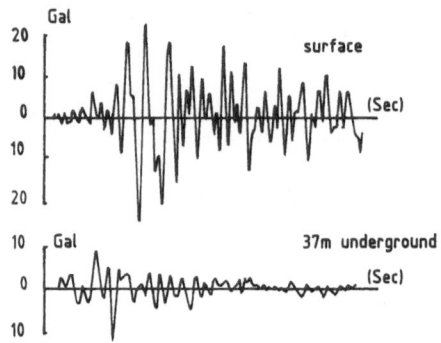

Figure 15. Accelerograms recorded at the surface and at 37 m depth at Urayama, Japan (from: Okamoto, 1969)

It can be assumed that for a repository depth of 1 km, except for major earthquakes (intensity 9-10), the damage will be negligible a few km from the source. Fig. 16 shows the attenuation of movement according to depth and distance of the fault.

Earthquakes may affect hydrogeology, modifying the water table or the discharge from springs (Fig. 17). Usually the effects can be felt over large areas (several 10,000 km²) and may be permanent, especially in unconsolidated sediments, due to compaction.

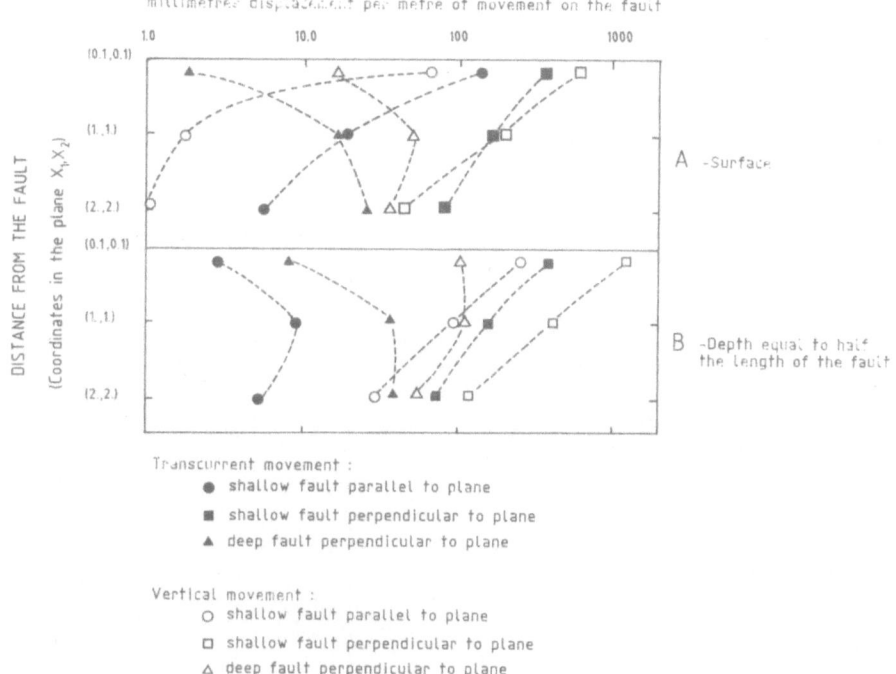

millimetres displacement per metre of movement on the fault

A – Surface

B – Depth equal to half the length of the fault

DISTANCE FROM THE FAULT
(Coordinates in the plane X_1, X_2)

Transcurrent movement :
- ● shallow fault parallel to plane
- ■ shallow fault perpendicular to plane
- ▲ deep fault perpendicular to plane

Vertical movement :
- ○ shallow fault parallel to plane
- □ shallow fault perpendicular to plane
- △ deep fault perpendicular to plane

Figure 16a. Displacement calculated at the surface and at depth 1.0, as a function of distance from a fault (from: Pratt et al., 1979)

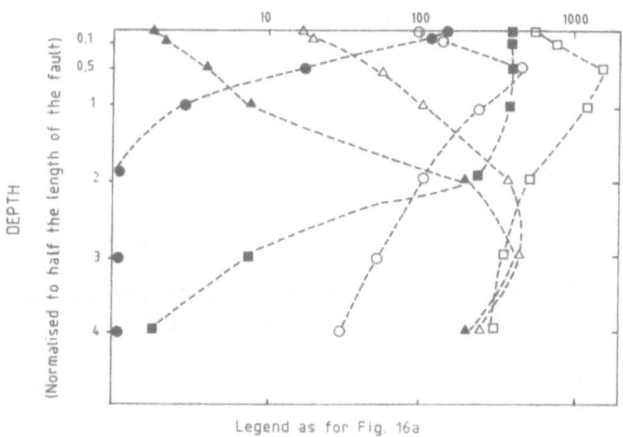

DEPTH
(Normalised to half the length of the fault)

Legend as for Fig. 16a

Figure 16b. Displacement calculated as a function of depth at the coordinates 0.2, 0.2 (from: Pratt et al., 1979)

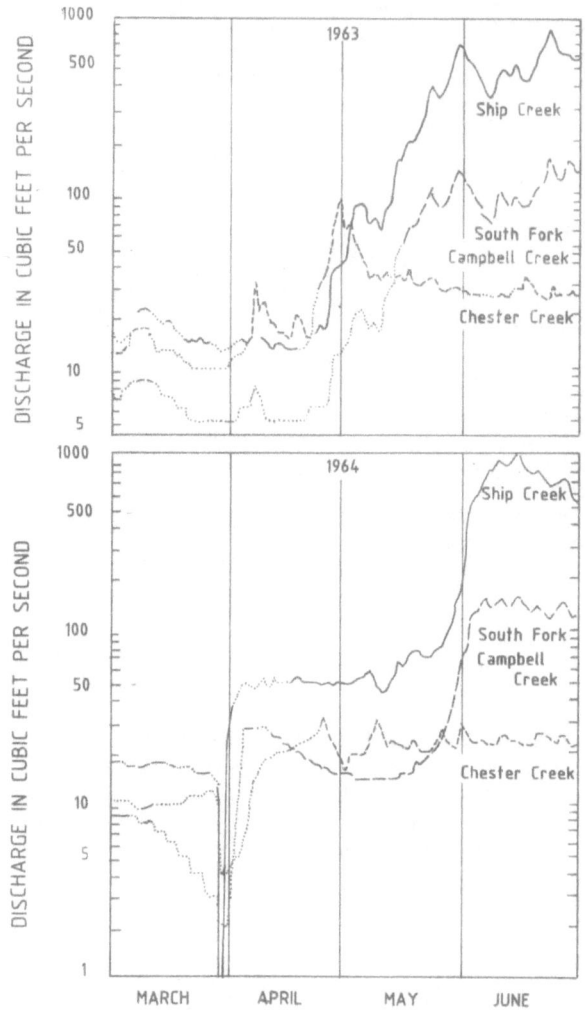

Figure 17. Examples of variation in flowrate of different streams,
recorded after the 1964 Anchorage earthquake in Alaska

Volcanic activity

The location of volcanoes is closely related to the geodynamic context, which may be subduction zones with andesitic lavas, hot-spots, intra-plate volcanoes linked to extentional zones, or lava flows in spreading zones such as the Djibouti area.

In Europe, we are mainly concerned with intra-plate volcanoes, such as the volcanoes of the Massif Central in France (Fig. 18), and subsidiary plate-edge volcanoes (Italy). The effects of volcanic activity can be two-fold:

- . direct intrusion into the repository: destruction with release of radionuclides is probable;
- . spreading of lava over the surface directly above the repository: no effect except further reinforcement of the geological barrier.

Prediction of the activity of a volcanically-active zone is possible if age, frequency and type of the emissions are known. For instance, in France the Chaîne des Puys is likely to erupt within 1,000 years, whereas it is unlikely to occur for 10,000 years in the Vivarrais and for 100,0 years at Mont Dôme. On the other hand, development of a **new** volcanic zone is practically impossible.

Horizontal and vertical movements

Horizontal movements occur under certain specific conditions:
- – global plate motion, with no effect on a repository;
- – strike-slip movements along faults;
- – continuous deformation of sediments during folding due to shortening.

The last two cases could directly influence a repository, except that sites are presumably chosen in areas where faults and folds **are already known.**

Vertical movements may occur under any conditions, whether the context is a basin, basement area, orogenic belt, or intra-plate graben. They may be positive as well as negative, and their rates can vary from 0.01 to 10 mm/year (Fig. 19), or 10 m to 1 km in 100,000 years.

Since collision is expected to last for several million years, there is no reason for the stress field to change significantly in the near future. However, it must be pointed out that unpredictable local variations of the stress field may occur, and that inversion of the two main stresses may occur. At present, the value of σ_1 is far greater than those of σ_2 and σ_3; nevertheless, though unlikely, an inversion between σ_1 and σ_2 is possible, which would change the possible movements along faults.

Diapirism

Movements due to the displacement and deformation of evaporitic rocks may be considerable. The factors controlling the displacements are mainly increases in temperature or in stress.

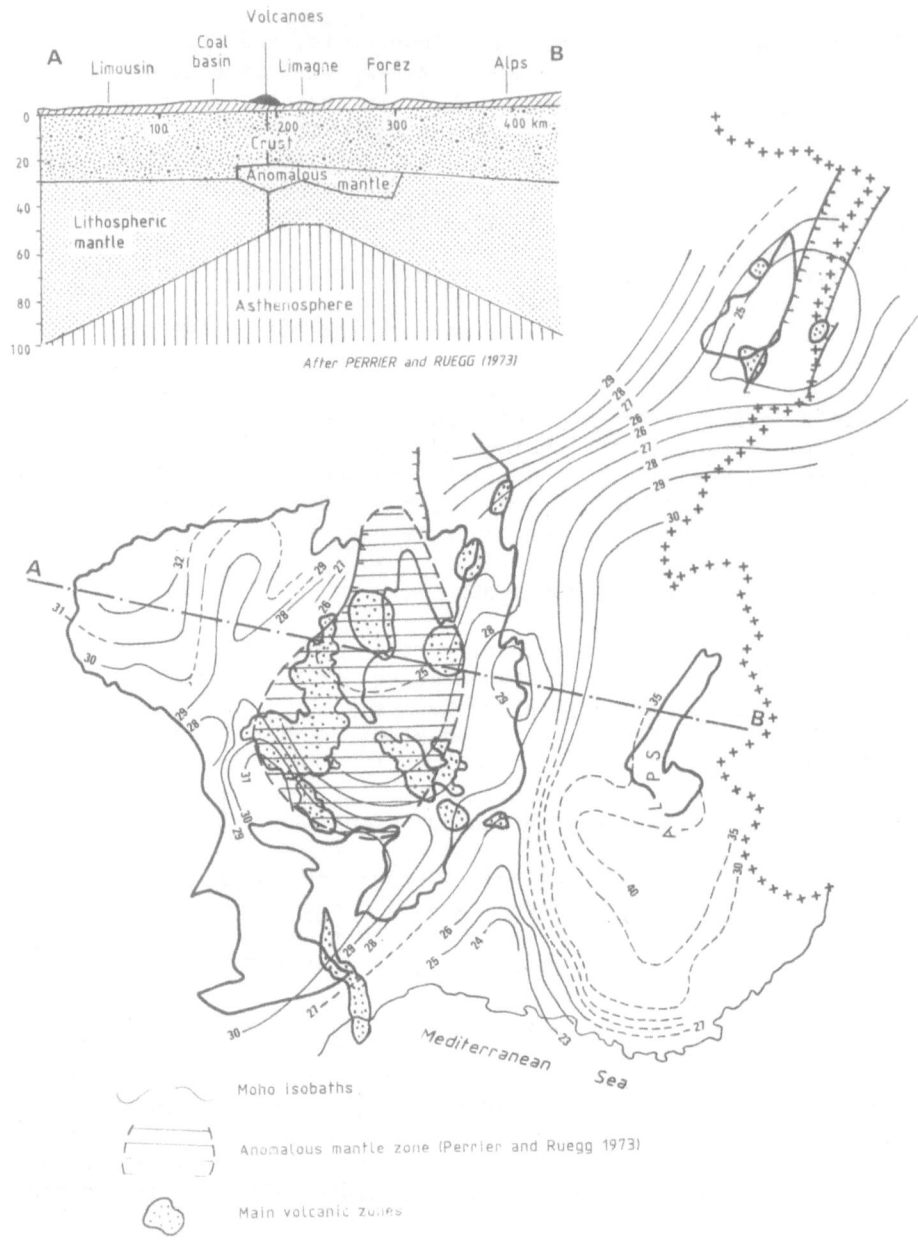

Figure 18. Structure of the crust and mantle beneath the volcanic zones in south-central France, showing crust-mantle transition iso-baths

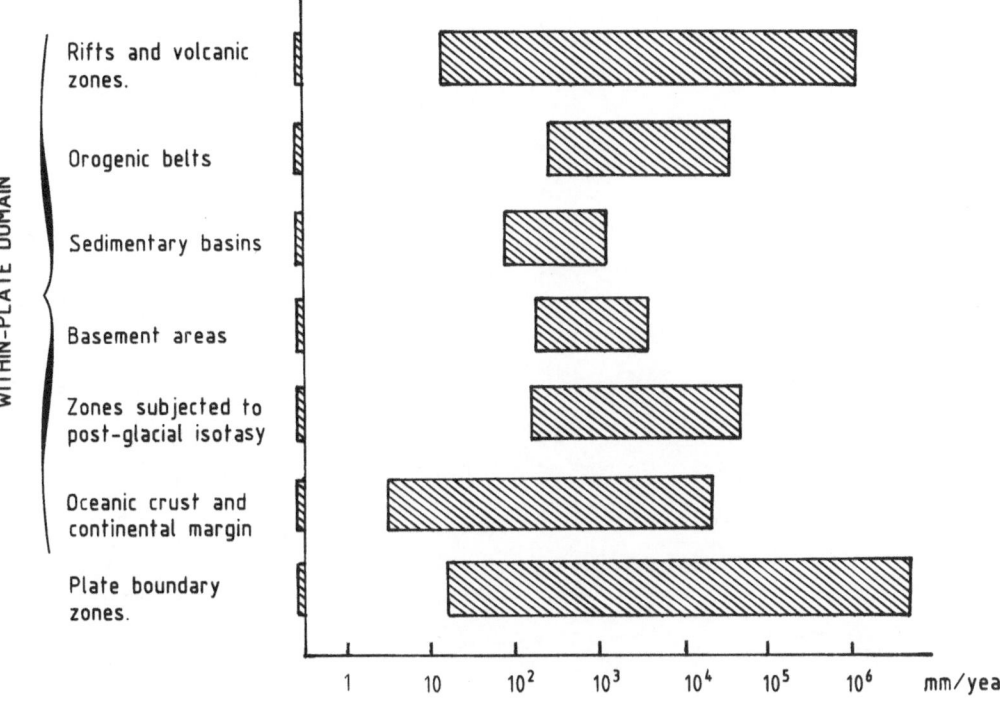

Figure 19. Comparison of the rates of vertical movement in selected geodynamic contexts

Generally speaking, once a diapir has begun to move, it is highly likely to move again under the effect of slight changes in the tectonic stress field, overload such as that caused by an ice-cap, or removal of sediment due to erosion. Even so, it is very unlikely that a **new** diapir would be created within 100,000 years, even if some changes of the stress field occur.

Heat flow

This parameter is less important. Provided the value is known (in this case 1.2 μcal/cm²/s), it can be assumed that the heat flow will not change much, even during the next 100,000 years, since the **very** reason for its existence is directly linked to global tectonics.

Tectonic activity

Information must be available on two main subjects:
1. The structural framework of the area, i.e. a map of all faults and structures, whatever their age;
2. The recent evolution and present state of the stress field. It is known that the stress field has changed during geological times

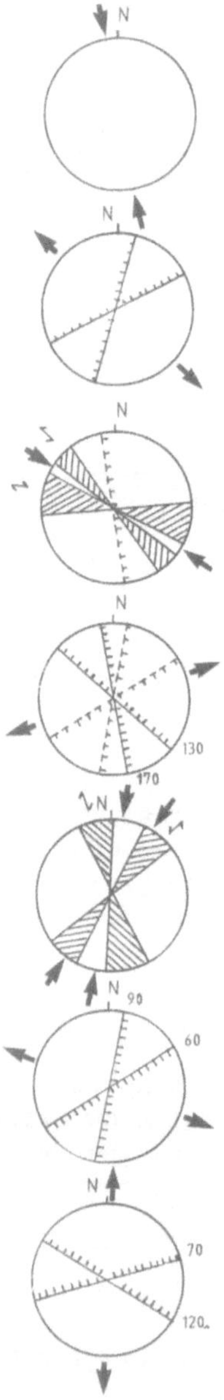

Orientations of the main phases of deformation	Ages and probable durations of the main phases of deformation	Ages of paroxysms of deformation with durations in millions of years
NNW-SSE compression	Quaternary (2Ma)	
NW-SE extension	Pliocene (4Ma)	
WNW-ESE compression (Alpine compression) (Transcurrent faulting)	Late Miocene (Alpine) (9Ma)	Messinian ("Pontian")
ENE-WSW extension (Normal faults)	Oligocene (10-15Ma)	Chattian Late Stampian Late Ludian 1-1.5Ma Early Ludian 1-1,5Ma
N-S compression (Pyrenean compression) (Transcurrent faulting)	Late Cretaceous- early Eocene.(Pyrenean) (25-30Ma)	Bartonian 4-5Ma Early Eocene (Post Rognacian pre-Lu-tetian)
ESE-WNW extension (Normal faults)	Early Jurassic (≃30Ma)	
N-S extension (Normal faults)	Permo-Triassic (≃40Ma)	

Figure 20. Phases of brittle deformation in the Paris Basin and its surroundings (excluding the Alpine domain)

(Fig. 20), but remained globally constant throughout the Quaternary; Western Europe is undergoing NNW-SSE oriented compression.

This situation is due to collision between the African and European plates, which accounts for all the (neo)tectonic activity.

5. CONCLUSIONS

The main conclusion of this study, is that the driving forces that are likely to act upon a waste repository are of external origin, except for some particularly strong vertical movements and the occurrence of very strong earthquakes close to the site.

6 - REFERENCES

Berger, A. & al. (1980) 'Long term variations of monthly insolation as related to climate changes', Int. A. Wegener Symposium, Berliner Geowissenschaftlische Abhandlange, Reihe A. Band 19, pp. 24-25.

Condie, K. (1982) 'Plate tectonics and crustal evolution', Pergamon Press Eds.

Duplessy, J.C. & al. (1976) 'Paléoclimatologie des temps quaternaires à l'aide des méthodes nucléaires', in La Préhistoire Française, t. 1, Ed. CNRS, pp. 352-361.

Lumley (de), H. (1976) 'La Préhistoire Française', IXe Congrès UISP, Nice, Ed. CNRS, 1459 p.

MacKenzie, D. (1972) 'Active tectonics of the Mediterranean region', Geophys. J. Astr. Soc., 30, 109-185.

Masure, P. (1982) 'Evacuation des déchets radioactifs à vie longue en formations géologiques continentales : point de vue du géologue', Note Technique BRGM-SGN/GEG, Orléans, n° 19/82.

Mörner, N.A. (1977) 'The fennoscandian uplift : geological data and their geodynamic implication', in Earth Rheology and Late Cenozoic Isostasic Movements, Int. Symp. Stockholm University, pp. 79-92.

Okando, S. (1969) 'Introduction to earthquake engineering', Univ. Tokyo Press, 571 p.

Perrier, G. & al. (1973) 'Structure profonde du Massif Central Français', Ann. Géophys., 29(4), 435-502.

Pratt, H.R. & al. (1979) 'Earthquake related displacement fields near underground facilities', Report DP 1533, Savannah River Lab., Aiken (SC).

IMPACT OF GLACIATION ON GEOLOGICAL DISPOSAL: THE PROBLEM OF PREDICTION
AND THE CHALLENGE OF MODELLING

M. D'ALESSANDRO*, A. BONNE**
* JRC Ispra
** SCK/CEN Mol

1. RATIONALE

One of the main concerns of the public, the Authorities and the nuclear
energy producers is the long-term safety of the disposal strategy of
the radioactive waste arising from the nuclear power production.

Since 1976 the JRC-Ispra has developed a methodology and various
models to assess the long-term risk related to the disposal of radioactive waste in geological formations. During these studies a co-operation with the CEN/SCK of Mol made it possible to check the validiy of
developed methodology by means of its application to a possible potential disposal site: i.e. a repository in the Boom clay formation underlying the nuclear establishment of Mol.

The risk assessment is performed in two steps: firstly, all of the
conceivable site evolution scenarios capable of provoking a failure of
the geological barrier are defined and ranked according their probability of occurrence. This is done by means of the application of the
Fault Tree Analysis [1]. The second step of the assessment requires the
development of a consequence analysis (at least for the most likely
processes) to quantify the effects of the release event.

As pointed out by the FTA, a scenario which is worth being analysed is a "glacial erosion scenario" concerning the formation of an
ice sheet over the disposal site and the related geomorphological processes, some of which seem capable of geopardizing the repository
integrity.

For the modelling of such a process affecting a nuclear waste repository in the case under study, the following approach was followed:
1) study of literature about glacial erosion evidences;
2) sampling of relevant evidences;
3) scenario development on basis of selected evidences and site specific repository conditions;

A. Saltelli et al. (eds.), Risk Analysis in Nuclear Waste Management, 201–227.
© 1989 ECSC, EEC, EAEC, Brussels and Luxembourg.

4) critical analysis of the developed scenario with experts of the glacial activity.

The main issues concerned in the Scenario description for the assessment of the related consequences were:
1) description of glacial dynamics;
2) description of time scaling;
3) description of the environment evolution on probabilistic basis.

A critical examination of these issues was expected from the consulted experts.

For the glacial erosion knowledgeable experts were available at Geological Survey of the Netherlands and constituted the body of the consulting team.

In the present paper a short description of a possible scenario for the glacial erosion of a nuclear waste repository is descripted as it was developed for the site of Mol. Emphasis is laid upon the dynamical, evolutionary and environmental aspects because these are the basis needed for modelling the phenomenon and evaluating the dispersion of radionuclides into the environment.

2. BACKGROUND

2.1 Repository site conditions

Within the Belgian programme of geological disposal of radioactive waste, clay formations have been chosen because of their low permeability, good ion exchange properties and for their capability of plastic deformation under tectonic stresses.

From among many clay thick levels lying within Belgian underground, Boom clay formation has been chosen as potential host rock because of its homogeneity, thickness and extent, as well as for lying at a mineble depth in the Mol region.

The Boom clay belongs to the rupelian age of the Oligocene and is covered essentially by miocene, pliocene and pleistocene sands. In the Mol region, the clay is 109 m thick and situated between -160 to -269 m below land surface. In our approach, we considered the clay layer homogeneous over a thickness of 90 m, not affected by any circulating ground water.

Neogene sands above the clay formation contain an aquifer which constitutes an important water resource for the region: the water table is shallow /1-2 m/. Below the clay, several aquifers exist, which are not exploited in the Mol region; however, further south, they are shallower and are utilized for industrial and domestic purposes.

The underground facilities of the repository consist of a horizontal network of galleries and storage holes excavated at a depth of about 200 m. Due to the huge pressure in a plastic clay body at such a depth, the waste will be separated from the geologic medium by a lining (cast iron plated with a cement) which resists the pressure.

The repository conceptual design (see Fig. 1) consists of a parallel set of galleries, 2.5 km in length, and about 3.5 m in diameter. The length and width of the designed repository are respectively 2.5 Km and about 600 m. The overall surface covered by the network of galleries can be calculated to be about 2 km^2, while the actual area occupied by the holes and galleries will account for only 10% of the total.

2.2 Evidences of glacial activities and dynamical aspects

From a general point of view an extensive reglaciation of the Northern Hemisphere would produce many important changes in continental areas with a complete upset of the pre-existing morphologic features and a drastic reduction of epicontinental seas. It is learned from the past that many of these changes have important consequences in the geological evolution of the affected regions. The importance of each process depends on the topographic features of the area, as well as on its position in respect of its distance from the glacier ablation centres.

The geomorphological features of the Mol site, as well as its geographic position are such that the consequences of a glacial cover should not be severe as far as its geologic stability is concerned, since the region is flat and far away of the possible glacial ablation centres.

However, in our case, the geological evidence shows something different: along the front of the Saalian ice cap in the Netherlands, a hundred kilometres from the potential site, five basins 130 m in depth are recognizable. Although the debate about the mechanism of their formation is still open, no doubt exists about their glacial origin.

Zagwijn [2] has suggested the hypothesis that a phenomenon called "glacial surge" was active to create these basins. At present this phenomenon is seen in polar regions, occurring when the instability of a glacial lobe causes the ice to flow rapidly, provoking important erosive effects. However it is not known whether it occurs at the edges of existing ice-caps. In facts no recent examples of the formation of glacial lobes at existing inland-ice fronts are known.

The depth reached by the glacial erosion on the sandy terrains of the Netherlands, about 125 m below the present sealevel, though not being sufficient for the repository to be affected, is such that we cannot exclude that the same phenomenon could reach 200 m, namely the

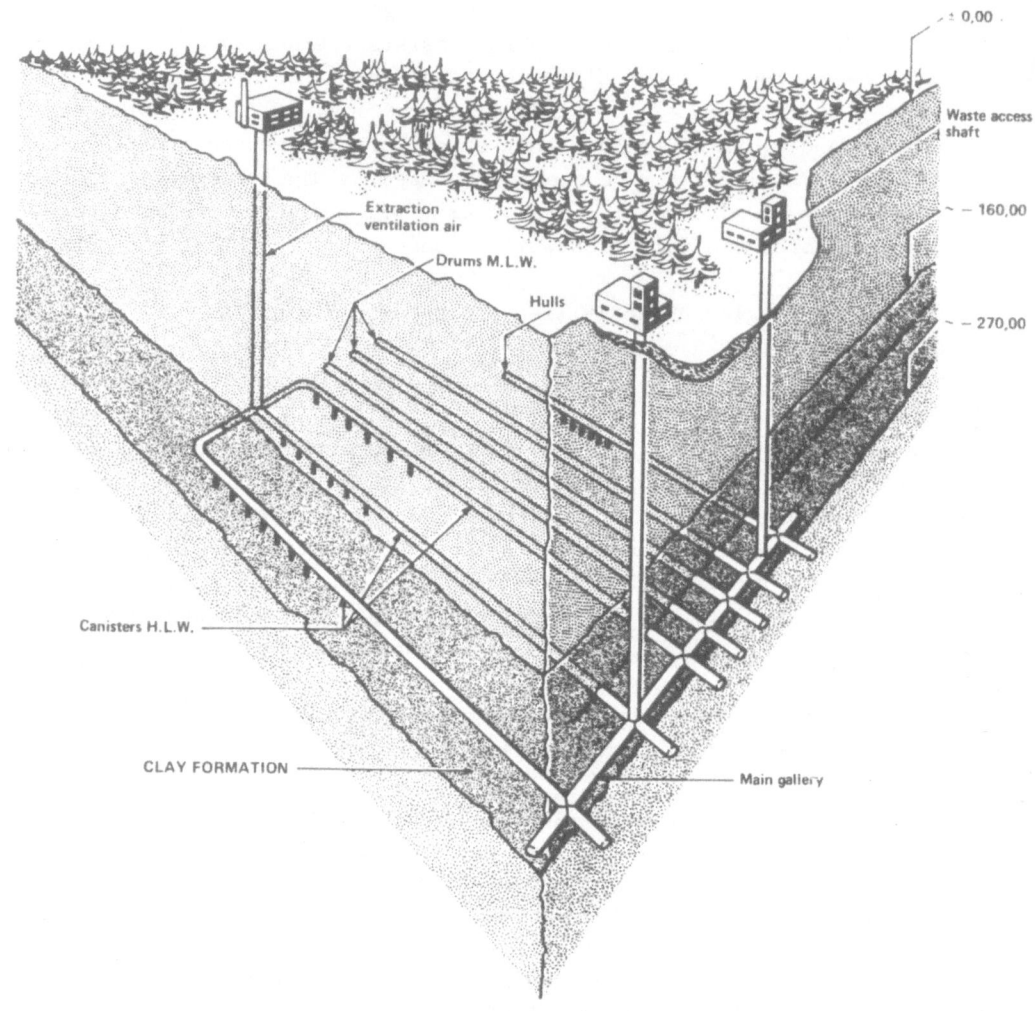

Figure 1. Conceptual repository design.

repository depth.

Moreover, in central and northeastern Lower Saxony, many channels due to glacial erosion cut across the prequaternary surface. Some channels can be traced over more than 100 km and can reach as much as 500 m in depth. The general aspect of such channels points to subglacial deepened structures. The erosion can possibly be ascribed to the activity of the subglacial melt-water during Elsterian glaciation [3].

Regardless of the mechanism of their formation, the presence of such large examples of glacial erosion denotes the importance of a phenomenon we cannot ignore. Some other statements underline the significance of the glacial scenario: "In future glaciations deep erosion by ice action will probably occur in those areas where deep erosion has already occurred" [4], where "those areas" refers to the "same environments" where the event occurred in the past. Thus, an ice cap, the southern borders of which will be hundred kilometres more expansive than the Saalian one, could cause the same effects as it did during Saalian glaciation, within the repository site.

The striking examples of glacial action mentioned above constitute two distinct scenarios, involving two mechanisms which are very different as fas as the vitrified waste dispersion and the final dose to man are concerned. Therefore, they cannot be represented by the same model, and the most important phenomenon must be chosen as guideline for scenario development.

The two kinds of erosion have more or less the same frequency per surface unit within the areas affected, although the comparison has been made on the basis of maps which perhaps present different levels of accuracy.

The glacial phenomena digging the Saalian basins are closer to the site and the mechanism of their formation is fairly understood, so that the terms of their possible development can be modeled fairly reliably. On the other side, the erosion channels, although being characterised by the same frequency, are less understood as far as their origin and development are concerned: the glacial sub-erosion is the most likely cause, but many aspects have still to be explained. For example, these specific erosion forms occurred mainly during the Elster glaciation, little traces of similar channels being recognizable among the effects of other glaciations.

For these reasons the Saalian glaciation were selected for the analysis. Thus the release model is developed by taking into account a repository disruption by the same mechanism which formed the Dutch basins during the Saalian glaciation, with the assumption that the entire repository has been affected by the process. The main goal of the model is to define the size and the distribution of waste after being affected by the erosion process.

All other ways of breaching, anyhow related to ice-cap development, the evidence of which is not recognizable near the potential site, will be disregarded in the present model.

Thus, permafrost, glaciofluvial erosion, changes in groundwater circulation patterns, although being considered as marginal aspects of the scenario, will not be considered as possible mechanisms of release. The isostatic adjustment is known to play an important role as trigger of basement movement along pre-existing fault plains. Such an occurrence, however, must be considered in the frame of faulting scenario, the glaciation being in this case only an indirect cause of release.

3. TIME SCALING OF A GLACIAL SCENARIO

Before entering the site specific evaluation, it could be useful to become familiar with the magnitude of the phenomenon as well as with the required time-spans.

The extensive glaciations of the Northern Hemisphere constitute a peculiarity of the most recent geologic times, the Quaternary Period, which is thought to have begun 1.5-2 million years ago.

A prolonged period of alternating glaciations (ice-age) develops in a time-span of a duration of 10^6-10^7 years [5]. Within such a "period", several glacial cycles can be distinguished, each of which covers about 10^5 years and comprises the following sequence:
1) Glacial Phase. This consists of the "cold" period of the glacial cycle, comprising the maximum expansion of the ice sheet and the first inversion of trend, when the glacial cap starts to melt and diminishes in area.
2) Interglacial Phase. This is a "warm" phase between two successive glacial expansions. We are at present in such an interglacial phase, which followed the Weichselian glaciation.

A slightly different classification of climatic cycles is used in alpine glaciology, where the terms "anaglacial" and "cataglacial" phase are widely employed [6]; the former term concerns the development of the glacial phenomenon (starting from the interglacial phase up to its greatest expansion) the latter refers to the decreasing phase of the phenomenon resulting in a new interglacial phase (Fig. 2).

Paleontological and palynological data indicate a sequence of 22 glacial cycles within the last two millions years, each of which covered a 80-100,000 year time span.

The disruptive scenario starts with the repository disruption occurring when the ice sheet is reaching its maximum extent, let us say 6-8.10^4 years after the start of the glaciation.

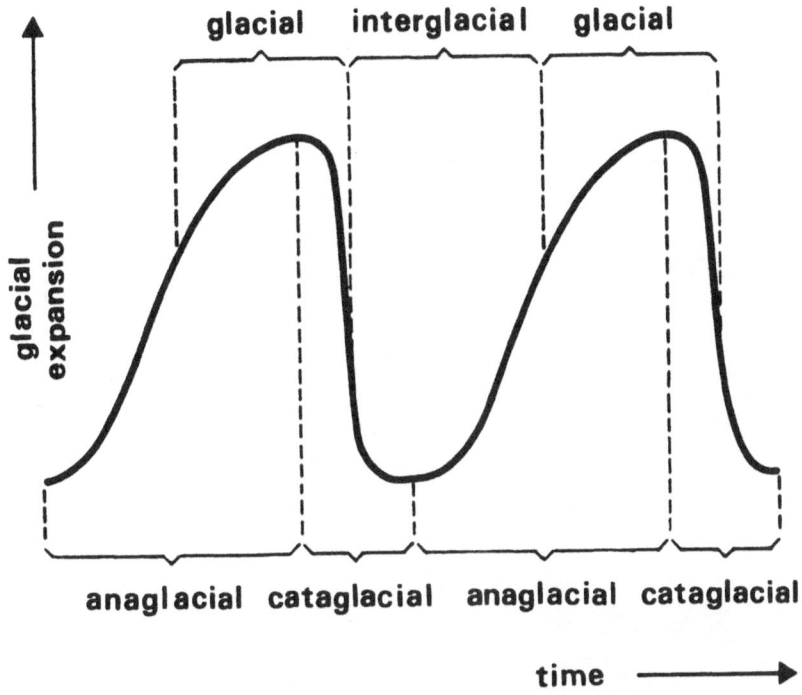

Figure 2. Climatic fluctuations during a glacial cycle.

The glacial action development must then be described by successive images of site evolution after glacier arrival and repository disruption: the quantitative data relevant for the fate of the released
radionuclides will be discussed in the next paragraph.

First phase: this could last about 10^4 years, from the repository disruption to the end of the anaglacial phase.
    The vitrified waste lies in pieces of broken glass within glacial
detritus in the ice-pushed ridges and ground moraine.
    The periglacial environment is still essentially dry (Fig. 3). In
classical alpine glaciology many traces suggest the presence of a
"lively" transport activity of the eroded materials during the anaglacial phase. However, in the present assumed erosion process the "glacial surge" is a completely dry phenomenon: no dispersion of accumulated drift materials occurs, as the undisturbed structures of the
Saalian ridges may indicate.

Second phase: glacier retreating because of cataglacial phase beginning: up to 15,000-20,000 years after repository disruption.
    The presence of an ice cap on the site will last no more than some

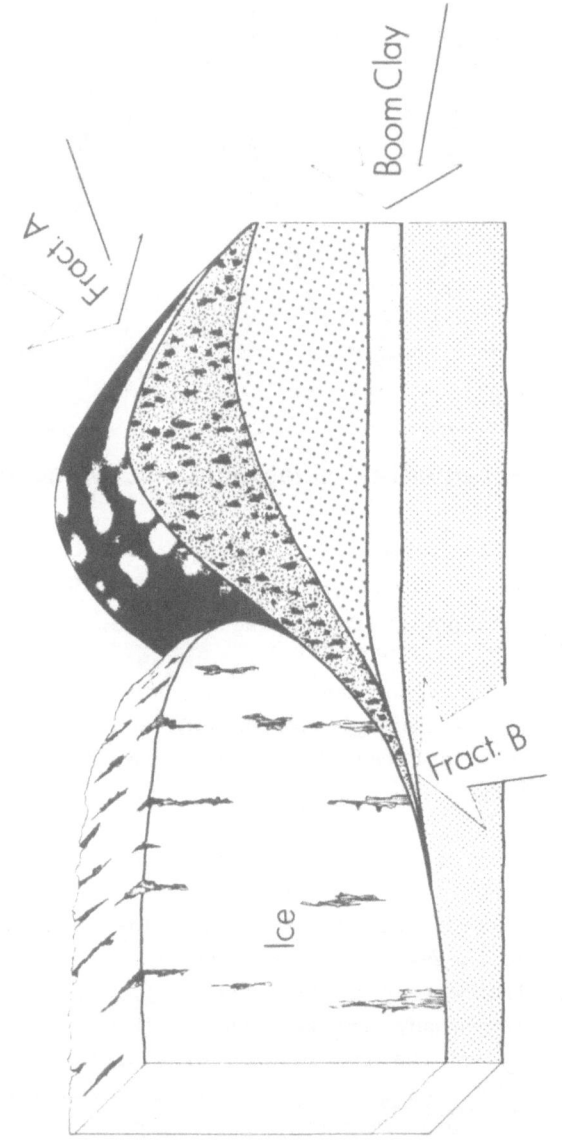

Figure 3. Formation of ice-pushed ridge by glacial action and dispersion of radioactive waste in periglacial environment.

thousands of years, let us say 10,000 years, also suggested by a 50 m deep old Rhine valley, the floor of which can be reached at an erosion rate of at least 5 mm/year. Afterwards, during the incoming cataglacial phase, an intense water circulation will affect the periglacial environment. The basin will be filled with meltwater and, after a while, it will be topped by clayey deposits as the five Saalian basins were.

Thus far, the region presents a characteristical tundra environment: though the glacier has retreated, the environment is still unsuitable for permanent human settlements: only a sparse nomadic hunter population could be imagined living in such an environment. Thus it is not reasonable to imagine any regular and continuous utilization of contaminated groundwater.

The situation during the anaglacial (or the incoming interglacial) phase could be represented as in Fig. 4. Due to the brevity of the anaglacial phase, such a situation could last no more than 5,000-10,000 years, i.e. up to about 15,000-20,000 years after repository disruption.

Third phase: 15,000-20,000 years after repository disruption. As the interglacial phase goes on, the site changes from erosion environment to sedimentation environment. What remains of till materials is partly covered by younger sediments, but the waste leaching continues.

Then, as the climate becomes warmer, man may settle down once again in or near the site, having obviously lost the memory or the knowledge of the repository. Thus he will be exposed to the radioactivity present in the ground and surface waters. Such a situation can be represented as in Fig. 5.

For the sake of completeness it could be added that a marine transgression will end the cycle: the terrestrial environment considered could therefore become a marine environment, forcing the possible human settlements to leave the site.

However, the time-spans required for such an occurrence, as well as the sequence of all the events concerned is so uncertain that it is not worth refining the model taking into account such a further development.

As a conclusion the vitrified waste disposed of in a deep permanent repository will be brought back or brought near to the surface, split into different fractions which are dispersed across different levels of depth.

4. ENVIRONMENTAL EVOLUTION MODEL

The previous analysis evolves through a scenario definition in which

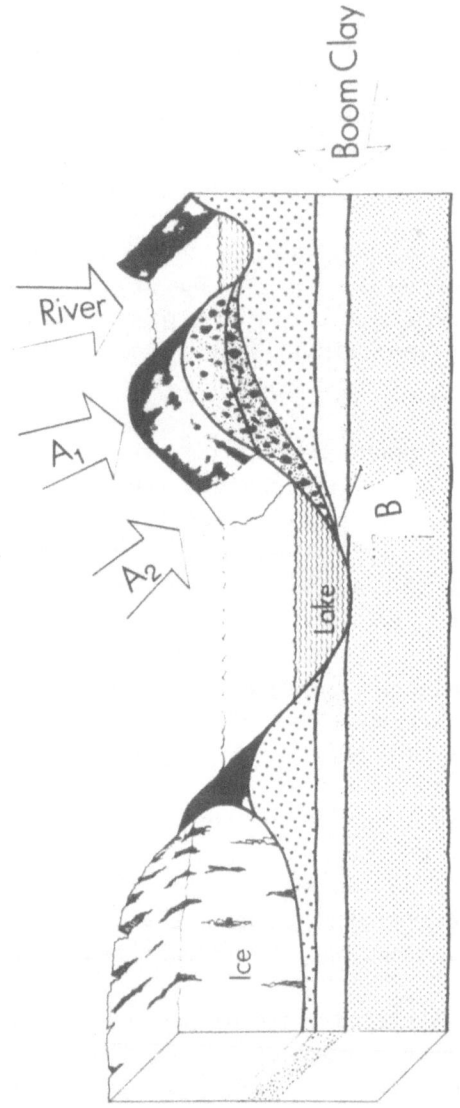

Figure 4. Evolution of periglacial environment during the first stage of glacial retreating.

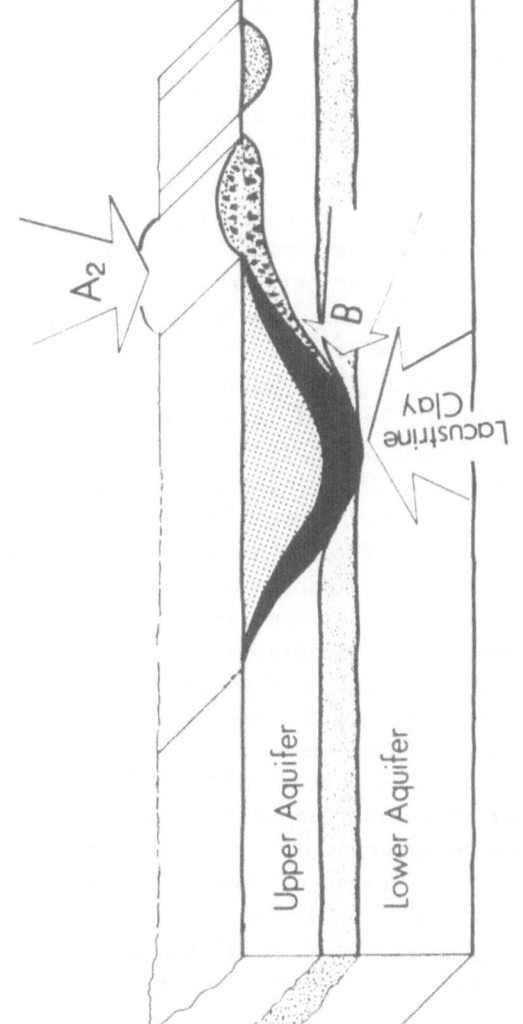

Figure 5. Interglacial phase showing the "final condition of the site

the possible developments of the glacial action was identified. In the present section it is investigated how during such evolution the possible fate of the radionuclides (waste) through the environment may be represented. The main goal of the study consists in defining a new radioactivity source term which is no longer confined to a deep repository, but dispersed in particular units of the environment.

During phase 1 a minor fraction which will be called fraction B, remains below the glacier embedded in the ground moraine. Its quantity can be reasonably evaluated to be about 30% of the total waste inventory: however, since this fraction constitutes the less important contamination source, a value as low as 10% can be assumed for conservative reasons. Such a fraction is leached out by pressurized meltwater coming from the glacier base. The ice thickness being at least 200 m [2] and the gradient being 0.09 bars/m, the surface water undergoes a pressure of ~18 bars which is quite sufficient to push the groundwater down, through the ground moraine up to the underlying aquifer.

The closest emergence point of the contaminated groundwater (although the flow direction cannot be defined) is about 16 km further South, near Veerle, where the lack of Boom Clay allows the underlying and overlying aquifers to come into contact within the Neogene sands. The path length would be such as to cause a very long delay of radioactivity migration due to retention processes.

Fraction A, 90% of the inventory, is embedded in the ice-pushed ridges which lies partly below and partly above the general topographic plain. The freely flowing surface streams are still far, further South.

For as long as the glacier maintains its presence on the "disrupted" repository, fraction A will be blocked in a permafrost so that no important leaching can act upon it: many fault plains are cutting the ice-pushed ridges of the Dutch basins: this fact could be explained by a frozen status of the pushed materials [2].

For phase 2 it may reasonably be assumed that fraction B in the ground moraine is still being exposed to the leaching. The ice disappearance has not changed the downward flow of leached out materials: the hydraulic equilibrium existing before the glacier's arrival has been re-established. (The ground moraine of the Saalian glaciation is still present at the base of the five glacial basins, chiefly near their borders).

The terminal ridge, where fraction A has assumed to lie, will be available to surface waters (meltwater, rain, run-off or streams). Its fate is not easy to define in detail, but some indications can be furnished by the present situation near the borders of the basins.

In the model, fraction A must be split further. After the permafrost melting begins, the top of the ridge is eroded and transported by river(s) which has (have) to be assumed to lap the southern side of the relief, taking into account the geographical frame of the zone. Such eroded materials will be transported to the sea or deposited along the river course, following the size of particles. The fraction discharged into the sea (fraction A,1) can be considered as leaving the environment. In fact, it will be transported at the same velocity as the river flow (0.5-2 m/sec), therefore its residence time within the river as well as its exposure to leaching will be negligible.

Eventually a marine model could be implemented describing a feedback to man of the radioactivity through a marine food chain. The fraction deposited along the river course (A1,2) shall be available for leaching and shall account for water contamination.

During phase 3, when the site is becoming a sedimentation environment, fraction B lies at a depth of about 200 m covered by impermeable clayey sediments. It is now in contact with the underlying aquifer which remains its receptor.

Fraction A1,2 continues to release radioactivity into the river water. The last main fraction A2 lying between 0 and -150 m, is partly covered by younger sediments. Its situation has not changed, and it is still contaminating the overlying aquifer. The lacustrine pelitic sediments sealed definitively the way between two aquifers, thus the receptor is still the upper aquifer, even if the previous downward gradient has been re-established.

As a conclusion, the vitrified waste disposed of in a deep permanent repository will be brought back or brought near to the surface, split into different fractions which are dispersed across different levels of depth.

Fraction B, embedded in the ground moraine at quite a deep level, will be practically cut-off from the shallow ground water flows.

Fraction 2 lies embedded in the ice-pushed ridge and is available to the leaching: it can either be considered as being completely immersed in the ground water (and undergoing a continuous leaching) or simply washed by percolating rainfall (undergoing a less thorough leaching rate). The more realistic model should take into account both possibilities: in such a model, however, other arbitrary assumptions would be necessary. Therefore, the former conservative hypothesis has been assumed, and fraction A2 is considered in permanent contact with the groundwater.

Fraction A1,1 leaves the terrestrial environment discharged into the sea.

Fraction A1,2 contaminates the river which collects all the streams flowing through the site (which was the periglacial environment).

The event sequence above described could be represented as in Fig. 6.

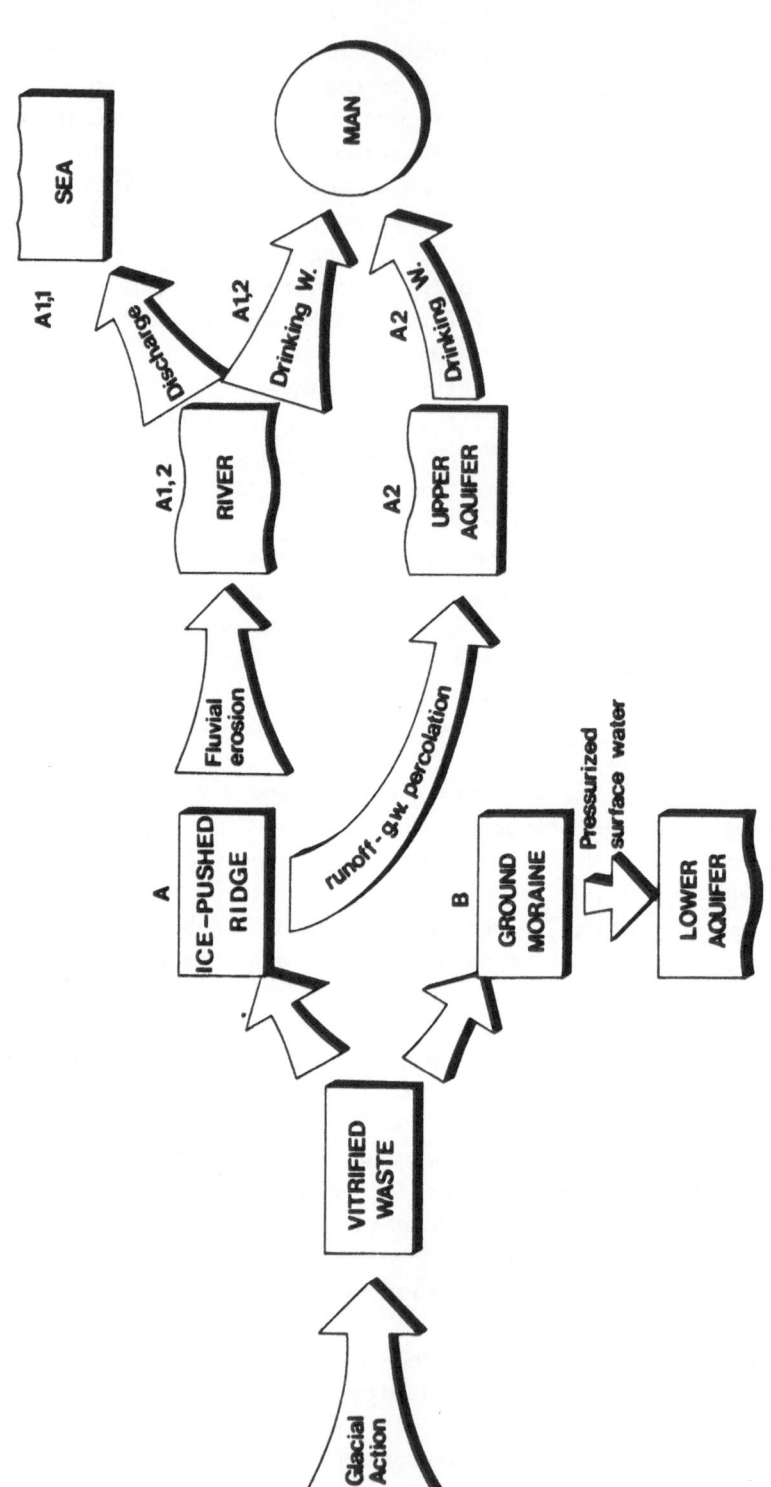

Figure 6. Glacial Scenario: model for consequence analysis.

5. QUANTITATIVE ASSESSMENT OF SOME SPECIFIC PARAMETERS

In order to assess the dose to population and thus to achieve a final
risk associated to the scenario under study, several parameters specific of the described model have to be taken into account:
1) the split of radioactive waste into different fractions;
2) the specific surface of the waste pieces reworked by glacier;
3) the time during which the waste undergo leaching processes, etc.

Among these, the split into different fractions and the grain size
of reworked glasses are of extreme importance in governing the risk:
more over their values are strictly dependent on the developed model.

Since the uncertainty affecting most of the parameters is very
important, the consequence assessment has been developed in such a way
to take into account probability distribution functions instead of
single values. Thus in the quantitative assessment the figures discussed within the next paragraphs will be spread over a range the width
of which reflects the uncertainty of the datum.

5.1 Specific surface of vitrified high level waste (VHLW) undergoing leaching

For the present model the definition of a granulometry for VHLW reworked by glacial erosion and transport is a matter of prime importance: the glass grain size, in fact, is among the most important factors governing the scenario's consequences in terms of radiation dose
to population, as it is learned from a first evaluation.

In a first approach it would seem reasonable to assume for the reworked glass pieces a granulometric distribution similar to that of a
generic till. These sediments, usually show a great preponderance of a
fine-grained and structureless matrix in which cobbles and boulders are
sparsely distributed [7].

However the problem of glass granulometry cannot be treated as a
natural phenomenon, having the geologic evidence as the reference
point, thus the process must be foreseen on a speculative basis.

The glacial attack on the underlying terrains can be split into
two main actions: quarrying and abrasion; the former consists in
opening the bedrock beneath and carrying away the loosened blocks; the
latter is responsible for the milling of the eroded materials. The
abrasion, therefore, produces the finest component of tills, but for
the action to develop, the ice base must be armed with a bed of rather
coarse detritus, acting as an abrasive paste.

In the present case the quarrying, obviously, is not of concern,
since in the site the "bedrock" is constituted by unconsolidated sedi-

ments. The effectiveness of abrasion should be also very poor, since the only tool available are the glass blocks themselves: the lack of coarse detritus within the glacial deposits of Dutch terrains is well known; unless the coarse detritus coming from the reworked man-made structures is taken into account.

The big towns, for instance, (if they will still exist at the moment of glacier arrival) could produce a huge amount of detritus. In the present assessment such a possibility will not be considered. Therefore, omitting the last consideration, the finest fraction on vitrified waste will be due to the fraction of glass blocks among themselves or to the splintering of blocks splitting into smaller pieces. Thus the starting point is the granulometry which was assumed for the undisturbed waste blocks in a previous report [8], where the glass was considered to be fractured into pieces having some cm in diameter. Thus the glacial action is assumed to cause the removal of this material and the production of very small splinters. The reworking by the glacier will produce other fine particles, probably still splinter shaped.

In conclusion after the removal and dispersion the waste will have granulometry showing a bimodal distribution: a coarse fraction in the range of gravel reaching 80% of the waste, and a fine fraction in the range of sand covering the remaining 20%. Taking into account finer granulometries in the range of silt-clay, would not be realistic since such very small glass particles would be leached out very quickly, and would not constitute an important radioactivity source for the environment. The consequent distribution of the grain size could be the following: a first coarse fraction centered around a diameter of 1.5-2 cm, covering a range between 1 and 3 cm, while the fine fraction could vary between 0.03 to 0.1 cm, being centered on 0.05 cm.

Within the above assumed distributions an airborne fraction is lacking. Although such loess-like materials would represent a serious threat because of inhalation pathway, on the basis of starting assumptions the finest component of reworked glass pieces exceeds the limits of airborne size. On the other hand no element is available for a quantitative assessment of such a fine fraction (if any).

5.2 Waste fraction present in the various forms of glacial drift

The method used to model the fate of vitrified waste after repository disruption by glacier bulldozing, consists of reconstructing the way of sediments eroded from basins during the late Saalian phases. A first simplifying assumption is needed, according to which the vitrified waste is evenly distributed in the glacial sediments, namely the ice-pushed ridge and the ground moraine.

Thus, in order to define the waste fractions present in various

forms of glacial drift, the volume of the ice pushed ridges still present around the Saalian basins has been compared with the volume of
basins themselves. A petrological analysis of the ice-pushed materials
has suggested that they came from the area now occupied by the basins,
so that the comparison seems correct. The ratio between the two volumes
(roughly assessed on the basis of the available geological maps) seems
to demonstrate that the sediments coming from the basins and still
present around the depressions account for not more 40% of the removed
materials; the same figure indicates the waste fraction A2, remaining
"in situ" after the erosion of the frontal ridge; thus for the probabilistic assessment the distribution of values for A2 can spread over
an interval ranging from 0.2 to 0.8 of the entire waste inventory
having the peak centered on 0.4.

The ramaining 60% of sediments coming from basins can be split
into a 50% which left the glacial environment, having been transported
away by streams (A1), and a 10% brought into the ground moraine (B).
The value has been chosen on the basis of author personal belief with a
conservative attitude. In fact, fraction B, will be a radioactivity
source of minor importance, being confined in deep levels and accounting for contamination of a poorly permeable aquifer, the emergence
point of which is very far; the figures mentioned have been assessed on
the basis of data furnished by the Dutch Geological Survey. A more accurate assessment could be performed once more detailed geological maps
will be available. At present the uncertainty affecting all these figures has been partly overcome by using probability distribution functions for each of them.

For further split of A (fraction worn away by strams) into A1,1
(fraction discharged into sea) and A1,2 (fraction deposited along the
river course and leached out) can be evaluated on the basis of particle
size and river velocity; therefore some basic assumptions on future
hydrologic patterns of the zone must be done.

The paleogeographic reconstruction of the North Sea and Channel
region shows that the old coast-line development could be indicated by
the present -200 isobath, which is not less than 500 kilometres from
the site. At such a distance from its base level the river which will
have eroded the end moraine (or which will have collected materials
washed from the periglacial zone) will have a velocity not lower than
0.5 m/sec.

Such a velocity is sufficient to transport all particles smaller
than 1 mm in diameter, according to the Hjulstrom diagram [10].

Thus the finer component of fraction A (10%) will be transported
at the same water velocity and discharged into sea in a time span lower
than 1 year. The coarser component (larger than 1 cm) will remain much
longer on the river bed before being dragged up to the sea, and it will

undergo leaching processes contaminating the river-water. The part lost on the way (A1,2) must be considered as a contamination source for the river. This quantity can be assessed by the Stamberg's law giving the reduction of pebbles dragged by rivers for a certain distance.

The quantity thus calculated of undissolved materials must be summed up to the 10% of fine grained materials, so that the final value for A1,1 (fraction discharged into sea) can be represented through a uniform distribution ranging from 10 to 15%, having 12% as its mean value.

The amount of A1,2 derives immediately from A1,1 being complementary to this latter in respect to the 50% of A1, namely 35-40% having 38% as mean value.

The splitting of the whole waste inventory into different fractions can be summarized in the scheme of Fig. 7.

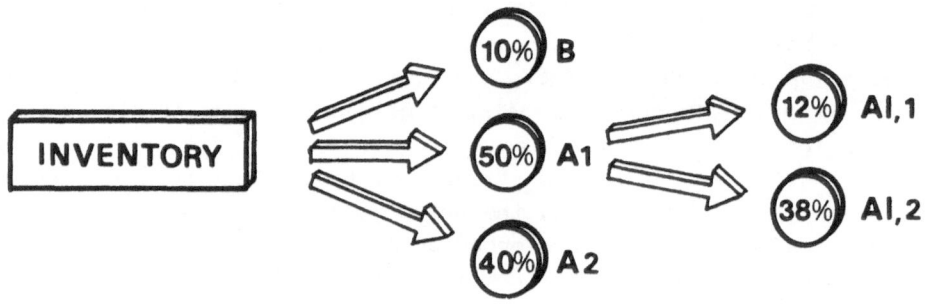

Figure 7.   Splitting of waste inventory into different fractions.

It should be noted that a certain fraction of materials excavated from Saalian basins and accumulated in the frontal ridge has been eroded and transported by the marine transgression during the Saal-weichsel interglacial (Eemian). Therefore the amount of waste discharged into sea could be greater than the value here assumed. However, for the usual conservative reasons a refinement of the model is not worth.

6. CONSEQUENCE ANALYSIS

6.1 Introduction

In the following section the consequence of the glacial erosion scenario are analysed, based upon a previous study conducted on the accidental scenarios for the Mol site [11].

Consequence are assessed in terms of dose rate of the most exposed individual of a critical group, and the risk is assessed by combining exposures with the corresponding probability of occurrence (scenario probability).

Since the treatment involves a considerable degree of uncertainty, a suitable computing tool is needed, able to handle probability distribution functions. The code LISA, developed at the JRC, was used for this purpose. As a result, a distribution of the output is given, which reflects the uncertainties affecting the input parametres.

The physical model developed in the previous section and the evaluation of the related parametres are the basis for the dose assessment.

It must be emphasized that the mathematical model requires drastic semplification to be made in the described scenario, due to the preliminary character of the exercise. Then many aspects which have been carefully described in the physical model will be neglected in the dose calculation.

6.2 Dose assessment model

After the repository disruption, the wastes are assumed to be dispersed throughout the various transport and accumulation glacial forms and to be available for leaching and erosion by groundwater and surface waters.

The radioactive wastes lie in what remains of the ice-pushed ridge, confined at various depths, in a semi-circular form. Water captation from wells, for domestic and agricultural purposes causes the groundwater to move, and radioelements to be brought to the biosphere. Another source of radioactive contamination is linked to the utilization of the river water. The contaminated ridge is supposed to be semi-circular in shape (Fig. 8). The ring radius and width are sampled from the corresponding distributions (see Table 3). It is assumed that the entire content of the repository is homogeneously spread in the ridge, where it is shared into an upper fraction A_1^o (which undergoes both erosion and leaching by surface waters) and a lower fraction A_2^o (which undergoes only leaching by groundwaters as it is confined below the topographic level).

Both fractions are also assumed to consist of "coarse" and "fine" components, φ_1 and φ_2, likewise materials of glacial origin. In order to assess the distance R between the contamination source and the abstraction point, the following procedure is adopted:
- first, a number N of wells are rendomly generated in the region, according to the assumed wells density: a random-sampled water abstraction rate is assigned to each well;
- the well's position are then checked to ascertain whether any of them falls within the ridge; if not, the distance between the closest well

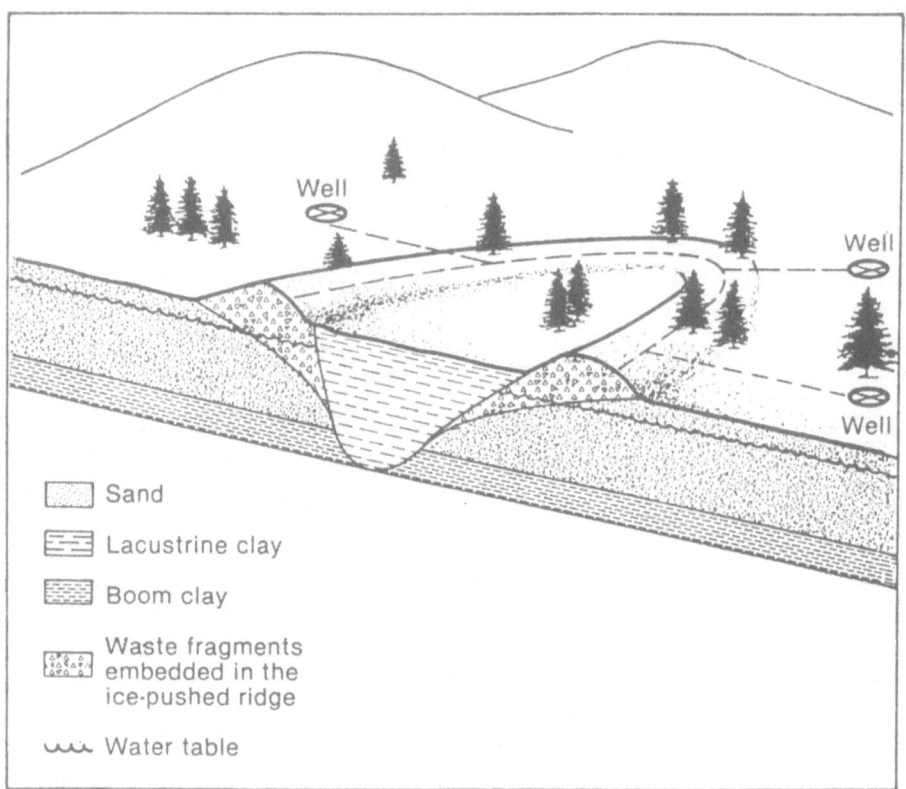

	Sand
	Lacustrine clay
	Boom clay
	Waste fragments embedded in the ice-pushed ridge
	Water table

Figure 8. Model for consequence analysis.

and the ridge median is assumed as pathlength R, and the radionuclide migration through the geosphere is modelled;
- if a well falls within the ridge, the geosphere model is bypassed.

The fraction of waste intercepted by each well is supposed to be proportional to the water extraction rate P_i, and inversely proportional to its squared distance from the ridge.

This assumption implies that the whole waste inventory is supposed to be intercepted by the N wells; however, only the doses due to the closest well are utilized for the assessment.

The above procedure is repeated for each run. The main equations utilized for mathematical treatment of the model are listed in Table 1 the symbols are explained in Table 2. In Table 3 the distributions of most important parametres are given.

It may be seen from Table 1 that all the radioelement released by erosion are assumed to collect into the river, while only the fraction $\emptyset$ intercepted by the closest well contributes to the leaching source

TABLE 1.

Submodel	Algorithms
Radionuclide inventory	Bateman's equations
Leaching source term to geosphere	$R_\ell = Q\phi\ell(A_1^o e^{-kt} + A_2^o)(\sigma_1\varphi_1 e^{-\ell\sigma_1 t} + \sigma_2\varphi_2 e^{-\ell\sigma_2 t})$
Erosion source term to surface water body	$R_z = QA_1^o e^{-kt} k(\varphi_1 e^{-\ell\sigma_1 t} + \varphi_2 e^{-\ell\sigma_2 t})$
Dissolution term to surface water body	$R_w = \ell QA_1^o(1-e^{-kt})(\sigma_1\varphi_1 e^{-\ell\sigma_1 t} + \sigma_2\varphi_2 e^{-\ell\sigma_2 t})$
Groundwater hydrology	$\bar{v} = (P.\ell n\ R/R_o)/\left[2\pi h(R-R_o)\right]$
Radionuclide transport through subsoil	$k_i \dfrac{\partial C_i}{\partial t} = D_i \dfrac{\partial^2 C_i}{\partial x^2} - v\dfrac{\partial C_i}{\partial x} - K_i\lambda_i C_i + K_i\lambda_j C_j$
Biosphere	$C_w = \dfrac{xi}{P}$; $C_s = \dfrac{C_w . R'(1-e^{-k't})}{d.\rho.K'}$
Dosimetry	$H = D_f . I_r . C_i$

term to the geosphere pathway.

The biosphere model then takes into account four pathways:
- well water (ingestion);
- well water (irrigation, resuspension, inhalation);
- river water (ingestion);
- river water (irrigation, resuspension, inhalation).

The ingested and inhaled activity is finally converted to dose rate through the use of the available dose conversion factors.

6.3 Results

The frequency distributions of annual doses on a log scale are shown in Fig. 9 for the four pathways considered. It may be seen that the doses

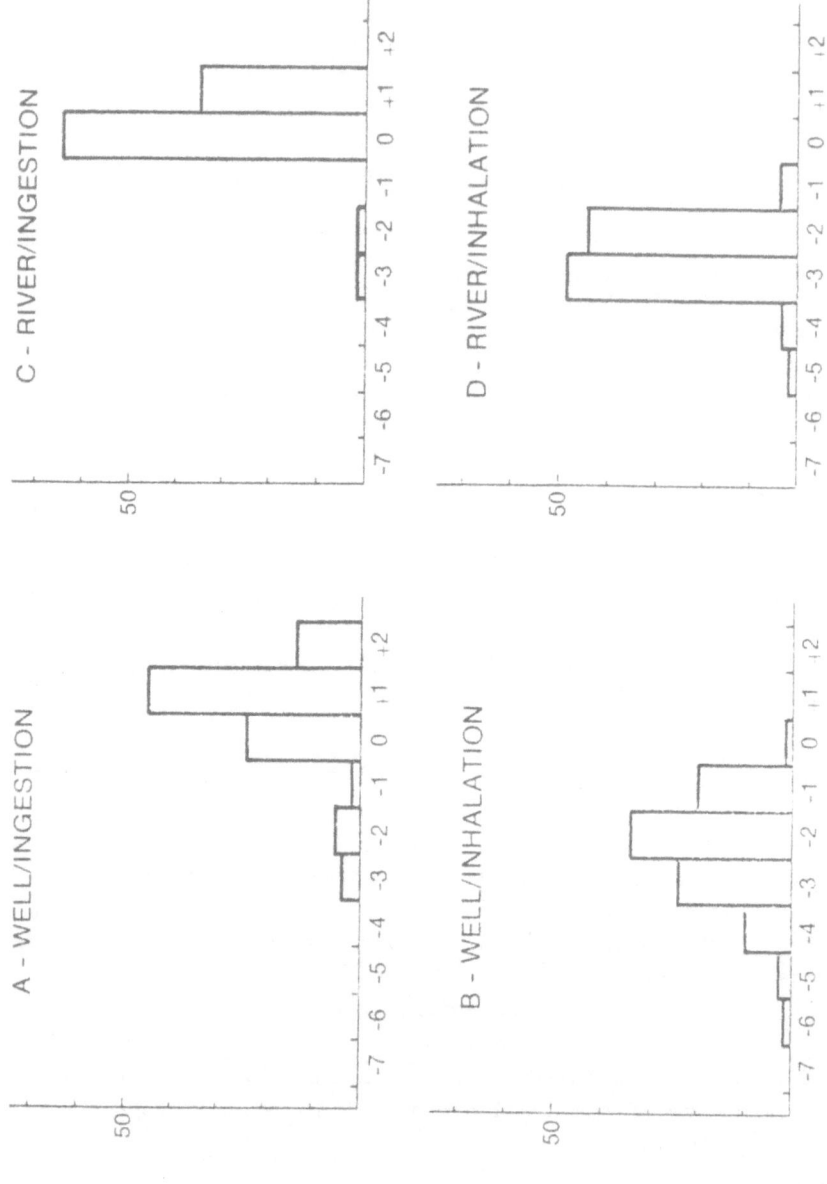

Figure 9. Frequency distributions (%, ordinate) of peak doses rate (rem/year, abscissa) for differ-ent pathways.

TABLE 2.

Q	= total waste inventory
ϕ	= waste fraction intercepted by the closest well
ℓ	= waste leaching rate
K	= waste erosion rate
σ_1	= specific surface area of the coarse fraction
σ_2	= specific surface area of the fraction
A_1^o	= waste fraction undergoing erosion and leaching by surface water
A_2^o	= waste fraction undergoing leaching by ground water
φ_1	= coarse fraction of waste
φ_2	= fine fraction of waste = $1-\varphi_1$
$\bar{v}$	= average groundwater velocity
P	= water extraction rate from well/or river flow rate
R	= pathlength in the subsoil
R_o, h	= well radius, water depth
K_i, λ_i	= retardation coefficients, decay constants
C_w, C_s	= concentration in water, in soil
K'	= transfer rate
d, ρ	= root zone depth, soil density
R'	= irrigation rate
H, D_f	= annual dose equivalent, dose conversion factor
I_r, C_i	= inhalation or ingestion rate, isotope i concentration

can be very large; there are three main reasons for this:
i)   the whole waste inventory is involved in the process;
ii) an important waste fraction is assumed to have been reduced to a fine granulometry by the glacial action;
iii) the pathlength in the subsoil can be very short and even nil.

    As it was expected, the ingestion of polluted water extracted from a well in the neighbourhood of the disrupted repository constitutes the most important pathway, and can result in doses as high as tens of rem.

TABLE 3.

Parameter	Distribution	Mean value (log)	St. Deviation (log)
Contaminated-ridge radius	Log-normal	1	0.1
Contaminated-ridge width	Log-normal	0.5	0.1
Time of cataglacial phase	Log-normal	4	0.05
Half-life time for erosion process	Log-normal	3.4	0.05
Fraction of waste A_2^o	Log-normal	-0.4	0.1
Sp. surface of coarse fraction	Log-normal	-4.12	0.1
Sp. surface of fine fraction	Log-normal	-2.7	0.1
Fine fraction of waste	Log-normal	-0.7	0.1
River flow rate	Log-normal	10	0.5

The ingestion pathway is dominated by Np-237; inhalation, is governed by Pu-239, Tc-99 and Th-229.

The consequences of this scenario could be very serious exceeding 100 rem/year in some cases (2% of the runs).

In such a situation the risk R must be quantified taking into account both the probability of the scenario occurrence (P_S)* and the probability that a health effect will arise (P_H).

$$R = P_S \times P_H \qquad (1)$$

It generally agreed that, for exposures lower than 100 rem P_H is proportional to the dose D, thus eq. (1) becames:

$$R = P_S \times D \times F \qquad (2)$$

where F is a conversion factor being equal to $2 \times 10^{-4} \text{ rem}^{-1}$.

Above 100 rem the health effects are no more stochastic [12], so that:

$$R = P_S \qquad (3)$$

(*) For the scenario probability (P_S) a value of 5×10^{-5} over a time span of 250.000 years has been evaluated. The assessment has been discussed in previous papers (1,11).

Thus the whole dose-rate histogram of Fig. 9 is to be elaborated to generate a histogram of the risk; this is made by multiplying each dose value on the abscissa by the product $P_s \times F$, following eq. (2), for doses less than 100 rem, while for the non-stochastic region eq. (3) is used.

A new histogram is thus obtained (Fig. 10) which gives the distribution of the individual risk linked to the scenario considered. In this histogram, 4% of the occurrences fall at a risk level of 5×10^{-5} yr^{-1}; taking into account the corresponding frequency the final risk can be assessed in the order of $10^{-6}yr^{-1}$.

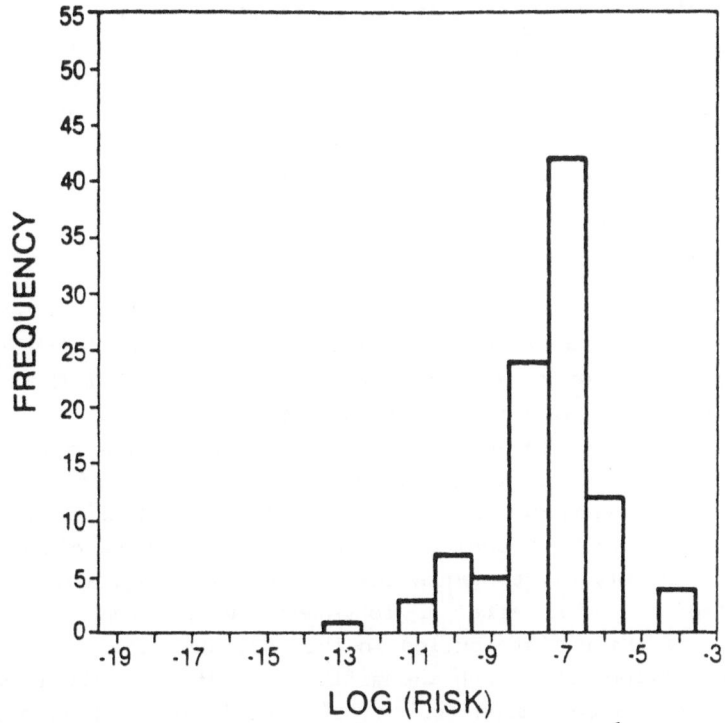

Figure 10. Distribuzion of the individual risk (yr^{-1}) linked to the glacial erosion scenario.

Such a risk level should be considered quite acceptable, in comparison with present practics. This case constitutes a good example of an occurrence which, if it were to happen, could result in serious exposures.

Therefore, on the mere basis of the consequences, the scenario should be considered unacceptable. If, however, the related probabilities are considered, the corresponding risk appears to fall at a reasonably acceptable level.

6.4 Sensitivity analysis

The radiation exposures associated to the glacial erosion scenario exhibit a different sensitivity to the various parametres. The analysis,
shows that the doses are mainly governed by the geosphere pathlength
and by the leaching rate, while the time of the scenario occurrence
does not play an important role. This behaviour can be easily explained: the scenario is characterized by very short pathlengths in the
subsoil, which require migration times correspondingly short, so that
even values for the time of occurrence near to the cut-off time permit
the radioactivity entrance into the biosphere.

The importance of pathlength is due not only to the retention
processes, but to the fact that the radioactivity fraction intercepted
by the well is inversely proportional to R^2. Unfortunately the subsoil
paghlength is intrinsically unknowable; this introduces in the analysis
an uncertainty level which cannot be eliminated.

7. CONCLUSIONS

The present study constitutes a preliminary approach to the elaboration
of a glacial erosion model needed for the assessment of the risk linked
to a possible glacial action over the repository site.
1) The assumed scenario appears reasonable, on the basis of the actual
   geologic evidence derived from pleistocenic sediments of Dutch terrains: some readjustments of numerical values and a refinement of
   the model should not contradict the validity of the content.
2) A large uncertainty concerns the chemical and physical state of the
   conditioned wastes after a long-term interaction with geological
   media, and their distribution in the environment as a result of the
   glacial action. The rough assumptions utilized in the present model
   to define the pollution source term should be refined according to
   experimental determination.
3) The consequences possibly caused by the glacial erosion scenario can
   be of relevant importance. However, the doses which will be assessed
   must be considered in connection with the corresponding occurrence
   probability in order to allow a correct comparison between overall
   risk linked to glacial scenarios and that due to other scenarios.

ACKNOWLEDGMENTS

The glaciological aspects here described have been discussed and
reviewed in a meeting held in Haarlem on June 26th, 1982, organized at

the National Geological Survey of the Netherlands on request of the authors. The authors gratefully acknowledge the participants:

B. Hageman   (Director of Geological Survey of the Netherlands, observer)

W. Zagwijn   (Staff member of Geological Survey of the Netherlands, Quaternary expert)

S. Jelgersma   (Staff member of Geological Survey of the Netherlands, Quaternary expert)

N. Vandenberghe (Staff member of Belgian Geological Survey, observer)

Mr. Beaufays   (CEN/SCK).

REFERENCES

[1] D'Alessandro, M. and Bonne, A. 'Radioactive waste disposal into a plastic clay formation (a site specific exercise of probabilistic assessment of geological containment)', Radioactive Waste Management, V. 2, Harwood Academic Publishers.

[2] Jelgersma, S. and Breeuwer, J.B. 'Toelichting bij de kaart glaciale verschijneselen gedurende het Saalien' from 'Toelichting bij Geologische Overzichtskaarten van Nederland', Rijks Geologische Dienst.

[3] Kuster, H. and Meyer, K.D. (1979) 'Eiszeitalter und Gegenwart' 29, pp. 135-156.

[4] Bull, C. (1980) 'Glaciological parametres of disruptive event analysis', A.E.G.I.S., PNL 2863.

[5] Vendenberghe, N. et al. (1980) 'Scénarios d'évolution lente, appliqués au site argileux de Mol (Belgique)', Réunion de travail sur les Scénarios de Libération des Radionucléides à partir de Dépôts situés dans des Formations Géologiques, OECD/AEN, 8-12 Septembre 1980, Paris.

[6] Trevisan, E. Tongiorgi (1958) 'La Terra', UTET, Torino.

[7] Flint, R.F. 'Glacial and Quaternary Geology', John Wiley and Sons, Inc.

[8] Bertozzi, G. and D'Alessandro, M. (1981) 'Radioactive waste disposal into a plastic clay formation: a site specific exercise of consequence analysis', JRC Ispra, Technical Note 1.07.03.81.66.

[9] Pettijohn, F.J. 'Sedimentary rocks', Harper and Brothers, N.Y.

[10] Vatan, A. (1967) 'Manuel de Sédimentologie, Ed. Technip., Paris.

[11] Bertozzi, G., D'Alessandro, M. and Saltelli, A. (1985) 'Waste disposal into a plastic clay formation: risk analysis for probabilistic scenarios', European Applied Research Reports, Vol. 6, No. 4, pp. 631-1002.

[12] ICRP Publication 26, Ann. ICRP 1, 1977.

Section 5 - Some worked examples

The first paper in this section, from Dr. Thompson of the UK HMIP, reviews the present-day state of the art in the use of Probabilistic System Assessment (PSA) codes in the UK. The need for a new generation of PSA codes able to incorporate climate changes and quantitative geomorphology is stressed with reference to the ongoing UK development of the VANDAL code. A different approach is presented in the subsequent paper, from Dr. Lewi of the French IPSN. Lewi demonstrates the use of a purely deterministic methodology on two examples (surface land repository and deep underground repository). The worked example presented by Dr. Stanners of the JRC draws upon his experience as member of the NEA Seabed Working Group, and as one of the authors of the feasibility study for a deep ocean nuclear waste repository under the seabed. Although this paper is in the "examples" section frequent reference is made throughout the paper to methodological issues. The combined use of deterministic and probabilistic methodologies and the description of the verification procedure adopted in the assessment make this contribution particularly interesting, providing the reader with a critical interpretation of some crucial methodological points. The closing paper of this section (Dr. Girardi, JRC and Dr. Cadelli, CEC) summarises the conclusions of the PAGIS (Performance Assessment of Geological Isolation Systems) study. Six years of work, several thousand pages of reports covering four selected geological media and a variety of disposal concepts make this study one of the most complete analysis of the waste disposal practice. The study was coordinated by the Commission of the European Communities in the framework of the Community Plan of Action in the field of radioactive waste and considered clay, salt and granite as continental geological options plus the sub-seabed option mentioned above. PAGIS is characterised by a strong effort in harmonising the assessment methodologies (a hard task when the number of participating countries and organisations is considered). The conclusion section of this paper is interesting for the critical stand taken by the authors, who revise objectively achievements ans limitations of PAGIS.

THE TIME DIMENSION IN RISK ANALYSIS
Examples from recent work in the United Kingdom

B. G. J. THOMPSON
Her Majesty's Inspectorate of Pollution
UK Government Department of Environment
43 Marsham Street
London SW1
United Kingdom

ABSTRACT. The objective is to estimate risks from the disposal underground of radioactive wastes over the long-term post-closure period, taking account of the possible future changes that may occur in the natural environment and the possible effects of the repository, and of future human action unconnected with disposal as such.

The 'steady state' approach and the present-day use of scenarios are reviewed largely on the basis of trial assessments carried out within the UK Department of Environment research programme.

Monte Carlo simulation of possible environmental futures is described against the background of the rapidly growing knowledge of climate change and of quantitative geomorphology. Such simulation results will be used to drive a 'second generation' p.r.a. code, VANDAL, to give risk estimates for deep disposal that account for temporal changes and associated uncertainties.

Early results for deep disposal indicate the 'steady state' assumption is a poor approximation to the full dose-rate time history for the particular conditions examined.

1. Introduction

1.1. OBJECTIVE

The objective is to estimate risks from the disposal underground of radioactive wastes over the long-term post-closure period, taking account of the possible future changes that may occur in the natural environment and the possible effects of the repository, and of future human action unconnected with disposal as such.

1.2. TEMPORAL CHANGE

There are two methods at present of accounting for temporal change and uncertainty associated with temporal change:

1.2.1. *The 'Steady State' Approach.* Here, the first generation system variability analysis codes, such as SYVAC, are used with defining variables whose values are drawn from probability distributions assumed invariant with time.

A. Saltelli et al. (eds.), Risk Analysis in Nuclear Waste Management, 231–262.

Two examples of the 'steady state' approach are outlined in Section 4 using using material taken from the Dry Run 1 [1] and Dry Run 2 [2] trial assessments in the UK. The UK DoE code SYVAC A/C [3] was used for the p.r.a. calculations.

The limitations of such an approach are discussed in Section 5 and referred to current controversy over the so-called 'scenario' method in risk assessment [47].

1.2.2. *Time-Dependent Analysis.* To overcome difficulties with the 'steady state' approach, the rapidly developing disciplines of quantitative climatology and geomorphology can be invoked to produce estimates of the variation with time of the probability distributions of the significant environmental parameters.

These estimates, together with the associated results of numerical modelling (of groundwater flow, transport and geochemical influences etc), can be used to provide inputs to a new time-dependent 'second generation' system variability analysis code, VANDAL, developed for the UK DOE [4]. This will enable estimates of individual risk and of other radiological performance variables to be made for periods of up to about one million years after closure for deep disposal facilities. See Section 5, below.

Preliminary scoping calculations for a hypothetical shallow engineered trench facility indicate the potential significance of climatic change on broad levels of risk, using the TIME2 code to generate the samples of possible future evolutions of the site during the current interglacial period of about 25,000 years (see Fig. 1).

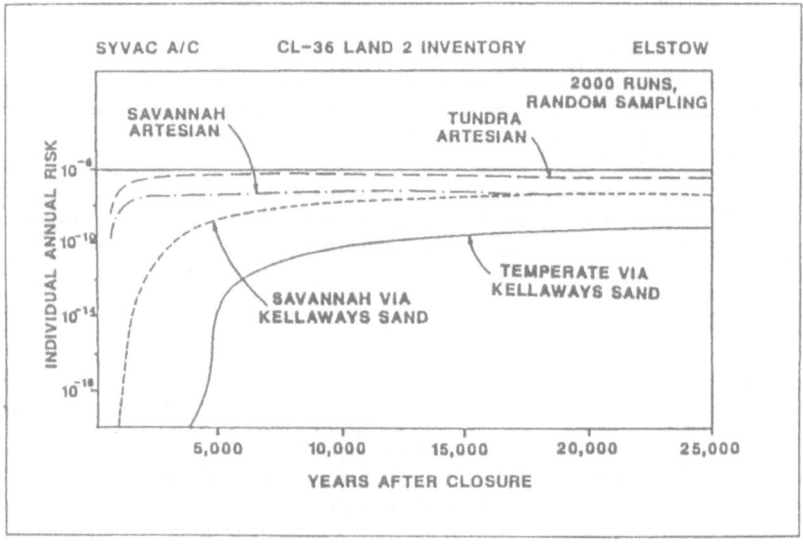

Figure 1   Effect of environmental change on risk from shallow disposal

For deep disposal, the 'steady state' approach is shown to lead to a poor approximation to the dose-rate time curve derived from the full time-dependent analysis using VANDAL.

A new environmental simulation code TIME4 is under development for use over about a one-million year assessment period. This requires models of multiple glacial cycles. The scientific basis for TIME2 and TIME4 is given by Boulton (ibid). Extensive references provide a good basis for a broad understanding of the international effort devoted to treating the problem of long-term future change in performance assessment.

2. Risk Estimation

2.1. BASIC PROCEDURE

The principal task is to estimate risk to individuals. In the UK, this is interpreted currently by UK DoE (HMIP) as a requirement to calculate the expectation value of risk to a representative member of the critical group at any specific time [6], and a comprehensive procedure has been published and rehearsed in a number of trial assessments of hypothetical facilities for both shallow burial of low-level wastes [7] and of deep disposal of a combined inventory of low and intermediate level wastes [1,2].

The procedure (see Fig. 2) introduces iterations between detailed models (of groundwater

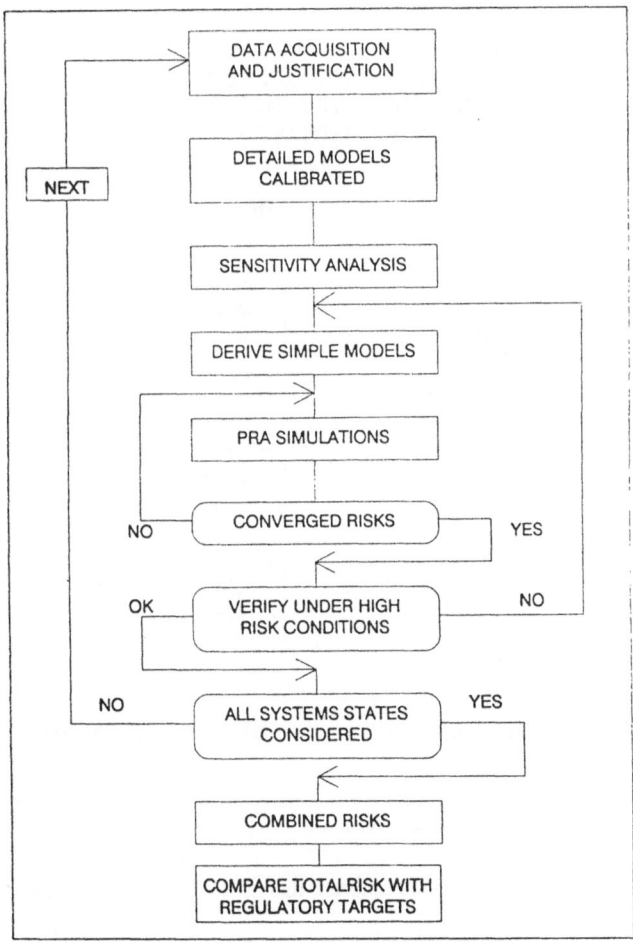

Figure 2   Outline of the UK DOE risk assessment procedure

– 233 –

flow and transport, geochemistry, etc.) and the use of the 'first generation' Monte Carlo p.r.a. simulators SYVAC A/C and D [3,8]. These examples, as with all currently published examples of p.s.a. calculations, from Canada [9] using SYVAC 1,2 or 3, CEC [10] using LISA, or studies by Sandia [11] in the USA, are limited to parameter values and probability distributions which generally do not vary over time. However, the logic of the approach and the general method of risk calculation remains valid in the presence of temporal variations and of uncertainty about these variations. Fig. 3 compares the 'steady state scenario' based present-day risk assessment procedure with the fully time-dependent approach under development.

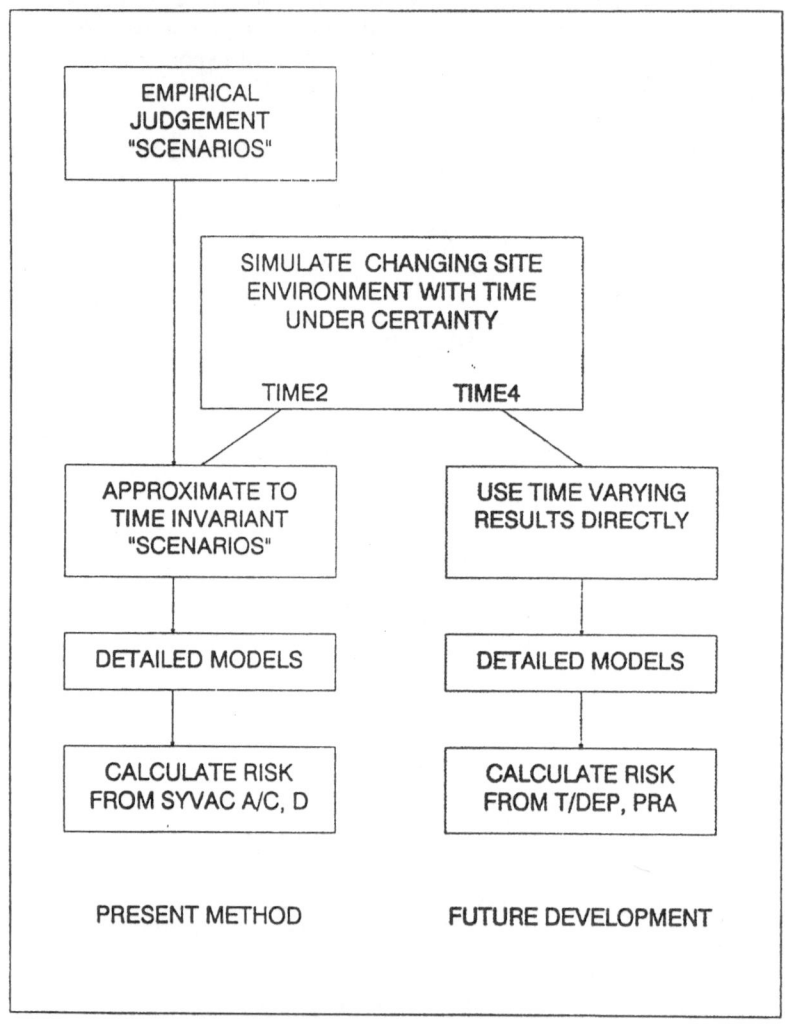

Figure 3   Comparison of different risk assessment procedures

2.2. RISK ESTIMATION IN THE PRESENCE OF TEMPORAL VARIATION

The mean risk $\overline{R}(t)$ at time t can be calculated for low doses and dose rates to which the linear dose-risk relationship adopted by the ICRP [12] applies, using:

$$\overline{R}(t) = \gamma \int_{x} p(x).H(x,t).dx \qquad (1)$$

where $p(x)$ is the joint probability distribution function defined over the entire multi-parameter domain of the problem, x is the vector of parameter values sampled and includes those describing the temporal changes. H is committed effective dose equivalent and γ is the ICRP risk factor.

In order to determine the risk for comparison with a regulatory 'target' figure, the number of simulations must be large enough to provide a statistically justifiable estimate. That is, it must converge and have confidence bounds that can be justified in terms of the underlying statistical assumptions (the statistical 'model').

Hence, a means of generating a statistically justifiable sample of the complete set of entire possible futures over the assessment period (T) is required to act as input to the p.r.a. simulator.

During 1986, two developments were accordingly undertaken:

(a) The development of a Monte Carlo simulation model (TIME2 [13,14]) of environmental change, for shallow facilities and;

(b) The development of a 'second generation' time-dependent Monte Carlo pra code (VANDAL [4]).

During 1988 the development of a second environmental simulation model called TIME4 has begun. This is to be used in the assessment of deep disposal facilities involving longer timescales up to about one-million years.

3. Environmental System and its Representation

3.1. INTRODUCTION

In this section, the nature of the environment is outlined and appropriate ways are discussed showing how it may be represented for the purpose of undertaking performance assessment of repositories for radioactive wastes.

3.2. THE MAGNITUDE OF CHANGES IN THE ENVIRONMENT OF NORTH-WEST EUROPE AND THE UNITED KINGDOM OVER RECENT GEOLOGICAL TIMES

The past 2 million years of geological history have been dominated by alternating sequences of glacial and interglacial climatic conditions in North-West Europe. On a number of occasions, large ice sheets have developed over much of Northern Europe, centred on Scandinavia, Scotland and the Alpine regions. The most recent retreat of ice is illustrated in Fig. 4 During the 25,000 years of this retreat, northern Britain experienced a change in conditions from complete ice cover to warm temperate conditions, involving an increase in average annual surface temperature of around 10°C.

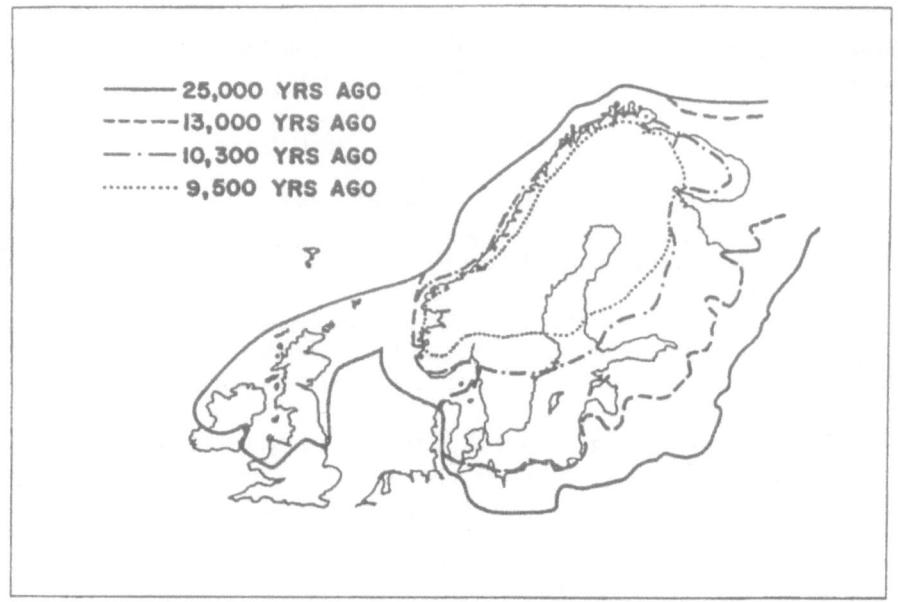

Figure 4   Retreat of ice sheets after the last glaciation in Europe

The equivalent change in parts of continental Europe was as much as 20°C. The magnitude of these climatic changes in North-West Europe is enhanced by the presence and action of the North Atlantic Gulf Stream which has tended to amplify the effect of global changes in climate. In warm periods, temperate conditions are extended to high latitudes because of the Gulf Stream, but in cold periods its absence results in very rapid southward propagation of the ice front.

At the height of these repeated glacial stages, marked erosion of both superficial unconsolidated sediments and bedrock occurred, such that on land the glacial stages are essentially erosive in character. The difficulty in reconstructing the on land glacial record before the most recent (Devensian) glaciation is largely due to the removal of previous glacial deposits by subsequent glaciations (consequently, the earlier record is largely reconstructed using data from ocean cores). Erosion rates varied considerably. More extreme occurrences include:

- Glacial meltwater channels cut several tens of metres into bedrock (eg: the 60m deep buried channel off the Humber Estuary, near Killingholme [15]).

- The removal, by ice erosion, of a 100m chalk escarpment in East Anglia [16].

The volume of eroded material in Britain during the last glaciation has been estimated at 2000 km^3 [16]. Such extensive erosion resulted in correspondingly extensive deposition. Much of this occurred offshore, but, on land, deposits of till, moraine and outwash sands and gravels are abundant and can attain thicknesses of several tens of metres.

Modification of the surface by glacial erosion and deposition is further enhanced by the effects of isostatic depression and rebound of the earth's crust in response to loading and unloading of ice. Since the last glacial maximum, isostatic rebound has reached as much as 200m in Britain and 800m in Scandinavia.

Tundra climatic conditions precede and follow glacial stages and prevail over large areas to the south of regions of ice cover. The effects of tundra conditions mainly relate to the presence of frozen ground (permafrost) which causes perturbation of groundwater flow patterns and mechanical break-up of rock and soil. Permafrost may occur to depths of several hundred metres. The harsh environment   results in sparse vegetation cover which, together with the mechanical break-up of the soil, allows rapid erosion to occur, especially by the agent of wind transport. In addition to this, reduced infiltration due to frozen ground results in enhanced surface flow and fluvial erosion.

The peninsular nature of NW Europe has heightened the importance and effect of sea level changes which have accompanied global climate changes. Substantial changes in the position of shorelines have resulted in important modifications to the surface. Widespread submerged platforms exist offshore of the United Kingdom at depths of 30m to 150m [17] indicative of exposure of much of the continental shelf during glacial times. Such dramatic changes in the position of the shoreline will have had correspondingly large effects on the onland groundwater and surface water flow systems.

3.3. MODELLING ENVIRONMENTAL PROCESSES

3.3.1. *Introduction.* The environmental system is undoubtedly multifaceted and any attempt to represent it in terms of models must account for a large number of processes (Fig. 5). These

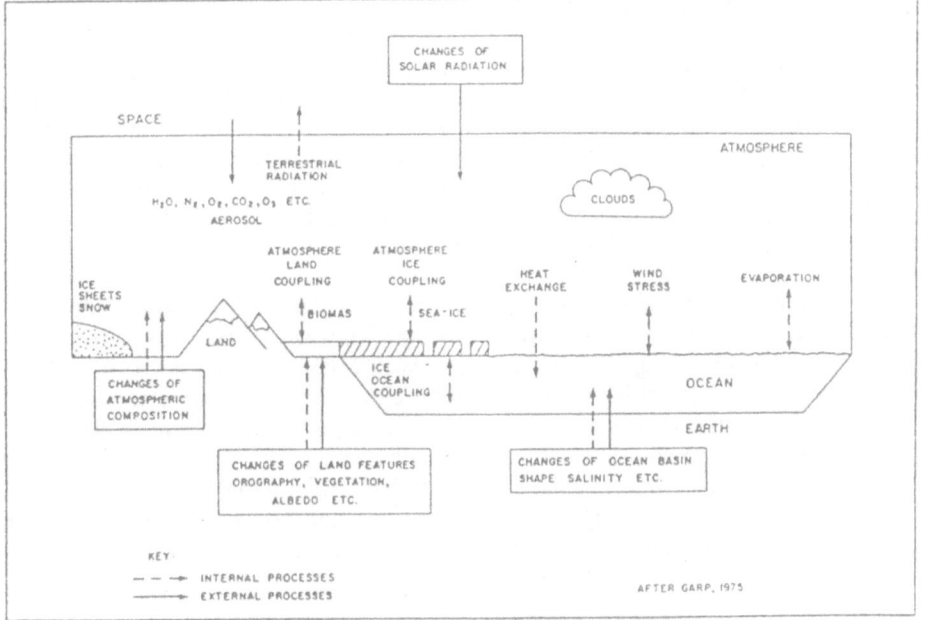

Figure 5   Components of the climate system

processes cannot be considered in isolation, but they must be considered in the context of the system as a whole.

Our ability to compose environmental system models [18] has only been relatively recently realised with the advent of sophisticated computer technology and rapid progress in the earth sciences.

An example of this recent development can be found in glaciation and ice-sheet modelling studies [19]. The decade of 1965-1975 saw extensive American, Soviet and European measurements of ice sheets (especially in Antarctica) and this was followed by a decade in which models of glacial systems proliferated [eg 20].

Similar developments have occured in most branches of earth science and, in recent years, attempts have been made to tackle integrated environmental system models on a global scale [21].

3.3.2. *Environmental Modelling for the Purposes of Radioactive Waste Management*.

Modelling for the purposes of radioactive waste management may be tackled with system models significantly less comprehensive than those which would be needed to represent the whole environmental system. However, it is important that this reduction does not involve liberal disregard of individual processes; the assumptions employed in assembling such models need to be carefully evaluated. Most importantly, processes should not be eliminated from a model without first considering their effect on the system as a whole.

For example, in composing a model, it may be decided to neglect the effects of earthquake activity (on the grounds that damage due to shaking would not be significant) whilst overlooking the fact that even minor earthquakes might affect fracture characteristics and groundwater flow conditions

3.3.3. *TIME2*.

The TIME2 model [13] uses a Monte Carlo simulation method, similar to previous models developed in the U.S.A. (GSM and FFSM) [22,23], but employs a deterministic environmental system model of climate change and glaciation in order to constrain time-dependant simulations to realistic potential futures. The model was designed for shallow disposal sites and considers periods of c.25,000 years (up to the first future glaciation). In addition to modelling the advent of future cold climatic conditions, the possible effects of a near-future (c.1000 years) rise in temperature due to the 'greenhouse effect' are included. In addition, the possible consequences of human intrusion are modelled.

The principal features of the TIME2 model are as follows:

(i)   Scientific basis: Climate change is modelled in terms of present-day worldwide data (assembled by the Climatic Research Unit of the University of East Anglia) and the history of climatic change preserved in the oxygen-isotope record from deep-ocean sediment samples (covering the last 700,000 years) [24]. Glacial processes were modelled using a deterministic generic model developed by Boulton and colleagues [25]. This model combines relationships describing the mass balance of an ice sheet, its non-linear viscous flow and the visco-elastic behaviour of the earth's crust in response to ice/water loading/unloading in order to model the advance and retreat of an ice-sheet and its effects on sea-level and land elevation.

(ii) The system modelled: This is shown in Fig 6. It is essentially a climate-driven model which is dominated by the effects of glaciation.

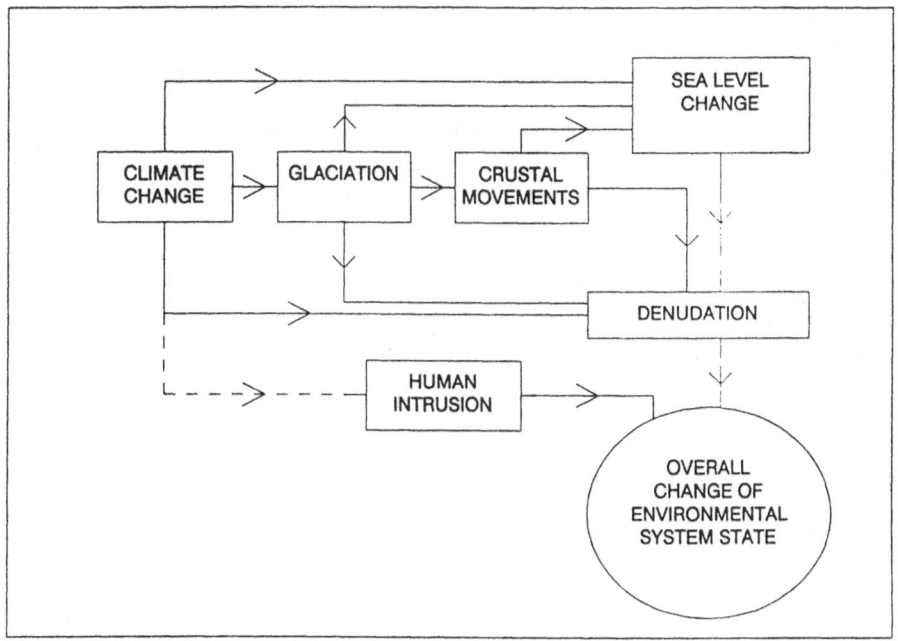

Figure 6   Environmental system represented in the TIME2 model

(iii) Model methodology: The record of the past climatic change preserved in deep sea sediment core samples is used to construct frequency distributions of the durations of climatic types for the location of interest. One of two standard sequences of climate states is used:

TEMPERATE	TEMPERATE
SAVANNAH	BOREAL
TEMPERATE	TUNDRA
BOREAL	GLACIAL
TUNDRA	
GLACIAL	

The durations and characteristics of these states are then sampled, to allow for uncertainty in modelling future change.

Research is currently under way to develop a model of successive glacial phases over longer timescales (c. 1 million years) with reference to deep underground disposal. This research programme [26] is aimed at describing the processes of permafrost development and decay and the influences of ice sheet loading/unloading and insulation effects on these processes (Fig 7). The objective is to better understand groundwater flow during repeated cycles of glaciation. This research involves theoretical modelling of ice and water flow using the continuum theory of mixtures [27].

A successor to TIME2 (to be called TIME4) is being developed as a Monte Carlo simulator of these environmental processes for the purpose of performance assessment of deep repositories. The model will be a simplified version of the detailed research models.

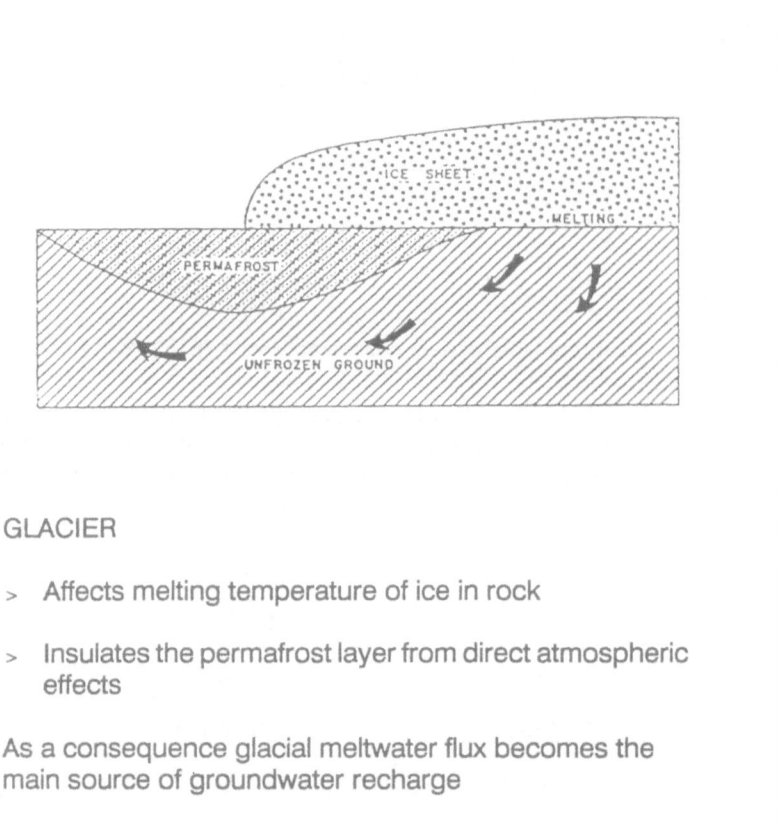

Figure 7   Ice sheet and permafrost interactions

3.3.4. *Typical Output from Environmental Simulation Using TIME2.* Predictions are obtained in the form of time series of values of output variables for each sample run and of probability distributions of values at each time from a complete simulation. Fig. 8 shows an example of TIME2 output in the form of a cumulative distribution function varying with time.

Typical outputs of relevance to the risk simulation are:

- Surface water flows and sediment fluxes;

- Infiltration rates;

- Point data on topography;

- Removal of material from the repository due to river meandering, human action (excavation, drilling etc) and general denudation;

- Time of arrival of ice sheet;

- Sea level.

These provide the basis for parameter values used in the models of radionuclide transport and radiological transport used in the p.r.a., supplemented, if necessary, by detailed modelling of the regional groundwater flow system.

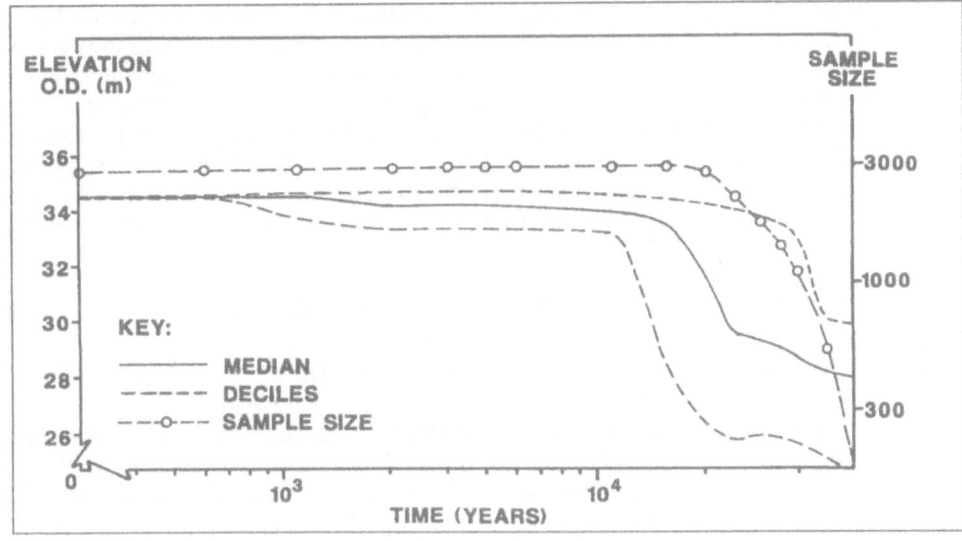

Figure 8   Changes in repository site evolution simulated by TIME2

4. The 'Steady State' Approximation

Until comparatively recently the only p.r.a. codes available were of a 'first generation' using mathematical representations of the radionuclide transport processes that did not account for temporal variations in the values of principal defining parameters. Hence in carrying out long-term performance assessments it was necessary to approximate the true behaviour of the changing environment by some assumed 'steady state'. In the first trial assessment, p.d.f.'s of sampled variables were somewhat intuitively chosen to cover a range of values that included some likely future conditions.

4.1. THE FIRST TRIAL ASSESSMENT: DRY RUN 1

The first attempt [1] to rehearse and demonstrate the proposed procedure outlined above, (see Fig. 2) was made during 1984-5 for a postulated disposal facility for intermediate level wastes situated at a depth of about 150 metres beneath the AERE Harwell site. A single groundwater transport scenario was considered with parameter values and uncertainties that were assumed invariant with time after closure.

A horizontal tunnel design concept was assumed for the repository consisting of a circular cross section with 34 steel drums encapsulated and grouted with cement inside a concrete liner. Consideration of activity, radiotoxicity and half-life allowed the inventory to be reduced to 23 nuclides.

The GEOHYDROTRAN finite element code was used in a two-dimensional vertical section to model the groundwater flow. This was calibrated to match observed heads at Harwell and near to the Thames. Sensitivity calculations were then used to derive the range of values of defining variables for a subsequent p.r.a. using a simplified one-dimensional nuclide transport model in the geosphere. Extreme values from individually specified ranges of the parameters values were used to estimate an upper bound on "risk" of $R_{max} = 1.3 \times 10^{-3}$ per annum at 54,000 years due to Np237 (see Fig. 9). This "risk" (R_{max}) would of course be associated in a full p.r.a. with a very low probability of occurence and hence contribute very little to the overall risk estimate $\bar{R}$. As R_{max} failed to satisfy the risk target, in accordance with procedure of Fig. 2, it was necessary to undertake a full p.r.a., using the UK DoE SYVAC A Monte Carlo simulator.

Risk was calculated for thirteen chains and, the only significant contributions obtained were from:

$$Tc^{99}, \ I^{129}, \ U^{234} \rightarrow TH^{230} = Ra^{226}, \ Pu^{239} \rightarrow U^{235} \rightarrow Pa^{231}, \text{ and}$$

$$Am^{241} \rightarrow Np^{237} \rightarrow U^{233} \rightarrow Th^{229}.$$

The overall risk rises to about 1.3×10^{-7} p.a. at 150,000 years and falls slowly to about 10^{-7} p.a. at one million years. In Fig. 9, the result of the p.r.a. simulation is compared with the "worst risk" and with a number of "best estimates". The latter appear to underestimate the the true risk and could lead to complacency in assessing a disposal facility.

4.1.1. *Discussion.* The logic of the assessment procedure was adequately demonstrated, but in a rather limited way. Only a single scenario was attempted with no explicit allowance for any future changes to the environment. Rather wide ranges of parameter values were assumed and (independently) sampled combinations of values resulted in physically unacceptable derived quantities (e.g. groundwater velocity). Post-hoc rejection of sampled values (e.g. permeability, as seen in Fig. 10) resulted in sampling from p.d.f.'s different from those assumed in the p.r.a. input.

Detailed modelling was restricted to groundwater flow and the derivation of simple models and data sets for p.r.a. lacked rigour as did the sensitivity analysis. These issues were at least partly resolved in the second Dry Run.

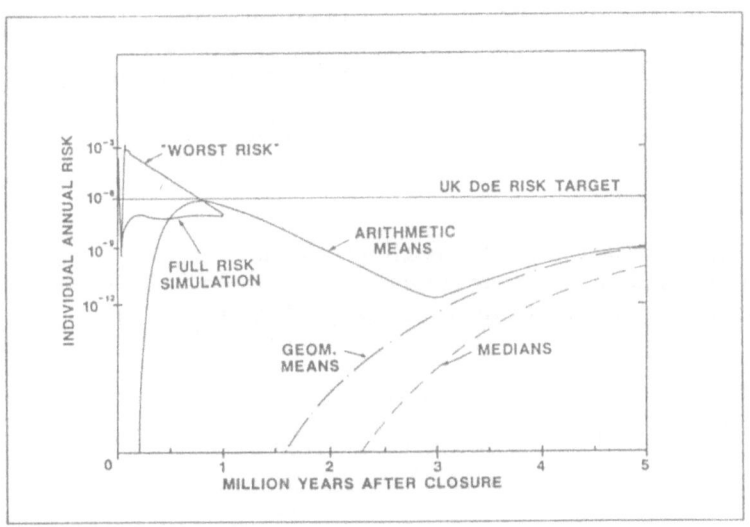

Figure 9   Comparison of different risk estimates. Dry Run 1

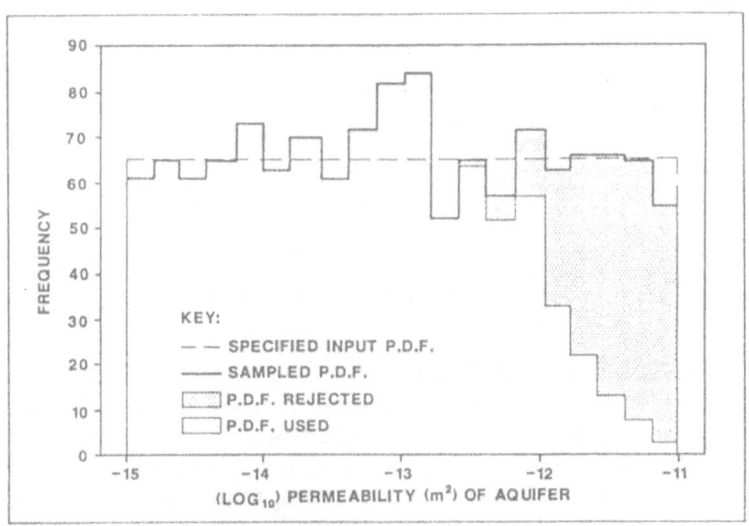

Figure 10   Modification of permeability p.d.f. due to sampling
infeasible groundwater velocities: Dry Run 1

4.2. THE SECOND TRIAL: DRY RUN 2

In the second Dry Run [2] during 1985, the Harwell regional and local hydrogeology was reanalysed and a different two-dimensional vertical section used for groundwater flow modelling. In addition to present day (reference release) conditions, five possible future conditions were explored as sketched in Fig. 11, to represent the effects of increased artesian head in the Corallian aquifer due to tectonic or climatic change, increased use of water resources -- especially pumping from the chalk, and exploration/exploitation of mineral or geothermal resources. Three of the limitations of Dry Run 1 were overcome:

(a)   The two-dimensional modelling was extended, using the TARGET finite difference code, to include calculations for migration of single nuclides such as Tc^{99} and I^{129} (see Fig. 12, for instance).

(b)   The use of several scenarios for specific possible future conditions each comprising fairly narrow ranges of parameter values, particularly for aquifer permeability, overcame the problem of physically unacceptable groundwater velocities.

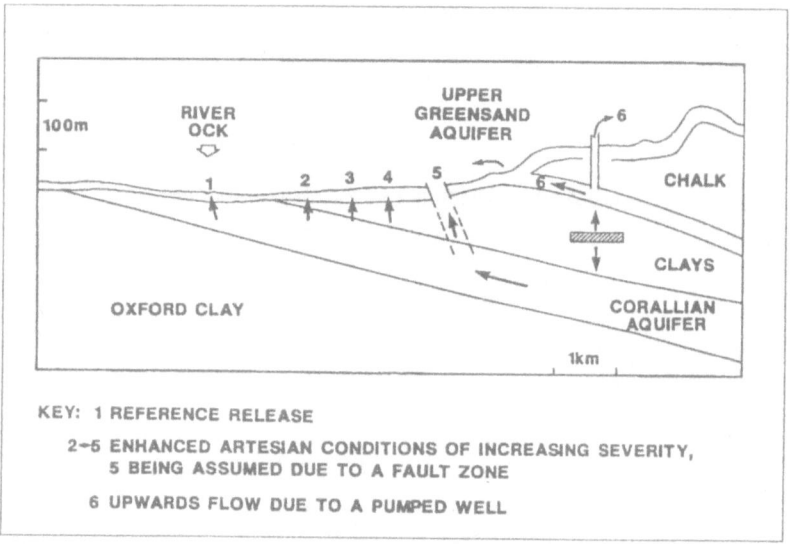

Figure 11   Groundwater pathways modelled in Dry Run 2

4.2.1. *Pr.a. Calculations.* SYVAC A was used to estimate risks for each scenario separately over a time period of five million years and the results are shown in Fig. 13 for random samples of 2000 runs. The only nuclides found to be significant were Tc^{99}, I^{129} and the $Am^{241} \rightarrow Np^{237}$ chain. The risks from the reference release (scenario 1) are very much lower than that for Dry Run 1, due to the changed hydraulic conditions leading to the different biosphere assumed. Only the severe faulted artesian condition (scenario 5) has risks comparable with the early result, peaking at about 2×10^{-7} p.a. at around 800,000 years.

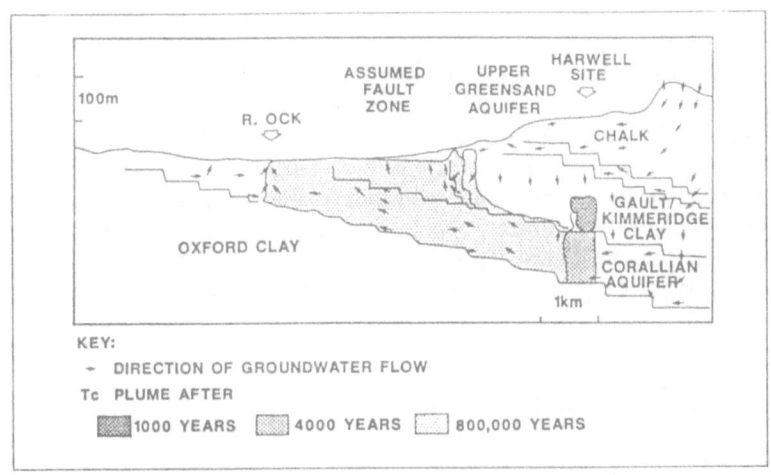

Figure 12   Predicted migration of I^{129} in two dimensional ground-water flow: Dry Run 2

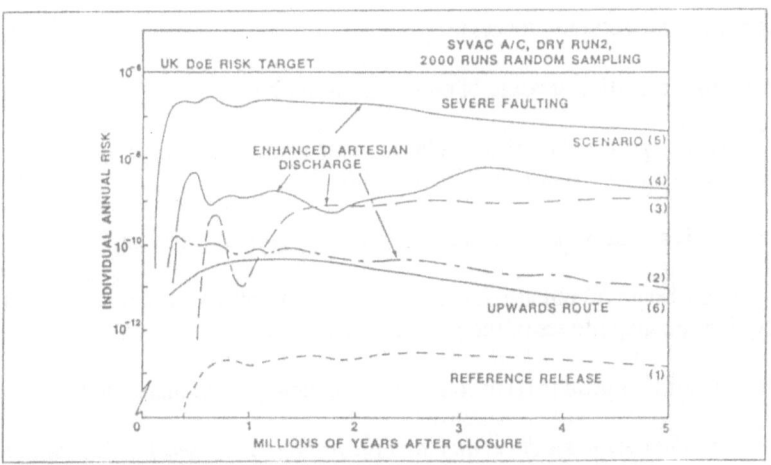

Figure 13   Variation of risk with time after closure for six different groundwater transport scenarios: Dry Run 2

4.2.2. Representation of Temporal Changes. In this particular case the risk-time curves for each scenario reach a plateau which is expected to be independent of the time of occurrence of the future state envisaged (pumping, faulting). Hence, explicit treatment of the time-dependent nature of those particular events and processes was not thought to be required although it was recognised that in general a p.r.a. code capable of explicitly representing the time variation of parameter p.d.f.s, ranges and values, as well as of the topology of the groundwater flow and transport pathways was required. Path switching and multiple releases to different biosphere locations could not be represented adequately in p.r.a. codes such as SYVAC, at that time.

4.2.3. *Combination of Risks from Different Scenarios.* Figure 13 shows that the risk is dominated by the fault condition (scenario 5) but even if this were assumed certain to exist over all time after closure the target level of 10^{-6} p.a. would not be reached. If, however, the risk from any scenario had exceeded 10^{-6} p.a., then the probabilities of each scenario would have had to be considered. This would have posed considerable difficulties in practice especially as sequences of different events and processes occur in a real future and their radiological consequences would depend upon their timing, magnitude and the conditions resulting from the cumulative effects of preceeding phenomena. It was recognised that a rigorous method of assessment must, if possible, incorporate a stochastic generator of futures sampled from the range of possibilities that might occur. TIME2 was at this stage being developed for this purpose.

5.   The Use and Limitations of Scenarios

5.1.   DIFFICULTY OF DEFINITION

Over recent years the subject literature has provided many interpretations of the term 'scenario'. For example:

(a)   A scenario is an isolated process [28].

(b)   A scenario is a particular route of contamination [29].

(c)   A scenario is a group of (logically related) concurrent processes comprising a 'state' of the environmental system at one time [30].

(d)   A scenario is a possible sequence through time of (related) system states [31].

(e)   A scenario of type (c) but also including the uncertainty of parameter values that determine the magnitudes of the processes involved [32].

We ourselves (see Thompson [47]) have taken the meaning of 'scenario' to be:

> "*The state of the system at any time (including the present) which state is then extracted (from the entire sequence or future) to allow calculation of performance to proceed with boundary conditions and modelling representations that are invariant with time over the entire analysis period.*"

This approximation of using a 'frozen' system state has been necessary because with the 'first generation' SYVAC-like pra codes the effect of changes in parameter values and/or the choices of models cannot be represented explicitly as functions of time, although of course this is not a fundamental restriction of the Monte Carlo approach.

The uncertainty in parameter values at the time the system state is frozen is included in the p.r.a. [5].

The diversity and lack of clarity of the definitions used leads to conflicts such as:

> How do we correlate scenarios? [34]

But on the other hand we cannot correlate because:

Scenarios are mutually exclusive [35]

In attempting to resolve the difficulties inherent in the approach we encounter a plethora of scenario types. For example:

Subscenarios [36]

Umbrella scenarios [36]

Normal evolution scenarios [37]

Altered evolution scenarios [37]

Disruptive scenarios [38,37]

It seems that the lack of a clear, unambiguous and economical definition for the scenario method is a strong indication that it has, at present, some scientific weaknesses although work is proceeding under the auspices of the OECD to clarify the situation [36].

5.2. THE USE OF AVAILABLE EVIDENCE

At any time in the development of science, the total body of knowledge and the individuals available to interpret it for use in any given application are the same, irrespective of the methodology adopted.

In particular the considerable records of geological history need to be reviewed and studied in order to provide a basis for assessing the range of possible futures and their likelihoods of occurrence. It is difficult to see how this can be achieved unless a modelling framework is adopted.

5.3. DEGREE OF REALISM

So far only very simple time sequences of system states have been identified by means of scenario-based methods. Often a 'base' or a 'present-day' scenario assumes no change with time in boundary conditions, overall geometry or the processes that are operating. 'Altered evolution' scenarios may invoke one or two major influences, such as a pumped well, a fault zone or a glacial ice sheet. These are usually assumed to operate over the entire assessment time period [2, 11, 29, 30, 37].

As part of the overall risk assessment procedure [1] we advocate the exploration of limiting conditions by means of scoping studies and sensitivity analysis. Simple scenarios can prove valuable for this purpose. However, it is difficult to determine the level of inaccuracy introduced by:

(a) Omitting the true complexity of the time sequence of climate-driven environmental conditions.

(b) Omitting the complex interactions between the different processes at any given time.

Finally, a geometrical framework by which the effects of geomorphological changes can be traced is not easily incorporated unless an environmental model is used.

5.4.   STATISTICAL CONSIDERATIONS

The use of scenarios with associated probabilities of occurrence is frequently put forward as a means of risk estimation, for instance in [39,2].
   However, the representation of the inherent uncertainty concerning possible future changes is inadequate statistically because:

(a)   Only a very small sample of possible futures is considered. For example, nine in [11], six in Dry Run 2 [2];

(b)   No explicit joint p.d.f. of parameters is specified;

(c)   The method of sampling is undefined;

(d)   There is a tendency to emphasise 'disruptive' or 'degraded' conditions [39, 38, 37] likely to lead to high doses, and this will bias the result.

As shown in Fig. 14, if the sample size is very small a considerable error in the risk estimate can occur. Also, the choice of sampling scheme has a large effect on the sample size required before a converged estimate of risk can be calculated [40].

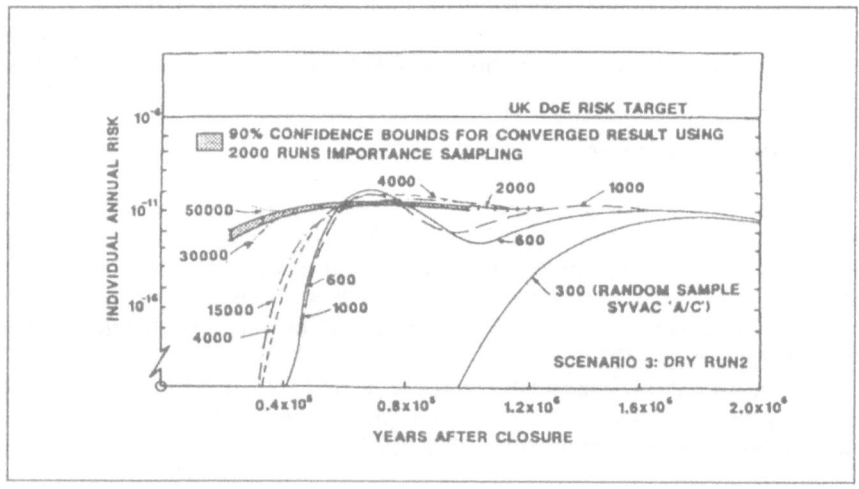

Figure 14   Effect of sampling size and sampling method on risk prediction

If there is no clear statistical model, almost any answer can be produced for 'risk' even though the same information (both 'hard' and 'soft') is used. This lack of constraint is clearly unacceptable in the general case.

Of course, it may be argued that the subjectivity of judgements over parameter values in the simulation approach is no more constrained than is the assignment of overall 'scenario' probabilities to entire futures.

It seems intuitively reasonable, however, that using past data to derive and calibrate a coherent statistical model is preferable to guessing a priori the probability of a complete time sequence of system states affected by a highly non-linear interaction of several concurrent processes over (say) 10^6 years.

6.   Risk Calculations Using the Results of Environmental Simulation

6.1.   USE OF ENVIRONMENTAL SIMULATION
RESULTS

The output from the TIME2 code or the planned TIME4 simulator can be used to generate a time series of values of some of the parameters needed as input for a single run of the pra code VANDAL, described below. Fig. 15 illustrates this, showing how a dose-time curve is roduced corresponding to each such time series. The overall expectation value of risk (R) calculated from the doses at any time (weighted by the appropriate run probabilities), using equation (1).

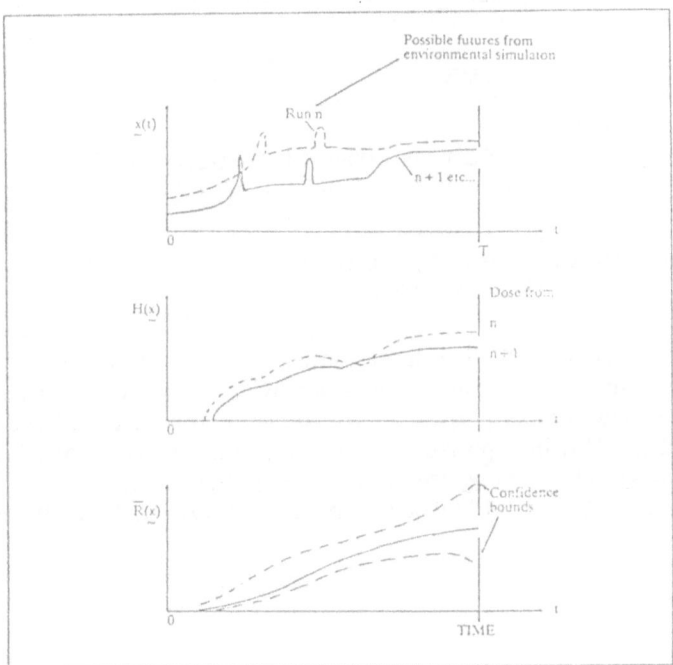

Figure 15   Calculation of risk from results of environmental
simulation

6.2.   TIME-DEPENDENT PRA: VANDAL

The new simulation code VANDAL has been developed from an earlier 'rapid prototype' code MAIA [41] which demonstrated the feasibility of overcoming several limitations of SYVAC by:

(a)   Incorporating a multi-dimensional network model for groundwater flow in the vault and the geosphere and for the associated radionuclide transport (see Fig. 16)..

(b)   Generalising the VERMIN nearfield model [42] to better represent the spatial arrangement of the repository and its contents. Also a simple representation of the post-closure influence of pH evolution in a cementitious environment on the near-field chemical system is included.

(c)   Adopting numerical methods throughout that permit geometric factors, bulk media properties and the boundary conditions to vary with time.

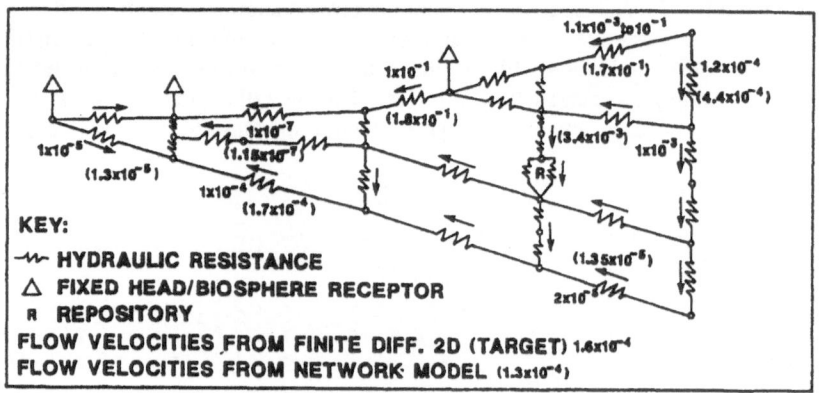

Figure 16   Initial calibration of a simplified groundwater flow model:
Dry Run 2, reference release scenario

The features of MAIA were then incorporated into the production code VANDAL by reprogramming under a formal QA plan involving commercial software engineering procedures. A new general purpose executive has been written [43] and an importance sampling package [40] incorporated. A new dynamic compartmental biosphere model DECOS [44] was similarly re-engineered and added to VANDAL.

Fig. 17 compares the overall functional structures of VANDAL and the earlier SYVAC simulators.

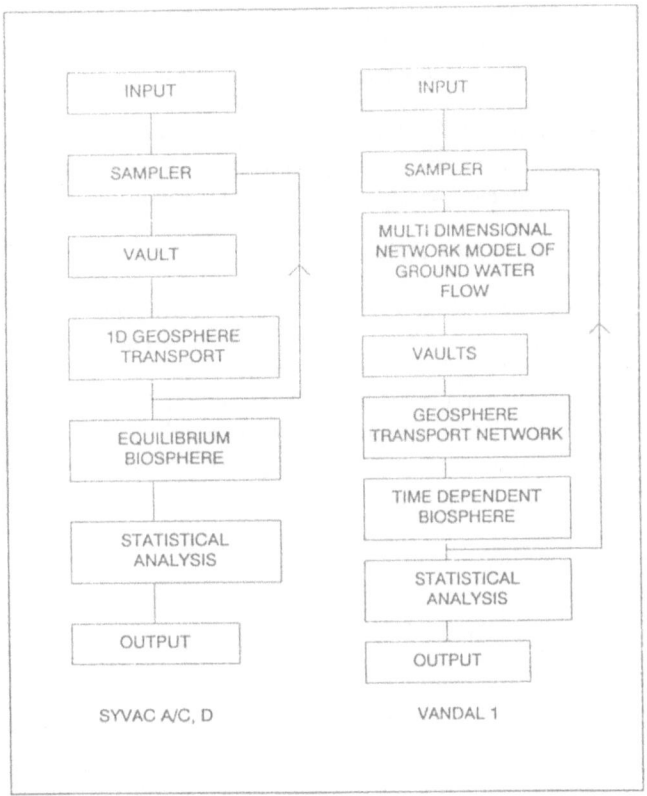

Figure 17   Comparison of main features of UK DOE p.r.a.
             simulation codes

6.3.   ILLUSTRATION FOR A SHALLOW SITE

During 1987, an exploratory study was made [5] for a hypothetical shallow disposal site at Elstow. Results from TIME2 and from professional judgement were combined to provide median estimates for two potential environmental system states:

(a)   For a Savannah climate at about 1,500 years after closure, due to 'greenhouse warming' and,

(b)   For a tundra climate at about 20,000 years from the present.

Biosphere dose-conversion factors were obtained on the basis of present-day analogues of such conditions [45, 46].
  These conditions were used to specify inputs to SYVAC 'A/C' risk calculations, assuming that these system states were 'frozen' over the entire assessment period of 25,000 years. Detailed analysis of groundwater flow and solute transport in two dimensions was undertaken to determine likely release and migration paths.

– 251 –

As Fig. 1 indicates, the broad levels of risk are much higher for potential future climatic conditions due to activation of shorter pathways to the surface.

The influence of erosion due to river meandering across the site and the effect of direct excavation of active material by human action was also examined.

As 'steady state' scenarios were derived from the interpretation of a stochastic simulation of a range of possible futures, this study represents an intermediate stage, as indicated in Fig. 3, in the move towards a fully time-dependent risk analysis procedure.

6.4.   ILLUSTRATION FOR DEEP DISPOSAL

A horizontal tunnel repository concept was assumed similar to that used previously in the trial assessments described earlier. The same inventory of combined low- and intermediate-level wastes was also used, although only the results of calculations for Tc^{99} are presented here. The site is illustrated in cross-section in Fig. 18, where it can be seen that the repository is situated in chalk at a depth of about 150m, with an overlying chalk aquifer and about 20m of glacial till assumed under present day conditions. Recharge occurs on the inland scarp and discharge to the coastal sediments.

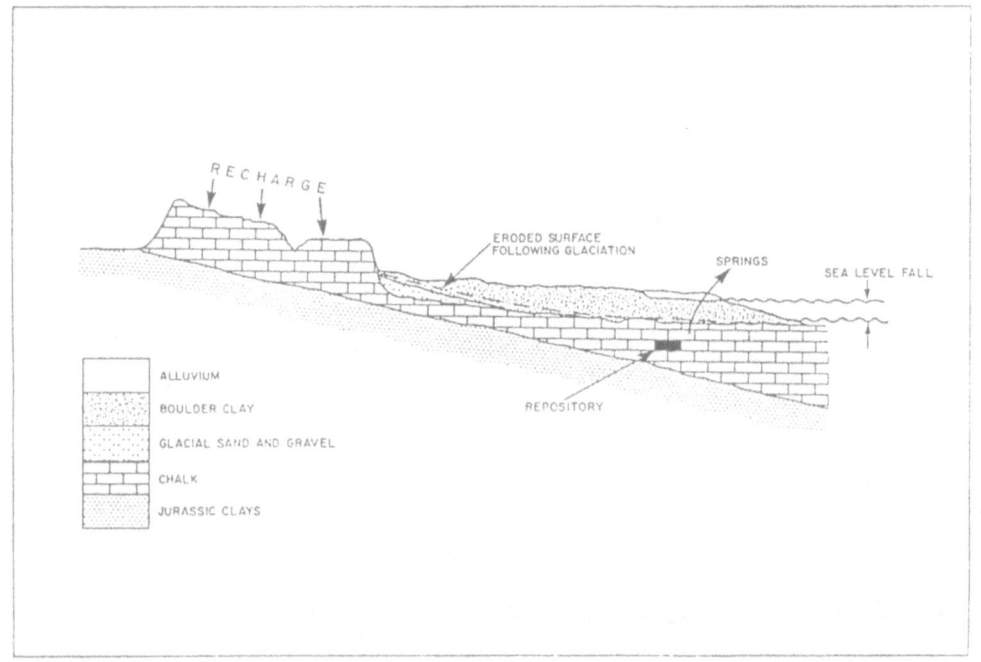

Figure 18   Schematic hydrogeological section through the hypothetical site

As TIME 4 is still under development a sequence of future climatic changes was derived from the oxygen isotope ratio data taken from deep ocean cores, as already mentioned in Section 3. The pathways for exposure to man are:

(a)   Upward flow through the chalk and the till into a spring used for drinking water.

(b)   Upward flow through the chalk into the aquifer and to the present seabed. For the purposes of demonstration, this pathway was considered negligible, except when a fall in sea level would allow cultivation of the exposed sediments.

A network model of flow and transport was derived for the site.
  The climate state sequence used initially was:

(i)   Present-day temperate,

(ii) Tundra, with about 10m of till eroded,

(iii) Reversion to temperate with a shorter path to the spring due to the erosion.

  During the tundra state, the sea level falls by about 80m and leads to an assumed agricultural pathway due to cultivation on the exposed sediments.
  Biosphere dose conversion factors are again taken from present-day analogues.
  Fig. 19 shows the variation of dose with time over 100,000 years when the conditions associated with each climate state are 'frozen' over the entire calculation.
  The full time dependent calculation is compared also with these three 'scenario' results in Fig.19 . The shape of the dose curve is very different from any individual scenario. Over the period between 15,000 to 40,000 years doses are only about 2% to 3% of the doses estimated from the 'present day' scenario and a peak dose of about 40% of the 'eroded temperature' scenario is found, but at a very different time (54,000 years instead of about 12,000 years).

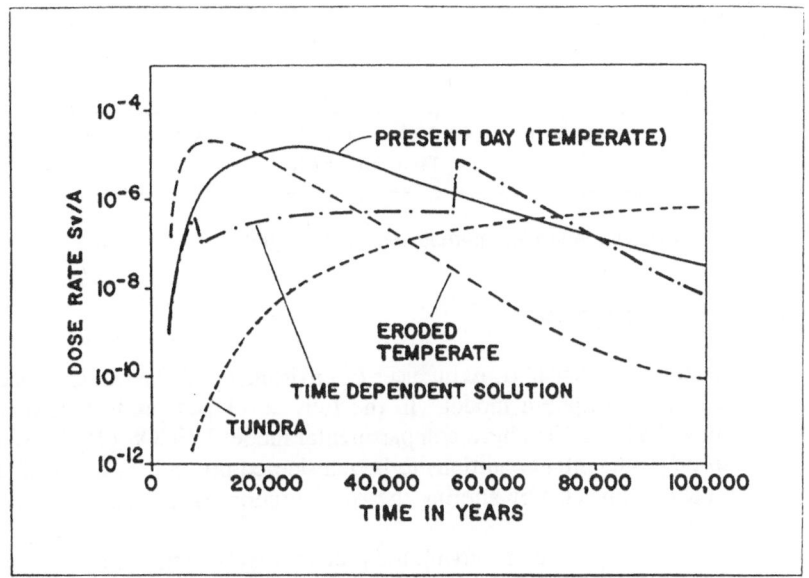

Figure 19   Comparison of full time-dependent calculation with results from each steady-state climate "scenario"

Fig. 20 compares two more time histories of dose in which the climate state sequences used were more complex and involved boreal periods before and after the tundra episodes. The effect of uncertainty over the duration of tundra on the overall doses is clear. It is intended to progress systematically to predictions over several glacial cycles, and to extend the analysis to multi-member decay chains.

These initial calculations are for a single nuclide (Tc^{99}) and further calculations for multi-member decay chains are in progress.

During the tundra climate, doses are much lower, because the reduced transport properties of the frozen layers and the lower infiltration adopted in the calculations more than compensate for the increased hydraulic gradient due to the erosion and fall in sea level.

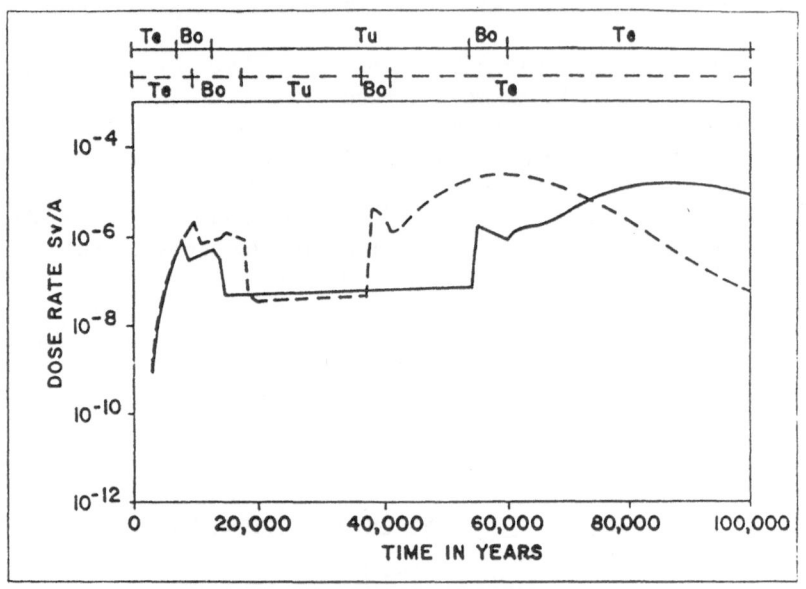

Figure 20   Dose variation for alternative climate sequence

6.5.   FUTURE HUMAN ACTIONS

The illustration of the radiological consequences of environmental change described above employed a very simple biosphere model. In the fully developed production versions of VANDAL, the new dynamic biosphere compartmental model DECOS [44] is used. This allows changes in environmental conditions to be considered not only in terms of changed dose-conversion factors, but also by altering the configurations of compartments and their interconnections.

For example, the effect of changes in agricultural practice may be estimated for a coastal site. At present, it is assumed that there is release of radioactivity via groundwater into the subsoil and by interflow to drainage ditches and then to an estuary. Irrigation of the topsoil is also practised.

After 1,000 years it is assumed that the land falls into disuse and reverts rapidly to marshy conditions.

The activity levels of Ra ^{226} in selected compartments for a rate of release of 1MBq per annum to the biosphere are shown in Fig. 21, where the effects of changes are compared with the time-invariant 'present-day' scenario.

The present-day assumptions lead to a considerable underestimate of activity in the topsoil but overestimates the activity in the estuarine sediments, for example.

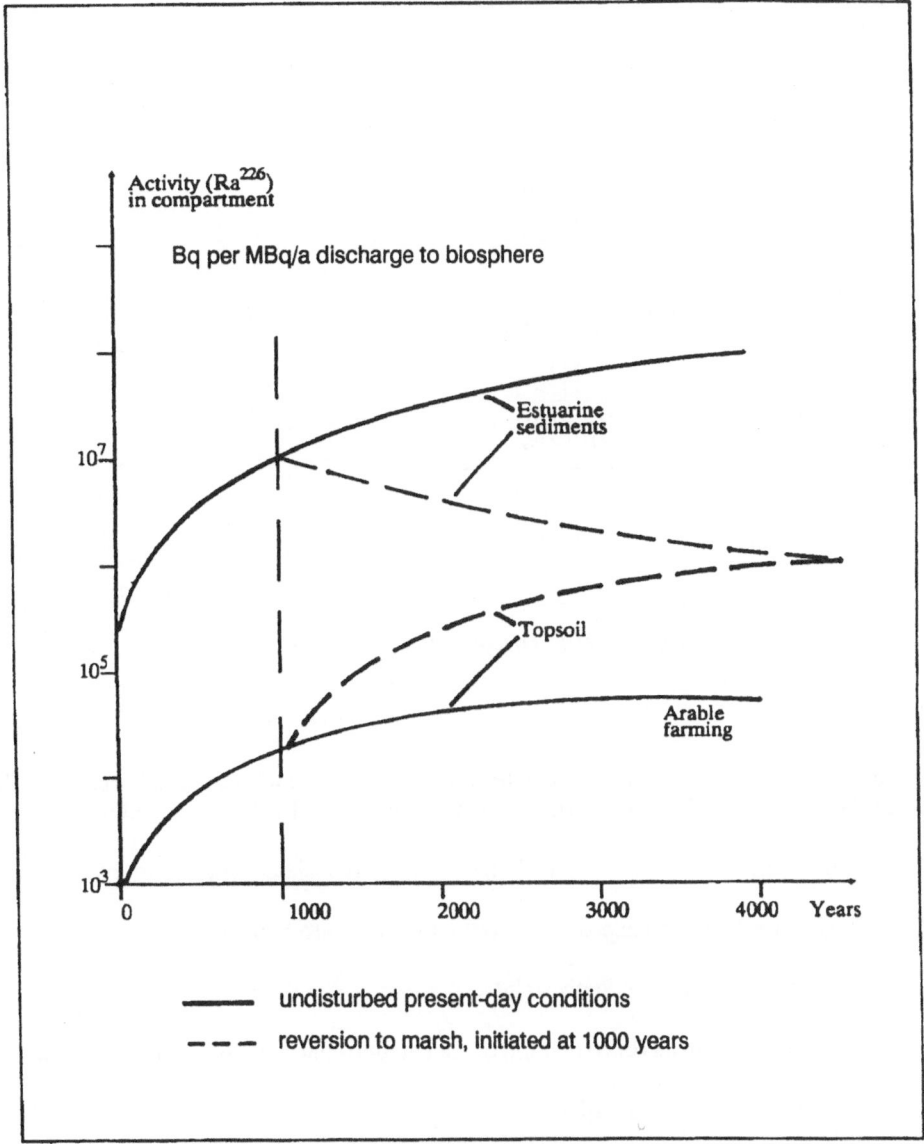

Figure 21   Human action - effect of change in land management practice

6.6. FUTURE DEVELOPMENTS AND TRIALS

Ultimately, it is hoped to link together all aspects of the simulation as shown schematically in Fig. 22. A full rehearsal of the new procedure will be necesssary before a real assessment is undertaken. This rehearsal (called Dry Run 3) is scheduled for 1989.

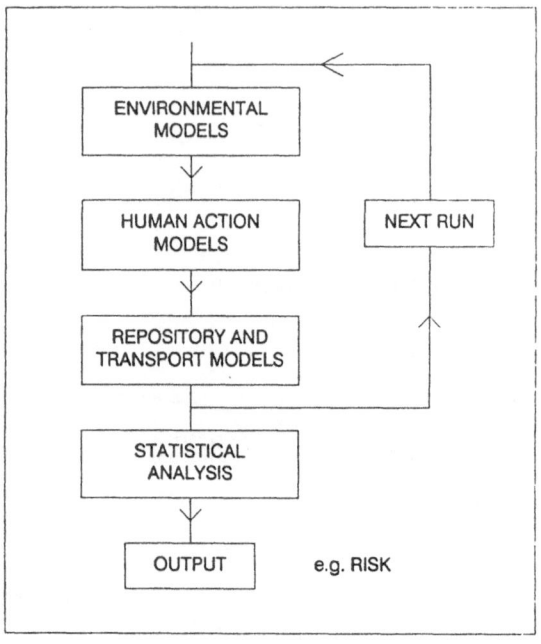

Figure 22   Overall environmental simulation

7.   Discussion

In addition to the issues considered here, many others of importance in regard to the present-day practice of risk analysis should be examined critically. These include:

- The influence of chemical and other changes associated with the repository.

- Other release and exposure pathways due to future human action.

- Spatial variability and uncertainty.

- Systematic errors (bias) and the overall audit of uncertainty in risk estimation.

- Verification and validation.

- The treatment of subjectivity: For example, probability elicitation, and the use of Bayesian methods.

8. Conclusions

The examples shown above reveal that the broad levels of risk from shallow engineered trench disposal of low-level wastes could be affected significantly by future likely climatic changes over the current interglacial.

For deep disposal, a full risk calculation incorporating temporal change and its uncertainties is not yet possible as the TIME4 simulator is not yet fully developed. The 'steady state' approach is, however, shown to be a poor approximation to the dose-rate time curve derived from a full time-dependent analysis using the VANDAL p.r.a. code, for the particular example considered here.

Verification and more especially validation remain a fundamental difficulty for such long-time frames, although the long-term environmental simulation approach may offer advantages if natural analogues are to be interpreted meaningfully. Boulton deals with the scientific basis of the subject in another paper in these proceedings.

9. Acknowledgements

The author is particularly grateful for advice and discussions with Mr C. Frizelle of ITEC, Dr P. Ringrose of Dames & Moore International, Dr M. C. Thorne of Electrowatt Engineering Services (UK) Ltd, Mr D. Nicholls now with R M Consultants Ltd, and with Professor G. Boulton of Edinburgh University.

HMIP are thanked for permission to present these views as part of the JRC ISPRA Advanced Seminar on Risk Analysis.

This paper may be used in the formulation of UK Government Policy but does not at this stage necessarily represent UK Government Policy.

10. Abbreviations

AECL	Atomic Energy of Canada Ltd.
HMIP	Her Majesty's Inspectorate of Pollution
MAIA	Migration Analysis for Initial Assessment
NEA	Nuclear Energy Agency of the Organisation for Economic Co-operation and Development (OECD)
p.d.f.	probability density function
p.r.a.	probabilistic risk analysis
p.s.a.	probabilistic systems assessment
SYVAC	SYstem Variability Analysis Code
UK DoE	United Kingdom Department of the Environment
US DoE	United States Department of Energy
VANDAL	Variability ANalysis of Disposal ALternatives
PAAG	Performance Assessment Advisory Group of the NEA

11.   References

1.   Thompson, B.G.J.   A probabilistic risk assessment of an underground radioactive waste disposal facility.
Proc. Second Int. Conf. on Radioactive Waste Management
Can. Nucl. Soc., Winnipeg, Canada (Sept 1986).

2.   Gralewski, Z.A.   Development of a methodology for post-closure
et al   radiological risk analysis of underground waste repositories.
pp 969-974, Proc. Conf. Prob. Safety Assessment and Risk
Management, PSA '87, Zurich (Aug 1987).

3.   SCICON   UK, SYVAC A/C Overview: Version 1.1, UK DoE Report
DoE/RW/87.014 (1987).

4.   SCICON   VANDAL version 1.1 User Guide, (Feb 1989) UK DoE
Disposal Assessment User Document UD-SCI-18 ed. 1 draft 2
(to be published).

5.   Dames & Moore   Evaluation of environmental change and its effects on the
radiological performance of a hypothetical shallow engineered
disposal facility at Elstow, Bedfordshire.
UK DoE Report No DoE/RW/87.124 (March 1988).

6.   Thorne, M.C.   Assessment of the radiological risks of underground disposal of
solid radioactive wastes.
UKDoE Report No DoE/RW/89.030 (Dec 1988).

7.   Nancarrow, D.J.   Preliminary radiological assessments of near-surface low-level
Sumerling, T.J.   radioactive waste repositories: Summary Report. UK DoE
Report DoE/RW/89.008 (Aug 1988)

8.   SCICON   UK SYVAC D version 2.0, Overview, UK DoE Disposal
Assessment Programmer Documents PD-SCI-39 to 42 (4 vols.)
(Feb 1988) to be published.

9.   Sherman, G.R.   SYVAC-2: A system variability analysis code for the assessment
et al   of nuclear fuel waste disposal concepts, AECL TR-317 (1985).

10.   Saltelli, A.   LISA, a code for safety assessment in nuclear waste
Bertozzi, G.   disposals, JRC Ispra. EEC Report EUR 9306EN (1984).
Stanner, D.A.

11.   Cranwell, R.M.   Risk Methodology for Geologic Disposal of Radioactive Waste:
Campbell, J.E.   Final Report, NUREG/CR-2452 SAND81-2573 (Aug 1987).
Melton, J.C.
Iman, R.L. et al

12. ICRP Publication 26: Recommendations of the International Commission of Radiological Protection, Annals of the ICRP, Vol. 1, No. 3, 1977

13. Dames and Moore Design of TIME2 code: time-dependent effects on land 2 type repositories.
UK DoE Report: No DoE/RW/85.076 (July 1985).

14. Dames and Moore Progress review: development of TIME2 code.
UK DOE Report: No DOE/RW/86.016 (Feb 1986).

15. IGS Hydrogeological map of north and east Lincolnshire. HMSO (1967).

16. Boulton,G.S
Jones, A.S.
Clayton, K.M.
Kenning, M.J. A British ice sheet model and pattern of glacial erosion and deposition in Britain. In: British Quaternary Studies, Ed. F.W. Shotton. Clarendon Press, Oxford (1977).

17. Sutherland, D.G. Submerged rock platforms on the continental shelf west of Sula Sgeir. Scott J. geol., 23,251-260. (1987)

18. GARP The physical basis of climate and climate modelling. Report of the International Study Conf., Stockholm (July 1984). GARP Publ. Ser.No. 16, World Meteorological org. Geneva.

19. Robin, G. de Q. Polar ice sheets: developments since Wegener. Geol.Rundshau., 70 (2), 648-663 (1981).

20. Morner, N.A.(Ed) Earth Rheology, Isostasy and Eustasy. Wiley and Sons, New York. (1980).

21. Price, R.A. Global change: geological processes, past and present. Episodes, 9 (2), 91-94. (1986).

22. INTERA GSM:Geologic Simulation Model for a hypothetical site in the Columbia plateau - large computer version. Tech. Report. ONW1-447. (1983).

23. INTERA FFSM:Far-Field State Model. Tech Report. ONW1-436 (1983).

24. Shackleton, N.J.
Oppdyke, N.D. Oxygen isotope and palaeomagnetic stratigraphy of equatorial pacific core V28-238: Oxygen isotope temperature and ice volumes on a 100,000 year and 1,000,000 year scale. Quaternary Research, 3. (1973).

25. Boulton, G.S. The reconstruction of former ice sheets and their mass balance
    Smith, G.D., characteristics using a non-linearly viscous flow model. Journal
    Morland, L.W. of Glaciology Vol.30, No.105, 140-152 (1984).

26. Dames & Moore A review of research into the effects of long-term environmental
    change on deep land disposal sites for radioactive waste. Draft
    report to the Commission of the European Communities and
    UKDoE Disposal Assessments Tech Report TR-D&M-12, ed. 1.

27. Morland, L.W. A theory of slow fluid flow through a porous thermoelastic
    matrix. Geoph. J. Roy. Astro. Soc., 55,393-410 (1978)

28. Van Luik, A.E. A prescriptive approach, the link between the performance
    Eisenberg, N.A. assessments of the US DoE projects and regulatory
    Alexander, D.H requirements of NRC and EPA.
    Proc. NEA Workshop 'System Performance Assessments for
    Radioactive Waste Disposal' pp 61-77, Paris (1986).

29. Nancarrow, D.J. Preliminary assessments of the post-closure radiological impact
    Sumerling, T.J. of disposal of low-level radioactive wastes in shallow engineered
    Ashton, J. facilities at four hypothetical sites.
    UK DoE Report DoE/RW/88.084 (June 1988).

30. Wuschke, D.M. Second interim assessment of the Canadian concept for nuclear
    et al fuel waste disposal. Vol 4: Post-closure assessment.
    AECL-8373-4. (1985).

31. Cranwell, R.M. Sources/Treatment of uncertainties in the
    Bonano, E.J. performanceoassessment of geologic radioactive waste
    repositories.Ref. 38 (ibid) pp 53-68.

32. Holmes, R.W. Uncertainty analysis for the long-term assessment of Uranium
    mill tailings. Ref 38 (ibid) pp 167-190.

33. Thompson, B.G.J. Notes to PAAG concerning the use of 'scenarios' in post-closure
    radiological risk assessment. (March 1987).

34. Brandstetter, A. The role of geostatistical, sensitivity and uncertainty analysis in
    Buxton, B. performance assessment.
    Ref 39 (ibid) pp - (Sept 1987).

35. Veneziano, D. Comments during final discussion session of US DoE/AECL '87
    Conf. Ref 39 (ibid) (Sept 1987).

36. Hodgkinson, D.P. Systematic approaches to scenario development. NEA (OECD)
    PAAG/DOC/(88)2. (January 1988).

37. Cadelli, N.P. PAGIS: A common Methodological Approach based upon European Data and Models: Summary Report of Phase 1, EEC Report EUR 9220EN (1984).

38. NEA(OECD) Uncertainty Analysis for Performance Assessments of Radioactive Waste Disposal Systems.Proc of NEA Workshop, Seattle, USA (Feb 1987).

39. Proceedings USDoE/AECL '87 Conference on Geostatistical, Sensitivity and Uncertainty Methods for Groundwater Flow and Radionuclide Transport Modelling. San Francisco, USA, (Sept 1987).

40. Johnson, K. The use of Importance Sampling in a time assessment to obtain
    Lucas, R. converged estimates of radiological risk.
    UK DoE Report No DoE/RW/87.080 May (1987).

41. Gralewski, Z.A MAIA: Time-dependent, multi-path p.r.a. demonstration code.
    Taylor, J.W. pp 1007-1012, Proc. Conf. on Prob. Safety Assessment and Risk Management, Zurich (Aug 1987).

42. Laurens, J.M. Validation of the vault computer model, VERMIN, for post-closure behaviour of repositories in geological formations. UK DoE Report DOE/RW/87.079 (Oct 1987).

43. Haydon, D. User Guide for the SEER executive.
    UK DoE Disposal Assessments Document UD-SCI-9, ed 1, (Dec 1986), to be published.

44. Nancarrow, D.J. Guide to the dynamic biosphere model DECOS.
    Mitchell, D. UK DoE Disposal Assessment Program Document PD-ANS-10 ed 1, (draft 2), (Aug 1988).

45. Jones, C.H. Basis of dose-conversion factors for a Savannah environment. ANS Report No 736-1, (Nov 1986), under UK DOE contract PECD 7/9/363.

46. Jones, C.H. Basis of dose-conversion factors for a tundra environment. ANS Report No. 736-2, (Nov 1986),under UK DOE contract PECD 7/9/363.

47. Thompson, B.G.J. A method of overcoming the limitations of conventional scenario-based risk assessments by using Monte-Carlo simulation of possible future environmental changes. UK DoE Disposal Assessments Tech. Report TR-DOE-12 (May 1988). NEA reference PAAG/DOC(88)11.

EVALUATION OF THE RISK ASSOCIATED WITH DISPOSAL OF RADIOACTIVE
WASTES :   THE DETERMINISTIC APPROACH

J.LEWI
CEA - Institut de Protection et de Sûreté Nucléaire
Département d'Analyse de Sûreté
B.P. n°6
92265 Fontenay-aux-Roses
France

ABSTRACT. Safety of radioactive wastes disposal depends on the inter-
position of "barriers" (waste package, engineered barriers, geological
formations) between the wastes and the biosphere : the combination of
these barriers must provide protection for man under all circumstances
considered plausible.

     Justification of the choice and of the characteristics of the
barriers, for the disposal of given quantities of radionuclides, is
provided by a risk assessment. Either a deterministic methodology or a
probabilistic one can be used to perform this assessment. This paper
describes the deterministic method currently used in France.

     After recalling the specific features of waste disposal options
and the meaning of an "acceptable risk", the deterministic approach is
described and it is shown how the results permit the acceptability of
the disposal to be reached.

     Two examples are presented, the first one dealing with a surface
land burial repository, the second one dealing with a deep repository.

Introduction

     Radioactive waste disposal facility safety depends on a certain
number of barriers being placed between the wastes and man. These
barriers, artificial (the waste package and engineered barriers) and
natural (geological formations), have characteristics suited to the
type of disposal facility (surface disposal or disposal in deep geological formations). The combination of these different barriers must
provide protection for man, under all circumstances considered plausible.

     Justification, for the disposal of given quantities of radionu-
clides, of the choice of the site, the artificial barriers and the
overall disposal architecture, is obtained by an evaluation of the

A. Saltelli et al. (eds.), Risk Analysis in Nuclear Waste Management, 263–284.

risk. This evaluation provides a basis for determining the acceptability of the disposal facility.

One of the following two methods is normally used for evaluation of the risk : the deterministic method and the probabilistic method. This report describes the deterministic method.

This method is employed in France for the safety analysis of the projects and works of ANDRA, the national agency responsible for the management of radioactive wastes. It should be remembered that in France, the La Manche surface disposal center for low and medium activity wastes has been in operation since 1969, close to the reprocessing plant at La Hague, and a second surface disposal center is to be commissioned around 1991 at Soulaines in central France (département de l'Aube). Furthermore, geological surveying of four sites located in geological formations consisting of granite, schist, clay and salt were begun in 1987 for the selection in about three years time of a site for the creation of an underground laboratory. This could later be transformed, if safety is demonstrated, into a deep disposal centre.

Specific features of radioactive waste disposal facilities

Waste disposal facilities differ from other nuclear installations essentially in :
. the extremely long periods (some thousands or hundreds of thousands of years) during which the potential risk remains significant ;
. the essential role played by the geological environment. Indeed, in view of the periods of time involved in disposal, it would not be realistic to assume that safety could depend on artificial barriers. These have a particular role to play during the operations preceding disposal (handling, storage and transport) and during the initial stages of disposal in which there is considerable beta and gamma activity. Artificial barriers can also, in certain cases, contribute to long-term safety, essentially by their degradation products (for instance in modifying the chemical properties of the disposal facility near-field).

It must also be pointed out that, in contrast to nuclear plants which benefit from the operating experience obtained from many hundreds of installations, for disposal only limited operating experience is available from the few surface disposal centres (the La Manche disposal site in France for example \1\) and "exercises" concerning deep centers (KBS studies \2\, \3\ the GARANTIE project \4\ and the PAGIS project \5\). Although limited, this experience is precious insofar as it enables the feasibility of the different methods proposed to be substantiated.

These features, which are specific to wastes disposal, result in considerable uncertainty with regard to the data necessary for risk evaluation. These uncertainties reflect :
. the difficulty of characterizing certain events liable to affect the integrity of the disposal facility in the future (natural phenomena or events of human origin),

. the problems involved in the validation of models for the time periods involved,
. the impossibility of extrapolating certain data into the future, particularly the biosphere and man data (life habits, diet, resistance to radiation),
. the spatial variability of the characteristics of certain geological media, and the difficulty of making representative measures in them.

Some of these uncertainties appear to be impossible to reduce (for example those concerning man). Others can be reduced by intensifying the effort concerning not only the characterization of the artificial and natural barriers but also studies performed in order to better understand the mechanisms involved in the evolution of these barriers and in the transfer of radioactivity to man.

The deterministic method

To determine the acceptability of waste disposal facilities, reference is normally made to the recommendations of publication No. 46 of the International Commission for Radiological Protection :
. for events which are certain or highly probable and in the case of prolonged exposure, the individual dose equivalence must be less than 1 mSv/year,
. for random events, acceptability of the disposal facility depends on an individual risk criterion. The individual risk is defined as the probability of effects on the health of an individual or his descendance. It is thus the product of the occurrence probability of an event and of the probability that the exposure associated with this event would have an effect on health. In view of the risk/exposure relationship which is in the order of 10^{-2} health effects/Sievert, the limit of 1 mSv/year for certain exposure corresponds to a risk of the order of 10^{-5}/year.

By extension of the dose limit system for certain events, the International Commission for Radiological Protection recommends that the individual risk associated with random events be less than 10^{-5}/year. It is however necessary to allow for all events liable to occur during one year and, therefore, to check that the sum of the risks associated with the events which can liable to occur simultaneously during the year is less than the 10^{-5}/year limit. For this reason the International Commission of Radiological Protection recommends use of a fraction of this limit for radioactive waste disposal.

It is possible to represent probability versus consequences for the acceptability of a waste disposal facility, as defined in Figure 1.

It should be noted that the International Commission for Radiological Protection also recommends optimization of radiological protection efforts, the objective being to limit exposure to the minimum reasonably attainable (ALARA principle). We shall not consider radiological optimization problems here.

The International Commission for Radiological Protection has not given any time limits for the recommendations concerning risk or dose limits. In certain countries, such the USA and Canada, it has been considered that evaluations should be limited in time to 10 000 years, in view of the decrease in the overall radiological toxicity of the waste disposed to levels comparable to those of natural substances, and also in view of the increasing uncertainty of evaluation with time. No official position has yet been adopted in France[1].

In the deterministic method, to ensure observance of the values recommended by the International Commission for Radiological Protection, a limited number of scenarios are considered which are representative of the different families of plausible situations and chosen in such a manner that the associated consequences are more severe than those of other situations of the same family. Each scenario is associated, generally approximately and in a subjective manner (expert opinion), with an occurrence probability. The most probable scenario, generally termed the normal scenario, is assigned a probability of one. For each scenario, the radiological consequences for an individual are evaluated as a function of time in a conservative manner. We shall later return to the methods which can be used to make such evaluations.

The disposal facility can be considered to be acceptable if the maximum radiological consequences are within the limits, for all situations taken into account, having regard to the probabilities of the situations. Thus, considering the probability/consequences domain, each scenario corresponds to a particular point ; acceptability of the disposal facility is deduced from a comparison of the position of these points with respect to the acceptability curve of the International Commission for Radiological Protection mentioned earlier.

The deterministic approach to the evaluation of the risk associated with radioactive waste disposal facilities is coherent with that used for calculation of the risk associated with the operation, under normal or accident conditions, of conventional nuclear installations : reactors or fuel cycle installations \7\. It differs from

---

[1]It should be noted however that the members of a working group \6\ set up by the French Ministry for Industry for proposing technical criteria for the choice of a deep site came to the conclusion that it was indispensable for the site choice to guarantee that the geological barrier would retain its confinement performance, subject to a certain margin of acceptable change, for 10 000 years in the absence of random events and that, for these first ten thousand years, the demonstration of safety should be based on accurate modelling, with realistic hypotheses regarding the effectiveness of the barriers. After 10 000 years, the safety demonstration should be based on a comparison between an estimate of the minimum residual effectiveness of the geological barrier and the evaluation of the effectiveness level necessary to attain the established radiological objectives.

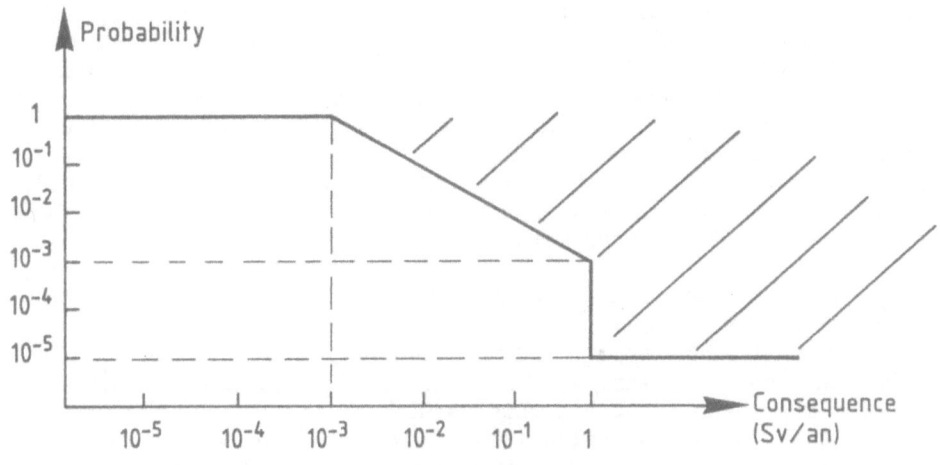

Fig. 1 - PROBABILITY CONSEQUENCES CURVES

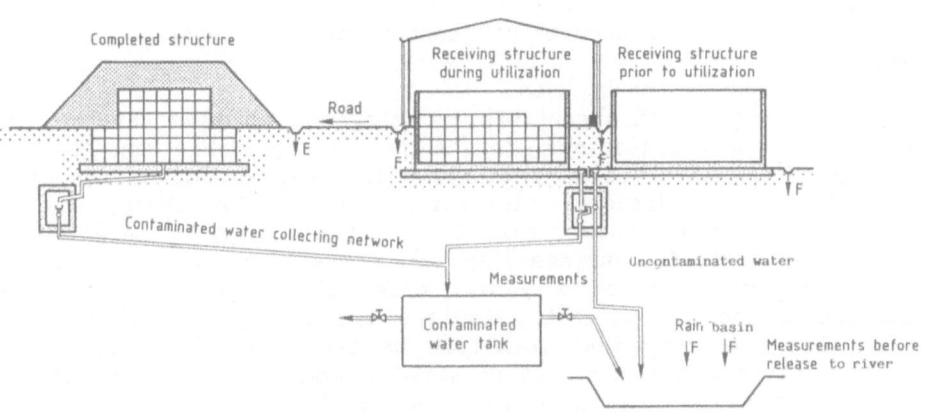

Fig. 2 - DESIGN OF DISPOSAL FACILITY AND RELEASED WATER NETWORKS

the probabilistic method insofar as each of the chosen scenarios "envelope" a family of events liable to affect the quality of the disposal facility. The scenarios cannot be construed as corresponding to the different representations of the disposal facility conceivable on the basis of the different values possible for the design parameters in view of their uncertainty. Indeed, a single representation of the disposal facility corresponds to reality. Uncertainty concerning this representation cannot be considered to be the origin of different situations to which a probability of occurrence can be attributed, and this applies even if the uncertainty concerning the basic data is determined by statistical methods.

The confidence in the results obtained in the deterministic approach implies that :
- the consequences of the chosen scenarios are more severe than any other of the same family. Preliminary calculations and sensitivity studies must help to obtain this result,
- all the families of plausible scenarios have been selected. This point is not specific to the deterministic methodology. A NEA working group has been set up in order to propose a methodology to exhaustively determine the scenarios to take into account in each case.

Conservative evaluation of the consequences

The conservative evaluation of the consequences associated with the disposal facilities, on which is based, for each of the scenarios considered, the deterministic approach, accounts, implicitly, or explicitly, for the uncertainty concerning the scenarios, models and data.
    Two methods can be used.
    The first method consists in using conservative data and hypotheses for the calculations. This method is applicable when the number of parameters involved is relatively limited, as is for example the case in surface disposal facilities. Indeed, for these, in French practice (see basic safety rule No. 1-2 \8\ concerning safety options and design basis for surface disposal centres), the choice of the site is to be such that not only area affected by disposal should be of limited extend and perfectly isolated in hydrogeological terms, but also that the geological formation be easily modelled, ie, that it only contains a few strata of relatively homogeneous properties.
    In the case of deep disposal facilities, the volume of the geological formation affected by the disposal facilities is extremely large (a number of km³ to a number of tens of km³). Furthermore, the natural heterogeneity of the formation, particularly in the case of crystalline rock, results in characterization involving a very large number of parameters. Under these conditions, to systematically take conservative values for all parameters would result in two types of problems : first it is never possible to justify the fact that truly conservative values have always been taken ; secondly, the fact that

conservative values have been systematically taken for all the parameters may lead to unnecessarly high consequences being predicted.

The characterization of geological media shows that it is generally possible to represent the parameters by statistical distribution
functions representative of both the intrinsic variability of the
parameters and the uncertainty inherent in the methods of measurement
(number of drillings necessarily limited and the representativeness of
the results).

To obtain a conservative evaluation of the radiological consequences, the method most usually used consists in calculating "best
estimate" consequences associated with the scenario under consideration and assigning them a reasonable estimate of uncertainty. The
"best estimate" consequences are obtained by giving reference values
to the parameters, generally central values of statistical distributions. It should be noted that the result thus obtained is not necessarily equivalent to the most probable dose, due to the non- linear
nature of the radionuclide transfer equations. Obtaining a reasonable
uncertainty value involves statistical techniques. The basic idea
underlying the use of these techniques is that uncertainty concerning
the result must reflect allowance for all possible uncertainty, both
that which would positively and negatively affect the reference
values.

To illustrate this situation, a description will first be given
of the application of the deterministic method to the evaluation of
the risk at the Aube surface disposal and secondly present certain
considerations relating to the application of this method to deep
repositories.

Aube center example

The Aube center \9\ is a surface disposal facility proposed by ANDRA
for wastes of low and medium activity and of short or medium half-
lives (less than about 30 years). Its operation, which is to start in
1991, is expected to last for 30 years. Authorization to create this
centre is expected by the beginning of 1989, on the basis of a preliminary safety analysis which reviews the possible radiological consequences associated with the principal radionuclides liable to be
contained in the wastes disposed of.

The disposal system chosen is shown schematically in Figure 2 :
the waste packages are placed in concrete disposal structures, which
themselves stand on a concrete raft. These structures are placed under
a cover designed to only allow an extremely small proportion of rain
water to infiltrate. A drainage network is provided for recovery of
rainwater infiltrating into the disposal facility under certain conditions.

The disposal facility is located on the side of a small hill on
the bank of a small river named "Les Noues d'Amance" (Figures 3 and
4). The disposal structures stand on a stratum of sandy clay approximately 10m thick above the highest level of the water table in the

Aptien sand layer

Albian

Upper Aptien sands

Lower Aptien clay

→ Flow direction

Repository

0   100   300m

N

Fig. 3 - GEOLOGICAL AND PIEZOMETRIC MAP OF THE AUBE SITE

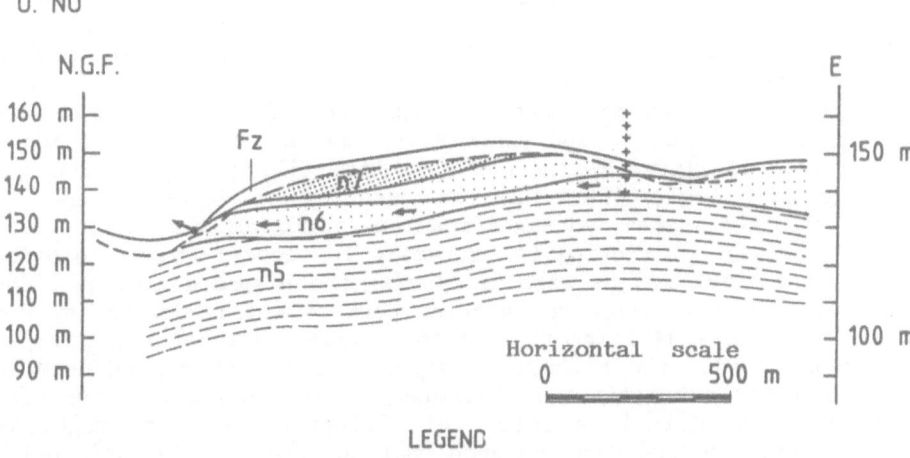

O. NO

N.G.F.

160 m	Fz	
150 m		150 m
140 m		
130 m	n6	
120 m	n5	
110 m		
100 m		100 m
90 m		

E

Horizontal scale
0   500 m

LEGEND

LITHOLOGY	HYDROGEOLOGICAL CHARACTERISTICS	STRATIGRAPH
Fz Limons et sables	. Semi-permeable to permeable	Quaternaires Formations superficielles
n7 Argiles ± sableuses	. Very low permeability	Albien inférieur
n6 Sables	. Permeable	Aptien supérieur
n5 Argiles	. Very low permeability	Aptien inférieur
↘ Emergence ／ Sens d'écoulement	⌒ Surface piézométrique Sept. 85	↕ Crête ↕ piézométrique

Fig. 4 - DIAGRAM OF HYDROGEOLOGICAL
SITUATION OF THE AUBE SITE

lower part of the sand. Thick strata of clay (a number of tens of metres) separate the ground water located immediately below the disposal facility from the deeper ground water. The studies carried out have shown that the eventuality of significant transfer of radioactivity between these ground water levels could be excluded, in view of the low permeability and high retention of the clay strata and the absence of signs of major faulting. Furthermore, piezometric data indicate that the "Noues d'Amance" river drains all the ground water affected by the disposal facility.

French practice \8\ specifies that the site must be monitored for a period limited to 300 years after the end of the operation of the disposal facility. Throughout this time, the artificial barriers play the major role (limiting the quantity of water coming into contact with the wastes and significant retention of the radioactivity by the package) restricting to extremely low levels any radiological consequences which may be associated with the disposed radionuclides. These barriers thus allow the greater part of the radioactivity with short and medium half-lifes to fall to extremely low levels. After the monitoring period, the site is derestricted, ie, it can be used for any purpose without limitation. Besides, it is assumed that the artificial barriers will have lost their integrity at this stage. The residual activity must then be such that no inadmissible radiological consequence be reached under all circumstances considered plausible.

Choice of the scenarios used for preliminary evaluation of the risk associated with the Aube centre allowed for all the preceeding conditions. Conservative analysis of the consequences gave the following results :

A) NORMAL SITUATION

A fraction of the rainwater percolates through the cover (during the monitoring phase) or through the soil (after derestriction). This water leaches those waste packages no longer protected by their casings. The water carries the dissolved radionuclides to the raft. It was assumed in the preliminary evaluation that, in view of the very low rate of infiltration through the cover up to 300 years, the water percolating through the raft and entering the water table would carry radionuclides with it. It is also assumed that after 300 years the cover will have lost its effectiveness and that the rate of rainwater infiltration through it will have reached that of the natural surroundings. The water then circulates in the water table where the radionuclides undergo physico-chemical reactions before flowing away at the discharges. The radionuclides are then ingested by an individual of the critical group in accordance with maximum water usage conditions (consumption of drinking water, ingestion of contamined plants or animals) which allows for the local diet.

These calculations have shown that the maximum doses reached are the following :
. in the monitoring phase : 5×10^{-6} Sv/year, 99 % of which is attribuable to tritium, three years after the beginning of the monitoring phase,

. after derestriction : 2.5 x 10^{-6} Sv/year, essentially attribuable to carbon 14, 3000 years after derestriction of the site.

B) LOW PROBABILITY SITUATIONS

During the monitoring phase, the only accident liable to occur is one involving loss of effectiveness of a fraction of the cover. Calculations have shown that, for collapse of the cover over 100 m² and delay of repair for one year, the radiological impact would not be more than 20 % greater than in the normal situation.

After derestriction, two types of low probability situations can occur :
. a situation leading to an increase, with respect to the normal situation, of the quantity of contamined water ingested : this would be the creation of a well in the immediate vicinity of the disposal facility. Under this hypothesis, man would short circuit part of the geosphere by drawing water directly from the water table in which the concentration per unit volume of radionuclides is higher.

Calculations have shown that, with the water used for drinking and irrigation of a vegetable garden, the maximum radiological consequences would be :
- 2.5 x 10^{-3} Sv/year, essentially attribuable to strontium 90, 50 years after derestriction of the site,
- 1 x 10^{-3} Sv/year, due to carbon 14, 2000 years after derestriction of the site.
. a situation causing radioactive dust to be placed in suspension and inhaled. Two types of situations have been considered :
- the construction of an expressway through the site : the effective dose at 50 years would be 1.8 x 10^{-3} Sv, mainly due to inhalation during the construction work of dust containing plutonium 239 and americium 241,
- the construction of a residential area : at the very beginning of the derestriction phase, external exposure, mainly due to caesium 137 would be largely preponderant and attain 3.3 x 10^{-3} Sv/year. This exposure rate would rapidly decrease with time (after 50 years, exposure would drop to 1.8 x 10^{-3} Sv/year) and tends towards a limit of about 4 x 10^{-4} Sv/year due to the presence of long half-life alpha emitter radionuclides.

The preceeding results, obtained by using conservative hypotheses, have made it possible to conclude that the Aube disposal centre would be acceptable for the quantities of radionuclides planned to be disposed of there :
. the doses associated with the normal scenario is well below the limit recommended by the International Commission for Radiological Protection (10^{-3} Sv/year),
. the doses associated with low probability situations are acceptable. Indeed, although it is not possible to accurately evaluate the probabilities of occurrence of the events considered, it would appear, subjectively, that at least in the first few centuries these probabilities are very low, in view of the memory that will remain (human memory and archives).

Plotting all the results in terms of probability against consequences as described in Figure 1, Figure 5 is obtained. For the purposes of this representation, a probability of 0.01 was assumed for
accidental events, which seems pessimistic. This representation illustrates how the criteria of the International Commission for Radiological Protection are applicable for the Aube disposal center.

It must be pointed out that the studies concerning evaluation of
the safety of the Aube center are continuing, in order, first to
reduce the uncertainty which persists concerning the behaviour of certain radionuclides (carbon 14, niobium 94, americium 241 etc).
Secondly to take into account the retention capacity of the concrete
of which the disposal structures are made, their filling material and
the raft ; and thirdly to make allowance for all the radionuclides
liable to be contained in the waste, and not only the major ones.
These studies should confirm that, throughout the monitoring phase,
the release of activity into the environment is extremely small.

The case of deep disposal facilities

A geological survey programme for the selection of a site for
creating an underground laboratory, which could be later transformed
into a deep disposal facility, has begun in four regions of France
\11\ : Deux-Sèvres (granite formation), Maine-et-Loire (schist formation), Aisne (clay formation) and Ain (salt formation). The choice of
the site will be made with reference to the technical criteria proposed by the working group headed by Professor Goguel \6\. Amongst
these criteria, the following are noteworthy :
. very low or zero water flow rates : this implies not only very low
permeability of the host rock but also very low hydraulic pressure
gradient ;
. high stability of the geological conditions.

The work conducted in the underground laboratory will provide
the data needed to confirm the choice of the site and perform the risk
assessment on the basis of which the decision to create a disposal
facility might be taken in the last years of the century.

This risk evaluation will be performed using a deterministic
approach for which models are currently developed in France by ANDRA
and IPSN. The MELODIE model \10\, under development since 1984 by
IPSN, is a complex model which associates sub-models representative of
the source term, the geosphere (flow and radionuclide transfer calculation) and the biosphere. In addition, it is possible to calculate
uncertainty and sensitivity to parameters by statistical methods
(Latin Hypercube parameter sampling, regression technique on the response area). The model has already been used for evaluating the radiological consequences of the granite option of the PAGIS project of the
european community commission \5\.

The scenarios to be taken into account for each specific site
are currently being determined. As for surface repositories, it is

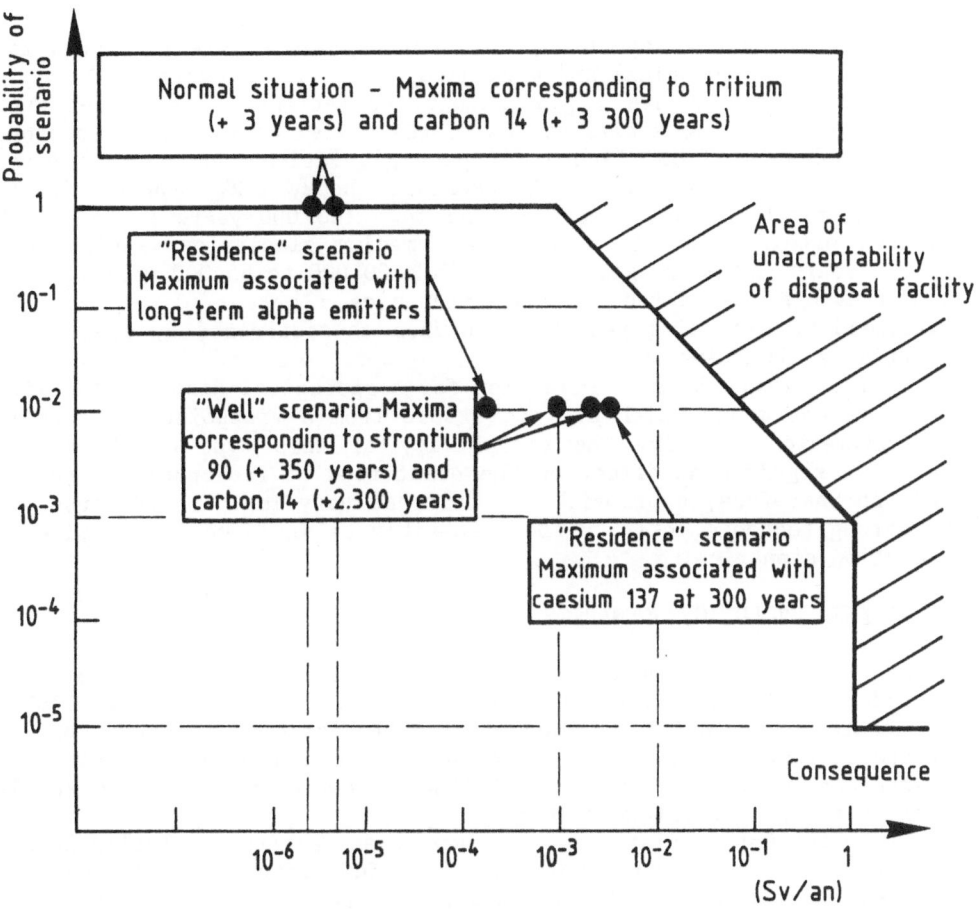

Fig. 5 - REPRESENTATION IN A PROBABILITY/CONSEQUENCE TERMS
OF THE RESULTS OF EVALUATION OF THE RISK
ASSOCIATED WITH THE AUBE CENTRE

possible to distinguish two main categories of families of scenarios in the following manner :

A) NORMAL EVOLUTION SCENARIO

For each geological formation, this corresponds to the extrapolation of current climatological and geological trends. It covers :
. climatological events : glaciations, variations of sea level etc. These events involve modifications of the relief of the site (erosion) and ground water flow. Records of past glaciations and current forecasting models, based on the Milhankovic theory \12\, would indicate the occurrence of glaciations in the next 100 000 years : minor glaciation within 5 000 to 6 000 year, medium glaciation within about 20 000 years and heavy glaciation within 60 000 to 65 000 years.
. the events associated with neo-tectonics : recent plio-quaternary deformations affecting recent structures characterized by plastic or brittle vertical movements.

The process of site selection for a geological disposal facility must exclude areas which are not stable from a viewpoint of tectonic and magmatic activity. The probability of this type of phenomenon having a significant effect on the quantities of radionuclides released can therefore, a priori, be considered very low. The consequences of strong tectonic activity could possibly be included in the altered evolution scenarios.

B) ALTERED EVOLUTION SCENARIOS

Amongst the altered evolution scenarios, a first category of families includes all defects in construction of the artificial barriers liable to create preferential pathways or locally increase the permeability of the surroundings : defect in shaft sealing, percentage of initially degradated packages, incorrect filling of engineered barriers resulting in a high percentage of voids.

In the deterministic methodology, these altered evolution scenarios must be considered during the first few centuries following closure of the disposal facility. During this period, the level of activity is greatest, the medium and short half-life beta and gamma radionuclides not having yet disappeared by radioactive decay. Also, for vitrified waste during this period, the thermal gradient is significant and the transfer of radionuclides could be facilited by thermoconvection phenomena or increased permeability on a large scale caused by thermomechanical phenomena. For these scenarios, the role of the waste package would be particularly important in limiting consequences.

A second category of families of altered evolution scenarios corresponds to certain natural events of low probability such as :
. important tectonic or seismic activity which could have the effect of modifying the permeability of the soil or creating preferential hydraulic pathways by the opening of new fractures,
. climatological changes greater than those allowed for in the geoprospective studies,

. the presence within the geological formation of undetected hetero-geneity of an undesirable nature (anhydrite or pockets of brine in salt),
. the reopening, for example due to thermomechanical effects associa-ted with the thermal phase, of faults not identified during the site survey,
. the fall of a meteorite.

A third category of altered evolution scenarios corresponds to human intrusions. Amongst the human activities liable to result in intrusion into a geological disposal facility, the following should a priori be taken into consideration :
. exploration or excavation of the rock for resource surveying. This includes all human activities leading to deterioration of the host rock, the creation of hydraulic passages, the extraction of minerals (for example exploratory drilling and mining work) which would have the effect of reducing the confinement properties of the repository ;
. the use of the surface of the land or the host in the vicinity of the repository. This includes the construction of dams (modifications of the flow characteristics of ground water), the use of the area for disposal of other materials (toxic waste, petroleum product etc.) and all types of activity liable to deteriorate the host rock and the hydraulic systems.

For each of the these altered evolution scenarios, characteris-tics depend strongly of the type of geological formation and of the site. Moreover, accurate probabilities are generally very difficult to determine. However, it should be remembered, in the deterministic methodology, orders of magnitude of the probabilities of the scenarios might be sufficient. Expert opinions are so used to provide those orders of magnitude.

To conclude this section, the sequence of calculations needed to evaluate, for a deep disposal facility, the deterministic consequences associated with a given scenario, can be illustrated by the following example, taken from the calculations carried out with the MELODIE model as part of the PAGIS project granite option. It concerns the normal scenario considered for the Auriat site and disposal con-cept A.

The first step consists in the choice of a "2 dimensional sec-tion" and the corresponding grid (figure 6). The second step is the evaluation of flow in this section (figures 7 and 8). A third step is the evaluation of the transfer of radionuclides and the consequences for man (figure 9). In the case under consideration, the maximum radiological consequences are reached after 3 million years and consi-dering the dose factors recommended by the International Commission for Radiological Protection in publication No. 48, are equal to appro-ximately 6×10^{-6} Sv/year, essentially attributable to neptunium 237 (52 %), thorium 229 (26 %), uranium 233 (11 %) and caesium 135 (7 %). Finally, the last step consists in allowing for uncertainty concerning the parameters ; this step results in roughly doubling the preceeding value.

This sequence of calculations allows the point associated with the normal evolution scenario (probability = 1) to be positioned in

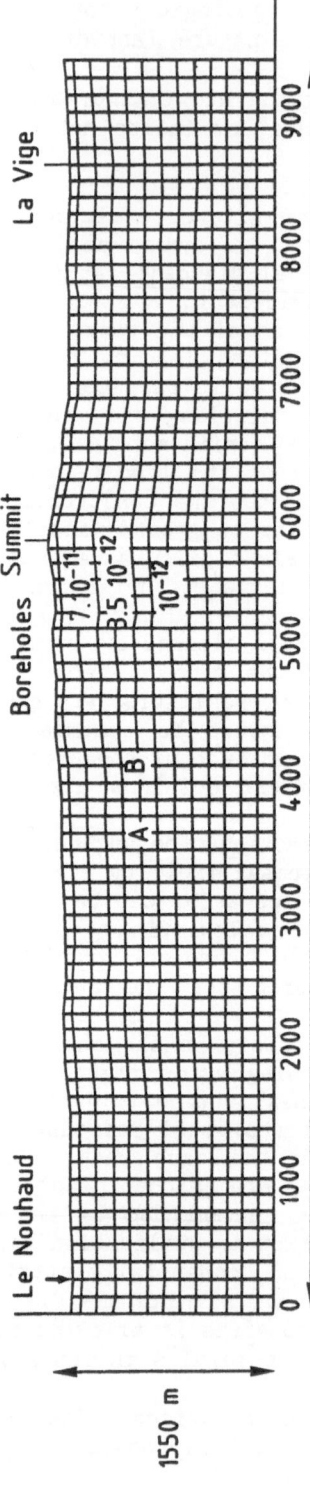

Fig. 6 – REPRESENTATION OF AURIAT SITE

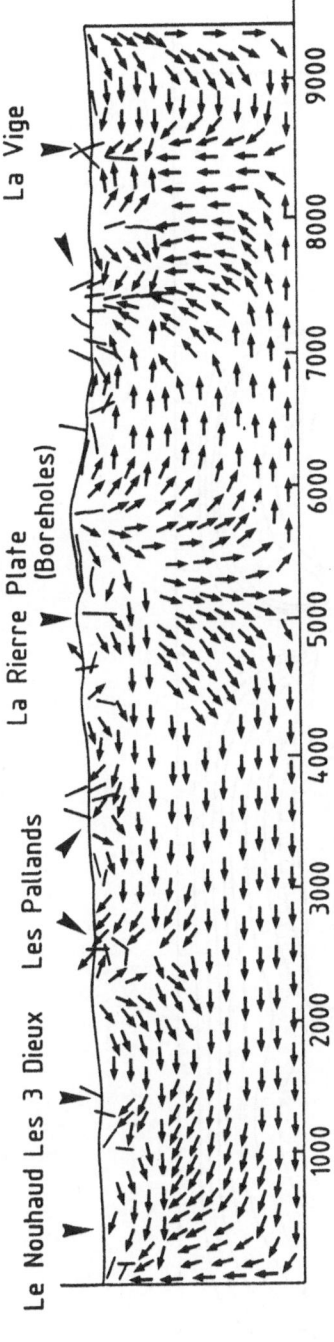

Fig. 7 – AURIAT : PORE VELOCITY FIELD

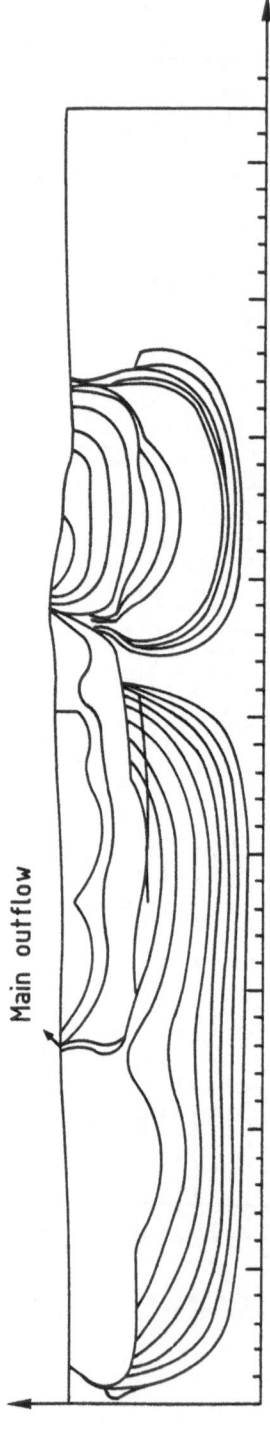

Fig. 8 – AURIAT – TRAJECTORY OF PARTICLES FROM SUMMIT MOVING
IN TWO–DIMENSION REPRESENTATION OF SITE

Main outflow

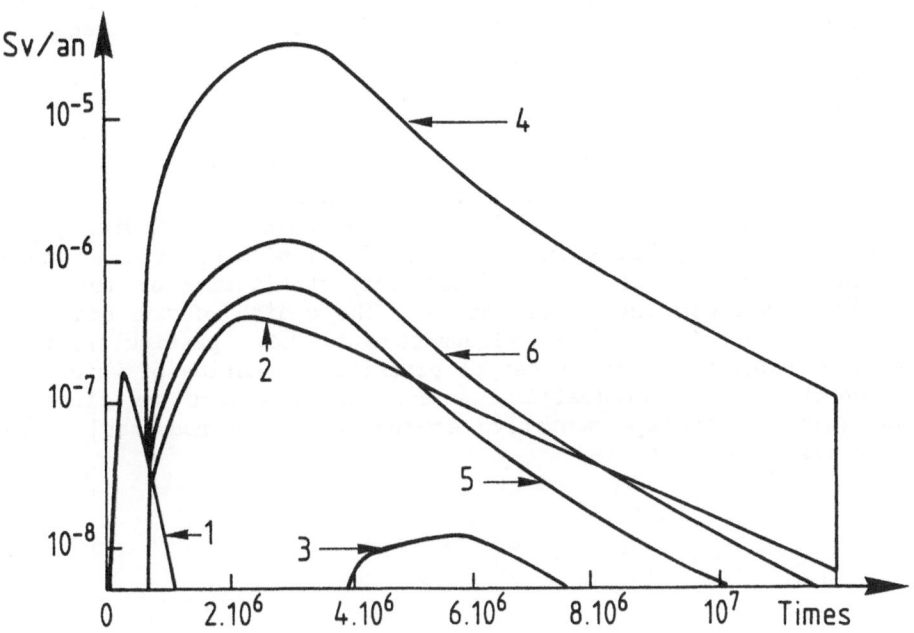

Les Pallands

Individual annual dose :

| 1 Tc^{99} | 3 Zr^{93} | 5 U^{233} |
| 2 Cs^{135} | 4 Np^{237} | 6 Th^{229} |

Fig. 9 - INDIVIDUAL ANNUAL DOSE AS A FUNCTION OF RADIONUCLIDE
(CIPR 30)

the probability/consequences plan. Altered evolution scenarios are subject to similar sequences of calculations. A human intrusion scenario has thus been considered in the granite option of the PAGIS project. Results of both the normal evolution and the human intrusion scenarios indicate that a repository in a granitic formation could be safely defined. Of course, this "generic" conclusion has to be confirmed by national authorities on the basis of a site specific and more exhaustive (regarding the scenarios) deterministic risk assessments.

Conclusion

The deterministic method allows the acceptability of a radioactive disposal facility to be assessed, with reference to criteria such as those recommended by the International Commission for Radiological Protection in its publication No. 46. The implementation of this method naturally necessitates thorough characterization of the barriers placed between the waste and man. The choice of the site and disposal concept, as well as allowance for their particularities, makes it possible to define a set of plausible scenarios of which the consequences require evaluation. A continuous effort to finalize models which are as representative of reality as is reasonably possible is also indispensable.

References

\1\ J. CHATOUX, A. BARTHOUX, A. FAUSSAT, F. VAN KOTE
Fifteen years of experience of shallow land disposal from site
selection to repository operation
4ème conférence nucléaire européenne ENC 86, Genève (Suisse)
1-6 juin 1986 - Proceedings published by the European Nuclear
Society (ENS), po Box 2613, CH-3001 Berne (Switzerland), vol. 4,
pp 851-856

\2\ Handling of expert nuclear fuel and final storage of vitrified
high-level reprocessing waste
Kärnbränslesäkerhet, KBS, November 1977

\3\ Final storage of spent nuclear fuel- KBS 3
SKBF/KBS May 1983

\4\ Project Gewähr 1985
Nuclear waste management in Switzerland : feasibility studies and
safety analysis
Project report NGB 85-09 (June 1985)

\5\ Programme PAGIS, option granite
Rapport EUR 11777 EN, 1988

\6\ Stockage des déchets radioactifs en formations géologiques
Rapport du groupe de travail présidé par le Professeur Goguel
Ministère de l'Industrie, des P et T et du Tourisme - Paris,
1987

\7\ P. TANGUY
The French approach to nuclear power safety
Nuclear Safety, vol. 24, n°5, september-october 1983

\8\ Basic safety rule n°I-2 - SIN 3210/84 (19 juin 1984)
Service Central de Sûreté des Installations Nucléaires
Ministère de l'Industrie et de la Recherche (France)

\9\ R. ANDRE-JEHAN - Le modèle ANDRA de stockage de surface : carac-
téristiques du site de l'Aube
Y. MARQUE, J. PUYCHEVRIER, J.Y. RAVACHOL - Le centre de stockage
de l'Aube
Revue Générale Nucléaire - Année 1988 - n°4 juillet-août

\10\ J. LEWI, M. ASSOULINE, J. BAREAU and P. RAIMBAULT
MELODIE : a global risk assessment model for radioactive waste
repositories - Proceedings of the workshop on mathematical modelling for radioactive waste repositories - December 10-12, 1986 -
Madrid (Spain), edited by ENRESA, paseo de la Castellana, 135 -
28048 Madrid, Spain

\11\ R. ANDRE-JEHAN
Le programme des caractérisations des sites pour un stockage en profondeur
Revue Général Nucléaire - année 1988 - n°4 juillet-août - p 327

\12\ S. COURBOULEIX
Etude géoprospective d'un site de stockage. Climatologie : évolution du climat et glaciations
Rapport BRGM n°83 SGN 143 GEO

THE SUB-SEABED DISPOSAL SAFETY ASSESSMENT: A WORKED EXAMPLE

D.A. STANNERS
Commission of the European Communities
Joint Research Centre
21020 Ispra (Varese)
Italy

ABSTRACT.   A worked example is presented concerning the safety analysis
of the sub-seabed option for the disposal of high-level radioactive
waste. This begins by asking what makes up such an assessment, outlining the major elements that are required to perform the analysis and
describing the methodology adopted. A number of important concepts essential to understanding the assessment are introduced with reference
to geological land based options. These include the development of
scenarios, and the definition of radiological pathways and cut-off
times. The way in which the seabed option was divided up for modelling
and analysis is then described. The many assumptions and simplifica-
ions required to do this are discussed with reference to each of the
sub-environments that the option was divided into, showing how the
various sub-modules were built up given the complex nature of the real
system. The performance of the final overall systems model is then
illustrated by following a release pulse through the system and
describing how the various barriers contribute to the overall retention
of migrating radionuclides. These preliminary simulations lead to a
number of simplifications which are introduced before proceeding with
the definitive assessment calculations. The results of the base case
assessment are then presented together with the results of the
stochastic sensitivity analysis. Where possible, the way in which this
data can be interpreted is discussed emphasising the importance of
intercomparisons and duplicate analyses conducted independently. Other
scenarios considered in the assessment concerned abnormal events and
transportation accidents which were were dealt with in some detail in
the assessment. The development of abnormal scenarios is discussed and
the manner in which potential events were interpreted into cases for
analysis is discussed together with the results. Conclusions are drawn
both from the point of view of the assessment results and the manner in
which the assessment was conducted.

A. Saltelli et al. (eds.), Risk Analysis in Nuclear Waste Management, 285–329.

1. INTRODUCTION

Conducting a risk or safety analysis when a real complex environmental system is involved is not a simple task, particularly when predictions far into the future are required. The purpose of this worked example is to describe how this has been done with a specific application in mind. This should help illustrate the generic methodological and statistical material, found elsewhere in this volume, and give some practical in-ights into carrying out such assessments. Although in the process spe-ific details of the example are entered into it is the general paral-lels that can be drawn with the performance of other assessments and the lessons learnt from this example, that are important in the context of the course from which this volume originated.

1.1. Historical Perspectives

The proposal to use the unlithified sediment strata under the deep oceans as a repository for high-level radioactive waste has received much attention over the last fifteen years (see, eg. Bishop and Hol-lister (1974); Anderson et al. (1987)). Studies have focussed upon sea-bed abyssal plains at ocean depths between 4 and 6 km. In 1977 the <u>Sea-bed Working Group (SWG)</u> was set up (under the auspices of the Nuclear Energy Agency of the OECD) with the purpose of investigating this op-tion on an international basis.

One major problem of all such studies involving 'real world' com-plex systems is the interdisciplinary nature of the work. The sub-seabed assessment involved scientists and engineers from among many different disciplines including, oceanography, geology, geochemistry, geophysics, marine biology, paleontology, radiation and health physics, mathematics and statistics, to name but a few.

Since it is unlikely that no single person is an expert in all these subjects, the problem arises how to divide up the system of in-terest and how to obtain and synthesize the information necessary for an assessment. The approach of the SWG was to form working '<u>task groups</u>' of international experts selected from member countries. Each group had its own area of interest and problems to solve. Eight groups were formed: the Radiological Assessment Task Group (RATG); the Near-Field Task Group (NFTG); the Sediment Barrier Task Group (SBTG); the Site Selection Task Group (SSTG); the Biology Task Group (BTG); the Physical Oceanography Task Group (POTG); the Engineering Studies Task Group (ESTG); and the Legal and Institutional Task Group (LITG).

The final outcome of the groups' deliberations was originally en-visaged as being a series of reports, one from each of the task groups, summarising the research work performed with conclusions concerning the safety and feasibility of the option pertinent to the group's field of interest. It soon became clear, however, that a modelling exercise was required which would have to draw on information obtained by the other task groups; the information would have to be assembled in a form able to be used for making predictions as to the future performance of a theoretical sub-seabed repository. This is clearly a non-trivial complex task. Such an exercise involves many aspects of which the calculation of consequences in terms of 'risks' or 'dose rates' to individuals or population groups is only a part. Hence, assessments were made of the overall feasibility (engineering and site identification) and safety (accidents during the disposal phase and the long term behaviour of the repository) of the option as a whole. Aspects of legal and social perception of the impact of such a repository were also considered. For full details of the overall feasibility exercise conducted by this international group see NEA (1988).

This worked example will concentrate on explaining how the radiological safety assessment was conducted by predictive modelling of the repository's behaviour.

1.2. Description of the Option

Two primary criteria were used for identifying appropriate disposal sites:
1) that the geological formations should be stable (minimal seismic activity) and predictable (life time integrity), and
2) that the geological medium should have the appropriate characteristics to make it an effective barrier to the release of radionuclides.

The abyssal plains were seen as the most promising areas containing zones conforming to these characteristics. After a considerable amount of field study investigating upward of 15 sites, the SSTG selected two on which to concentrate the assessment. These were Great Meteor East (GME) in the Madeira abyssal plain of the Northeast Atlantic and the South Nares Abyssal Plain (SNAP) in the Southwest North Atlantic (Figure 1).

Two concepts of disposal have been suggested; drilling and the use of free fall penetrators, where in situ tests have shown that sediment depths of over 30 metres can be obtained. Figure 2 shows an emplaced penetrator, but could equally apply to packages emplaced down backfilled drill holes where the sediment depth would be up to 200 metres. The abyssal plain sediments are unlithified material with a high water content (30 to 70% by weight) and therefore have excellent sealing properties. Moreover, they are composed of material (clays and oozes) which also makes them excellent 'sorbers' of elements and compounds of many types. This potential ability of the sediments to act as an im-

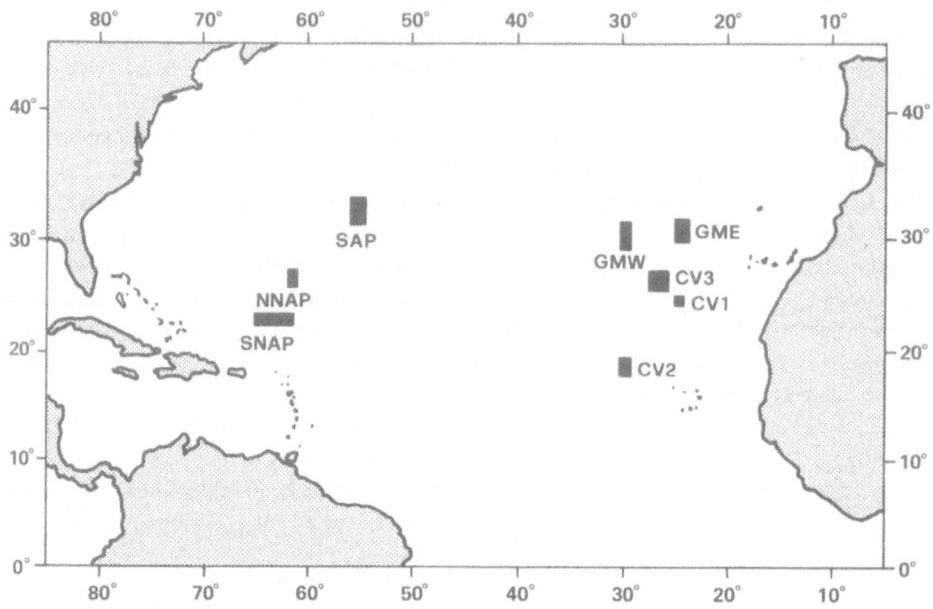

Figure 1: Location of candidate study areas in the Atlantic

SAP - Sohm Abyssal Plain;
SNAP - South Nares Abyssal Plain; NNAP - North Nares Abyssal Plain;
GME - Great Meteor East; GMW - Great Meteor West;
CV1, 2, 3 - Cape Verde sites (Madeira abyssal plain)

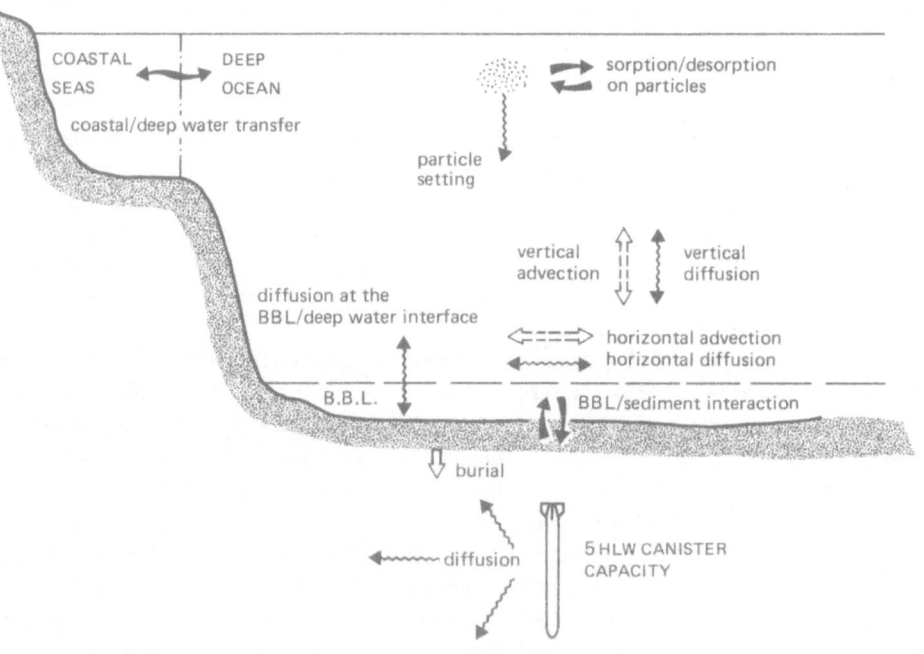

Figure 2: The seabed disposal concept

portant barrier to the return to the biosphere of disposed radionuclides is, together with the remoteness and stability of these regions,
one of the most important characteristics of this option. For migrating
radionuclides to return to the biosphere, they must first traverse the
sediment column and come into contact with the bottom boundary layer on
the seabed. Only when the nuclides have escaped from this zone can they
enter the main ocean waters to become involved with the complex mixing
and scavenging processes therein (Figure 2). The consumption of contaminated marine foodstuffs provides the major route by which the
radionuclides are able to come into contact with man.

2. SETTING UP THE ASSESSMENT

The goal in this example was to examine the potential radiological impact on man (and fauna) of the disposal of vitrified high-level radioactive waste within deep seabed sediments. Such an assessment is composed of many elements and these must now be investigated. First of all
information was required concerning the following points:

i)   Scenarios of emplacement of waste in the sediments and the sub-
      sequent evolution of the whole system. - These scenarios are based
      on judgement and experience and cover both the expected evolution
      and a broad spectrum of possible events or mishaps. The choice of
      parameters for these scenarios reflects the uncertainty or range
      of possible values.

ii)   Behaviour of the waste package in the medium. - This information
      comes from laboratory experiments, measurement and models devel-
      oped by the NFTG.

iii) Migration of radionuclides in the sediments. - The relevant in-
      formation and measurements were gathered by the SSTG and SBTG and
      models were developed to represent the on-going physical processes
      in the sediments.

iv)   Release, transport and dispersion of radionuclides in the ocean. -
      These mechanisms are represented in models developed by, and with
      input from, the POTG.

v)   Transfer of radionuclides from seawater or sediments to the bio-
      sphere considering all major pathways, including the food chain. -
      Data were provided by the BTG but other data already accepted in-
      ternationally, such as concentration and transfer factors, were
      also used.

vi)   Dose criteria. - Recommendations of the International Commission
      for Radiological Protection (ICRP) were used concerning the de-
      termination of the dose to man resulting from each pathway and
      yardsticks to evaluate the consequences of these doses.

The data needed for all the models and scenarios were obtained continuously throughout the course of the SWG work. A 'data base' was eventually formed with this information which proved to be an invaluable tool for two particular reasons:
i) it encouraged the the experts in each group to provide the best available definitive data on the subject, and
ii) it promoted compatibility, essential when analyses are being conducted by different groups.

It was of course essential that this data base did not contain errors or misunderstandings and it was therefore initially assembled in a draft form and circulated widely to enable ample time for corrections to be made.

The assessment was performed both stochastically and deterministically. The deterministic approach was used primarily because such calculations are more common and better understood than stochastic analyses. It was also felt, however, that the quality of the data, especially those concerning the probability functions of the input parameters, did not merit complete reliance on a stochastic analysis. Nonetheless, the results of the deterministic calculations should not be regarded as an alternative assessment but rather as one case (the 'best estimate' case) included in the stochastic calculations.

The assessment was conducted using the following methodology:
i) A series of models representing each of the major physical processes known to occur were assembled.
ii) A number of scenarios were selected representing normal, abnormal and accidental behaviour of the system.
iii) For each scenario a deterministic calculation was made over 10^6 to 10^7 years assigning 'best estimate' values to model parameters. Sensitivity studies were performed by 'bracketing studies'.
iv) Stochastic calculations were made for the normal behaviour or 'base case' scenario using, to a large degree, the same input data and models used for the deterministic assessments.
v) The results, scenario by scenario, were analysed and compared for both the deterministic and stochastic approaches.
vi) The stochastic sensitivity results were analysed and compared with the deterministic 'bracketing' studies.
vii) Conclusions were drawn regarding the radiological safety of the system analysed, comparing with ICRP radiological standards (ICRP, (1985)); suggestions for improvements in the analysis were made.

A number of important concepts, which have greatly influenced the methodology used for this assessment, need to be explained. These particularly concern the development of scenarios, radiological pathways to man and the determination of an appropriate cut-off date for the calculations.

2.1. Development of Scenarios

For the sub-seabed option four types of scenario were considered.

2.1.1. The Normal or Base Case Scenario. This corresponds to the normal emplacement of the waste package at its prescribed depth (50 $\pm$ 20 m) in the sediment. The path in the sediment between the sea bottom and this emplacement depth is assumed sealed and the environment is assumed stable. The emplacement method considered is that of free-fall penetrators. The base case assumes gradual corrosion of the penetrators eventually leaving the vitrified waste exposed to the sediment pore water. As the glass gradually dissolves into the interstitial water the radionuclides are released to migrate into the surrounding sediments by molecular diffusion. Advective pore water flow is assumed absent. Depending upon the individual chemical properties of the radionuclides and the redox conditions of the sediments, the migration of the radionuclides will be retarded to varying degrees. Sediment thickness is assumed constant despite the fact that the sediments are steadily accumulating with time. The upward diffusing radionuclides are eventually released into the water column after first contaminating the bottom boundary layer of the sediment-seawater interface. The nuclides then slowly disperse becoming involved in various mixing and scavenging processes in the oceans. Eventually they enter the food chain contaminating marine biota, where they become available to man through ingestion and inhalation. Throughout all these processes the radionuclides decay at their appropriate rates cases becoming transformed to daughter products as dictated by their decay schemes.

2.1.2. Altered Evolution Scenarios. The seabed geological environment and its evolution can be predicted with a higher degree of certainty than in the land-based options. Indeed, the mid-plate areas of the seabed are among the most stable parts of the earth for the selected time frame. Provided there is no erosion, most climatic or geological changes in the earth will result in little or no alteration in the sediments at the depth of emplacement. In fact, the changes most likely to occur should result in a deeper burial of the wastes by the deposition of new sediments on the seabed, thereby gradually increasing the safety of the repository. Three types of altered evolution of the seabed environment are nevertheless considered:
- Vertical pore water flow in the sediments resulting in an enhanced transport of radionuclides towards the ocean. The study of this phenomenon has received much attention by the SWG. They conclude that,

at present, the vertical pore water velocity, if any, is below detection limit ($\sim 10^{-3}$ m/a) and that at this level there should be no influence on the predicted radiological impact. Future enhancement of this velocity over periods in the order of 10^6 years cannot, however, be ruled out and thus it has been included as an abnormal case.
- Faulting of the sediments and their effects have been tentatively evaluated since numerous syn-sedimentary faults have been observed in both selected study areas.
- Unexpected erosion of the sediments leading to a reduction of the thickness of the sediment cover above the waste package. Such an event is extremely unlikely in the selected areas but was nonetheless considered.

2.1.3. Human Action. A special characteristic of sub-seabed disposal, compared with land-based geological repositories, is that human intrusion after disposal is extremely unlikely due to the remoteness of the site. Also, any intrusion would have very limited consequences. Indeed, part of the selection criteria stipulates that there are no known resources at the sites, so no direct exploitation is expected close to the disposal area. Future requirements, however, are difficult to foresee, and therefore, as an example of a possible future activity close to the site, the effects of manganese nodule exploitation in nearby ocean areas were studied.

2.1.4. Accidents or Human Errors. Although human intrusion is not considered a serious problem for a sub-seabed repository, there is, however, a specific series of human errors or accidents that must be analysed in the radiological assessment, which are generally not considered for a land-based repository. These are the following:
- Transportation accidents at sea from harbour to disposal site. This may involve the loss of a number of waste packages, damaged or undamaged, or even the loss of a whole transportation ship! Such accidents can happen anywhere in the sea (coastal waters, continental slopes or in the deep sea). Their analysis forms a major part of the assessment.
- Improper emplacement of the waste package in the sediments. This can result in shallower burial depths than prescribed or even the presence of waste packages on the seabed.
- Incomplete sealing of entry path of waste package between the sea bottom and emplacement depth. This is particularly associated with the penetrator method of emplacement and may be due to a malfunction during emplacement or to unexpected deviations in the natural properties of the sediments.

2.2. Radiological Pathways to Man

For the seabed option the potential pathways to man are much better defined and much less subject to changes with time, climate or eating habits than for a land-based repository. For this reason there is less uncertainty in defining potential pathways. Among the most important pathways are:
- The Marine food chain, including the consumption of seaweed or sea animals at different trophic levels;
- The Consumption of desalinated water or sea salt;
- Inhalation through airborne dispersal of radioactivity from the ocean to the atmosphere and back to land.

Since, for the seabed option, there is a more restricted number of possible pathways it is possible to maximize potential doses to represent future dietry habits, by defining diets based solely on one product.

2.3. Cut-off Time for Analysis

The time beyond which consequences are no longer computed is referred to as the 'cut-off time' (this should not be confused with the regulatory time limit which is the period of time for which the safety of the repository must be assessed). In some countries regulations prescribe a cut-off date for the radiological assessment. In other countries some preliminary assessments have used 10^5, 10^6 or 10^7 as cut-off dates even in the absence of specific regulations. Difficulties arise, however, in the interpretation of results calculated for extremely long time periods (eg, greater than 10^6 years) since it can be argued that such predictions are meaningless. In this context it should be kept in mind that the lifetime of our solar system is in the order of 10^9 years and that the next glaciation is expected to occur within the next 10^4 years. Although it is often suggested that the safe behaviour of a repository may only be shown over short time spans it is, nevertheless, an important tenet of radiological safety that future generations should be protected to the same degree as is acceptable today. Furthermore, despite the rough assumptions contained within the models predicting behaviour far into the future, it would be unreasonable to accept a disposal option which could forecast unacceptable doses after the regulatory time limit.

Taking all these aspects into account the RATG decided on a cut-off date from 10^6 to 10^7 years. The particular reasons for this selection are as follows:
- The quality of the confinement of the sub-seabed option for the base-

case is such that with a short cut-off date (eg. 10^4 years) almost no radionuclide release to the biosphere occurs. It would seem reasonable therefore to extend the prediction period to cover that over which the processes take place.

- The rationale for studying the seabed option is precisely its long-term stability, both geologically and in terms of transfers to man. This makes predictions far into the future more certain than for land-based options.
- The history of deep-sea sediments down to proposed disposal depths can be traced to at least one million years. An assessment into the future over a similar time span therefore seems reasonable.
- By 10^7 years nearly all the radionuclides will have decayed away to a very small percentage of their original activity. The two exceptions are I and Np which have half-lives greater than 10^6 years. These two nuclides will either be mixed or recombined in the environment (I-129), or still sorbed to sediments (Np-237). Extending the calculations beyond 10^7 years did not therefore seem useful.

2.4. Dose to Fauna

In land based repositories only the dose to man is generally calculated, since:
i) man is considered the most sensitive element to radiation in the biosphere, and
ii) radiological protection of fauna is intended to protect populations or species and not individuals.

These arguments may not hold for the sub-seabed option. Furthermore, since man does not live near to the potential release areas of radionuclides on the seafloor we may not be the most exposed species. Therefore, the possibility that deep-sea animals might be the most endangered species has to be analysed. The doses to fauna and their consequences were therefore evaluated using the results of calculations describing the transfer of disposed radionuclides to the biosphere.

3. CONVERSION OF THE 'REAL WORLD' INTO MODELS

For the purposes of research and modelling the seabed option is usually divided into four parts: the near-field zone, the far-field zone, the ocean and radiological pathways to man (dosimetry). The first two are normally regarded as being contained within the geosphere and the last two within the biosphere. These divisions are not very different from those made for other options; the contents simply differ. In fact, in any assessment a division of this nature is essential to help design appropriate models for the analysis of the effects of the major mechanisms controlling the transport of the radionuclides. Dividing up a real system of integrated processes is, however, fraught with many problems. These may not only be associated with the manner in which the divisions were made in the first place, but may also concern the way the pieces are reconstructed. This linking of sub-models can only be done correctly if proper interfaces are defined.

Our knowledge of the real world will always be incomplete and thus uncertainties will always be present in the models and data adopted for an assessment. For this reason, together with the necessity for simplification inherent in a modelling exercise, assumptions have to be made. Models written for the purpose of radiological assessment should obviously concentrate on the driving mechanisms controlling the transport of radionuclides from the repository to the biosphere. Any process that can be seen to alter the efficacy of such mechanisms, either acting on the nuclide to change its sensitivity to the mechanism in question, or acting on the mechanism itself, should be taken into consideration.

The question arises as to how detailed and realistic the models should be. Often, the paucity of information imposes a natural constraint on the complexity of the models employed, but then it must be ensured that the simplified model does not <u>underestimate</u> consequences. <u>Conservatism</u> is usually the solution; by taking conservative estimates and assumptions an assurance is built into the assessment that any calculated consequences will be on the high side. Conservatism can be applied either to compensate for the lack of data in a certain area or to simplify a model. To do this requires fore-knowledge of the models' behaviour, which is not always easy to determine, and leads to the use of the terms 'most favourable' and 'least favourable' for the data and assumptions used. Of course, conservatism must be balanced with realism so as not to exaggerate unnecessarily. This is particularly important when applying a stochastic analysis to the system in question, since it is the purpose of this approach to include, <u>as much as possible</u>, all uncertainties into the probability distribution functions used for the input data. Uncertainties should not just be taken up by conservative choices. In practice the correct balance is difficult to make; for the seabed assessment the appropriate decisions were made by experts confronted with the available data. These aspects will be discussed below for each of the system compartments individually.

Uncertainty is also associated with knowing whether account has been taken of all significant processes that might occur in each of the zones of interest. To help tackle this problem each of the task groups were encouraged to adopt what could be termed a 'devil's advocate' approach. In other words, an approach that would encourage a critical view of the 'base case' situation, and possibly result in the investigation of new processes. This was found very useful for exploratory purposes, especially during the development stage of the work and mainly resulted in the proposal of new abnormal scenarios.

The final point of general consideration concerns the aspects of verification and validation. Verification usually refers to whether the model performs the functions it is programmed to do; whilst validation refers to the extent to which the model reflects the real world. The former is normally established through independent studies of inter-comparison, and the latter by comparison with experimental or, even better. in situ tests. The models described here have been through ex-haustive tests of verification, but in most cases have not been thoroughly validated (this could be the aim future investigations).

A summary now follows of the way in which the above points of gen-eral interest were applied to seabed to model its various sub-systems. This will take a descriptive approach; for more detailed information and the mathematical formulations used refer to De Marsily et al. (1988) and Flebus and Stanners (1988).

3.1. The Near-Field Model

The purpose of the near-field model is to compute the rate of nuclide release from the glass as a function of time (eg, in moles/year). Re-lease takes place from the repository into the sediments due to the dissolution of the glass blocks contained in the penetrators (Figure 3) and this can only take place if the penetrators and canisters are breached by corrosion or cracking.

To convert these processes into a series of models it was first assumed that the thin (5 mm) stainless steel canister does not act as significant barrier and is therefore neglected. Thus, for the model, the thicker (75 mm) mild steel penetrator becomes the only barrier against glass attack by sediment pore waters and consequent radionu-clide release. Two consecutive sub-models are therefore used to describe the near-field; the container degradation and the radionuclide release models. The former computes the number of breached penetrator casings at each time point. Once this has occurred, the dissolution of the glass and release of the radionuclides into the sediments begins using the second sub-model.

3.1.1. Container Corrosion and Failure Rate. The lifetime of penetra-tors buried in ocean sediments depends upon many factors, the most im-portant being the design characteristics and the type of material from which they are composed. Experimental studies have shown that the cor-rosion of mild steel is highly temperature dependent and consequently it is very important to define the temperature history of the decaying glass waste contained within the penetrators. For a penetrator having a

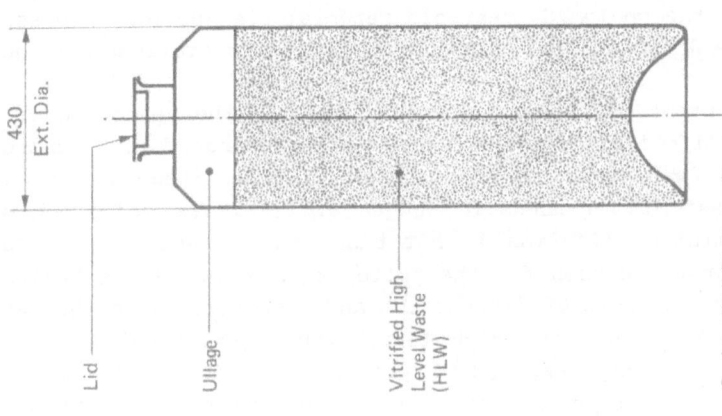

0.59 m

CORROSION ALLOWANCE OVERPACK

Overpack/
Penetrator Shell

430
Ext. Dia.

Lid

Ullage

Vitrified High
Level Waste
(HLW)

Canister Wall Tickness: 5 mm

HLW Volume: 0.150 m³

GLASS CONTAINER

Figure 3: Glass container and corrosion allowance overpack for penetrator

75 mm wall thickness loaded with 33 year-old waste, the calculated temperature history gives an expected lifetime of approximately 9800 years. If corrosion were to take place at 4°C the expected lifetime would be around 30000 years. Unfortunately, the corrosion rate factor used in these calculations is not yet accurately defined, moreover, it can be shown that the penetrator lifetime is very sensitive to this value; increasing it by 20% reduces the 9800 year lifetime quoted above to 4600 years.

The processes expected to occur in this sub-system are thus reasonably well known but our uncertainty of the exact corrosion rate, small may it be, causes a large uncertainty in our estimate of the container lifetime. What model and data can we choose, therefore, to describe these processes for the assessment? The <u>controlling factor</u> here is the failure rate so it is this that should be defined in the model in place of a specific description of the processes leading to canister corrosion. A value for corrosion rate is consequently not directly needed. It was therefore assumed that the lifetimes of the 14667 penetrators required for this option were distributed in a truncated Gaussian function having a mean of 5000 years, a minimum of 300 years, a maximum of 10000 years, and a standard deviation of 1500 years. In addition, to reflect the uncertainty on the corrosion rate, the mean value was taken as a variable parameter for the stochastic assessment with a uniform distribution between 3000 and 20000 years, adjusting the maximum as required to maintain mass balance. Thus, from a few experimental data and a theory to describe the temperature controlling behaviour of corrosion, a model is built-up to describe this complex part of the system. The result is deemed both conservative (based on 33 year-old waste and not the cooler 50 year-old material planned for disposal) and realistic (peak of failure corresponds with expectations but not exaggerated).

A final important point remains; this concerns the selection of a minimum canister lifetime. Given high engineering standards and control procedures it is reasonable to assume that for the 'base case' all canisters will last some reasonable length of time. The selection of 300 years as a minimum lifetime is not based on an estimate on what this might be, but was chosen for the following reasons. The behaviour of glass and the mechanisms of dissolution and dispersion into the sediments is not well known under high temperatures (eg. > 90°C) and in the presence of a strong temperature gradient. Consequently, if our model were to cause some of the glass to be exposed to sediment pore waters in the first few hundred years, the leach model would be highly compromised. The assumption that no canister is breached before 300 years is therefore imposed so that the interface temperature is below 25°C. This is a good example of the type of compromise that sometimes has to be reached between the model used and the present state of knowledge of a particular set of processes.

The boundary or interface between the near- and far-fields is usually defined as the sphere inside which the presence of a hot, radioactive canister is felt. In the near-field the sediments are altered in some way causing changes to their properties. In the far-field, however, the sediments are in their fully natural state. With this in mind it is useful to reflect that as part of the 'devil's advocate' approach adopted by each of the task groups, the NFTG concentrated in searching for processes that could cause the formation of mobile 'species' of radionuclides under the heat and radiation fields experienced in the near-field zone. The species of most importance would be those that could be transferred to the far-field without change, and, despite their expected lack of thermodynamic instability, migrate quickly to the biosphere. None has so far been identified.

3.1.2. Release model of radionuclides from the glass matrix. The degredation of the radionuclide containing glass matrix in the sea sediment environment depends upon many factors. These include, temperature, exposed surface area and dissolved silica concentration. Since the radionuclide concentrations in the glass are known values, the output fluxes from the near field can be easily calculated once the amount of glass dissolved in the sediments per unit time is computed. The question arises therefore, how do large glass blocks (cylinders approximately 1 m tall and 0.43 m in diameter) disintegrate and dissolve with time in contact with sediments and their pore waters?

One of the most important characteristics of the glass blocks is that they are fractured as the result of the casting and cooling processes during the blocks' early history. The effective surface area of the glass exposed to the pore waters is therefore considerably increased. An average fracturation factor of 11 is normally assumed. For modelling purposes the fracturing is assumed to result in the formation of a series of small glass spheres of equal size (at the start 2800 spheres with a radius of 0.023 m are calculated to be present per canister). From here it is a simple matter to calculate the release rate for one and therefore all the spheres as a function of time. Applying the appropriate formulae, one finds that under normal conditions, once a canister fails, all the glass is dissolved in 5000 years. For the stochastic assessment uncertainties are accepted for the fracturation factor (uniform distribution between 8 and 15) and the leach rate (uniform distribution ranging a factor of ten either side of the mean value).

Again, a complicated system is converted into a model which is simple and yet realistic. Note however that of the three factors mentioned at the start as having an influence on the glass degradation, only one has been included in the model, viz. surface area. Temperature is assumed constant and low which accords with the minimim cut-off time for canister breakthrough mentioned before (a leach rate for 4°C was taken for the base case). For abnormal scenarios, however, where canister failure occurs at a time when the wastes are still hot, the leach rate is taken as a function of temperature. The dissolved silica concentration, which will eventually cause saturation effects and re-precipitation of some complex silicates in the zone close to the glass surface, is neglected. This and the related effects of solubility limits for elements such as Am, Pu, Np, U and the Rare Earths, cause a reduction in the release rate of radionuclides from the glass, and are not included in the model mainly due to the lack of appropriate data. This is nonetheless a conservative assumption since these processes can only lead to a reduction of release rates from the glass.

3.2. The Far-Field Model

The concept of the sub-seabed disposal option supposes that by burying wastes deep within the sediments a major barrier to the migration of released radionuclides is introduced, (without this you simply have sea bottom dumping). The large retention capacity of sediments for a wide variety of substances, including radionuclides, is well known. This is linked to the large surface areas and the latent surface charges of many of the sediment particles. This assumption of an effective barrier therefore appears well founded.

In addition to radioactive decay and chain ingrowth effects which are included in all model compartments, three aspects have to be considered for the far field models; transport processes and physical and geochemical properties of the sediments.

- Transport process - Advection and dispersion are the two principal mechanisms responsible for mass transport in an isothermal fluid system. A major assumption for the base case in the calculation of radionuclide transport in the sediments, is the absence of advective movement of the interstitial waters (no pore water velocity). The evidence for this is strong but not conclusive mainly due to the difficulty in predicting future events, especially those of an episodic nature (eg. pore water venting from sediment slumping or events associated with faulting). Such events are considered under altered scenarios. With the 'no pore water velocity' assumption, the only remaining migration mechanism possible in the far-field is molecular diffusion. Emphasis is thus given to porosity values (which control the molecular diffusion coefficient) and to distribution coefficients (which control the adsorption and retardation of the radionuclides).
- Physical properties of the sediments - The two sites selected for investigation exhibit different physical and geochemical characteristics. In both cases the sediment column is made up of a series of

horizontally inter-layered deposits of pelagic and tubiditic origin
of varying thicknesses, porosities and mineralogies. The complexity
of this structure makes it extremely difficult to generalise even for
one particular site. Nevertheless, for modelling purposes an attempt
was made by the SSTG to define a standard sequence of sedimentological layers which resulted in the definitions of characteristic
modules for each site (Table 1).

TABLE 1. Characteristic sediment module sequences for the two
study sites.

GME SITE

SEDIMENT TYPE	THICKNESS (m)	POROSITY
Calcareous ooze (turbidites)	2.0	0.60–0.75
Silt turbidites (base)	0.1	–
Pelagic sediments	0.1	0.65–0.70

SNAP SITE

Turbidite clays	0.23	0.40–0.60
Fine grained silts	0.02	0.75–0.80

These modules were expected to repeated themselves down to several
hundred metres. The affect this layered structure has on porosity
values is different for each site. For GME the calcareous turbidites
clearly dominate and the difference between lithological types is not
extreme (10–15%). For SNAP, however, porosity is found to vary a
great deal (up to 30%) over small depth intervals. In addition, a
gradual reduction in porosity with depth is always evident due to
compaction processes; this is particularly important in the first few
metres of the sediment column.
- Geochemical properties of the sediments – The most significant geochemical characteristic of the sediments is the presence of an oxidized surface layer and then mildly reducing conditions down to the
depth of emplacement. At GME the oxidized zone has a thickness of 0.1
to 1 metre, while at SNAP this thickness is from 1 to 13 metres. Except for the influence of organic rich layers this redox layering is

largely independent of the sediment lithology and as such is effectively superimposed upon it.

Experimental studies show sorption reactions to be independent of the sedimentological nature of the different units (ie. silts, clays, or oozes). In contrast, the oxidation state of the sediments can have a significant influence on the behaviour of redox sensitive nuclides. It was thus decided to structure the base case model on the chemical redox layering neglecting the complex lithology. The model therefore becomes a simple two layered system representing the upper oxidizing zone and the underlying reducing region. Average porosity values were chosen with a uniform stochastic range large enough to cover the variations measured in the different lithological units (0.58 to 0.78 for GME and 0.46 to 0.67 for SNAP). The porosity-depth function was only considered for the deterministic analyses.

The problem now arises as to how to model the sediment interactions with the solid phases. Mention has already been made of sorption coefficients, and certainly these are one of the easiest ways in which to describe such reactions. Furthermore, models exist describing the sediment interactions of migrating species including, for example, reactions of colloidal formation and filtration, precipitation and solubilization at redox fronts and kinetic effects. However, the appropriate data is not available in sufficient detail, for all conditions and nuclides, to enable such models to be used. It was therefore decided to assume that radionuclides, migrating by diffusion in the sediment pore waters, are sorbed by the sediments according to an instantaneous reversible equilibrium and a linear isotherm. This is known as the 'constant distribution coefficient K_d approach' and is a simple and pragmatic solution for describing the complex processes known to occur in sediment pore waters.

Consider the significance of this decision. Geochemical experts are in agreement that the K_d approach is too simplistic for describing the 'sorption' processes that migrating chemical species undergo. In reality kinetics play an important role and the possible presence of colloidal species and the effects of micro-reducing environments and other localised surface phenomena might also contribute significantly to their behaviour. Such processes cannot be taken into account by single K_d values. There is, however, no practical alternative available at the present for far-field models when data is required for a large number of different species and models are needed to be interfaced with stochastic codes requiring small computer processing time.

Despite the simplicity of the K_d approach and the plethora of data in the literature, the selection of which K_d value to take in each case was not an easy task. The spread of experimental points is notorious and data appropriate to conditions within deep sea sediments are not always available (low temperatures 2-4°C, reducing and oxidizing conditions, and large hydrostatic pressures of 50-60 MPa). This could be a perfect situation for applying stochastic techniques, plotting the spread of experimental values and using this distribution as the input probability distribution function. This was rejected, however, due to the difficulty in making a proper comparison between the data. So many factors, not always reported by researchers, can intervene to influence the obtained result, that a distribution formed from such data could be completely misleading. Values were selected instead using conservative assumptions about the behaviour of each radionuclide in question. For the stochastic analysis, ranges about these values of one or more orders of magnitude were used.

K_d values are usually determined by static batch methods. An additional source of information from which K_d values can be derived is data concerning effective diffusion coefficients. Whilst K_d values provide information about the sorption capacity of sediments, effective diffusion coefficients give an integrated measure of the effect of sediment interactions, and other retarding mechanisms, on the movement of the radionuclides directly. Unfortunately, the data on effective diffusion coefficients is not as comprehensive as that for K_d values. However, by comparing K_d values derived from the available effective diffusion coefficient data (see eg, Li and Gregory, (1974)), with the spread of directly determined values, it was possible, for some of the nuclide elements, to make better estimates of the K_d parameter distribution functions to be used in the assessment. Thus, by a combination of conservative choices and the use of more precise information where available, a usable K_d data set was constructed for all the elements of interest (Table 2).

On completing this exercise it was noticed that despite the important distinction that had been made between the oxidizing and reducing layers, this was only reflected in differences in K_d values for four radionuclide elements (Pu, Ra, Tc and U). In other words, the range of uncertainty in the data was, in most cases larger than the difference in the sorptive capacities of the two zones! Furthermore, when time considerations were made for running the model in a stochastic mode, the space step chosen by the optimisation routine for minimising computer processing time, resulted in a value greater than the thickest possible oxidized layer for the GME site, (in other words, the space resolution was not fine enough to distinguish this surface zone). Hence for this site, only a single mildly reducing layer was considered. Since, for the nuclides in question, retention coefficients are smaller in oxidizing sediments, this is not a conservative assumption. However, when account is taken of the large spread of values already allowed by the defined distributions (Table 2) and the small amount of sediment represented by this surface layer (0.1 to 1 m thick)

TABLE 2. Data set of sediment K_ds (ml/g) - best estimate values and ranges - applicable to GME site for use in the far-field model.

ELEMENT	OXIDIZED	MILDLY REDUCED
Am	1×10^6 $(1 \times 10^5$ to $1 \times 10^7)$	1×10^6 $(1 \times 10^5$ to $1 \times 10^7)$
C	0 (0 to 1)	0 (0 to 1)
Cm	1×10^6 $(1 \times 10^5$ to $1 \times 10^7)$	1×10^6 $(1 \times 10^5$ to $1 \times 10^7)$
Cs	1×10^2 $(1 \times 10^1$ to $1 \times 10^3)$	1×10^2 $(1 \times 10^1$ to $1 \times 10^3)$
I	1 $(0$ to $1 \times 10^2)$	1 $(0$ to $1 \times 10^2)$
Np	2×10^3 $(1 \times 10^3$ to $5 \times 10^3)$	2×10^3 $(1 \times 10^3$ to $5 \times 10^3)$
Pd	5×10^3 $(0$ to $2 \times 10^6)$	5×10^3 $(0$ to $2 \times 10^6)$
Pu	2×10^2 $(1 \times 10^2$ to $3 \times 10^2)$	2×10^2 $(1 \times 10^2$ to $1 \times 10^5)$
Ra	1×10^4 $(4 \times 10^3$ to $4 \times 10^4)$	1×10^4 $(7 \times 10^3$ to $6 \times 10^4)$
Se	0 $(0$ to $1 \times 10^4)$	0 $(0$ to $1 \times 10^4)$
Sn	5×10^4 $(0$ to $5 \times 10^5)$	5×10^4 $(0$ to $5 \times 10^5)$
Tc	0 $(0$ to $1)$	0 $(0$ to $1 \times 10^2)$
Th	1×10^6 $(5 \times 10^5$ to $1 \times 10^7)$	1×10^6 $(5 \times 10^5$ to $1 \times 10^7)$
U	1×10^2 $(8 \times 10^1$ to $8 \times 10^2)$	1×10^2 $(1 \times 10^2$ to $1 \times 10^5)$
Zr	1×10^4 $(0$ to $5 \times 10^6)$	1×10^4 $(0$ to $5 \times 10^6)$

compared with the penetration depth (30 to 70 m), the effect of this is expected to be negligible.

As mentioned above K_d data relevant to deep sea sediment conditions is not often available and the assumption is made that temperature and pressure effects are negligible. Since the 'constant K_d' approach adopted for the far-field model requires as input data molecular diffusion coefficients for each radionuclide element, there is an opportunity here to make some corrections for these effects. Data for these coefficients were first derived from tracer and self-diffusion coefficients of appropriate ions at 'infinite' dilution. These values were then corrected for salinity, temperature and pressure due to their effects on viscosity.

The final model adopted and data used is thus a very simple representation of the real world. The route by which this was derived, however, gives us confidence in it even though many principal processes known to be active in this system are not directly modelled. In addition, before deciding on which particular model to use, several different codes were applied to the same transport equations. Some were analytical, others numerical, some one-dimensional, others two or three-dimensional with and without radial symmetry around the waste package. All these codes were compared and verified through intercomparison studies, and in their later versions they all gave similar answers for the flux of nuclides out of the far-field.

The interface between the sediments and the ocean is represented by the Benthic Boundary Layer (BBL). This is a biologically active zone, approximately represented by the top 10 cm of sediment and the bottom half metre of the water column. For modelling purposes this zone was regarded as part of the ocean, so that the output from the far-field model would be the input flux of radionuclides to the top 10 cm of the sediment column.

3.3. Transport in the Oceans

Under the guidance of the POTG, a number of models describing transport in the ocean were developed and tested (Shephard et al. (1988)). The eventual aim was to develop a credible model suitable for use in the radiological assessment. The result was the Mark-A box model. This was initially constructed (Robinson and Marietta (1985)) to serve as a sub-model of physical transport to be combined with other units describing, for example, biogeochemical and sedimentary processes, thus forming an overall systems model. Because the assessment involves stochastic calculations a very large number of computer runs are required, and therefore economics dictates that the models are simple. At the same time they must represent reality sufficiently well that reasonable confidence may be placed in the results.

Robinson and Marietta (1985) point out, however, that the available 'dynamical' models (eg. GCMs, General Circulation Models and REMs, Regional Eddy-resolving Models) are too large to be incorporated into an assessment model, due to their extensive computational requirements. For efficiency in systems studies, the complex calculations performed by 'dynamical' models are usually replaced by box model calculations. Despite the similarity of these two modelling approaches in the way that a continuum is treated by a discrete set of points, there are important differences between them (Robinson and Marietta (1985)). Not only is the resolution of box models far smaller, but also, with box modelling, a bulk circulation is imposed rather than being derived from 'dynamical' laws.

The approach adopted by the POTG was to develop a box model that adequately reproduced the coarser features of the general ocean circulation. By "relating the box model to the evolving hierarchy of more complex 'dynamical' models" (Robinson and Marietta, (1985)) the chances for misinterpretations were minimised. In this way the box model was able to be continually improved during the development stages, with reference to updated results from the more realistic continually evolving 'dynamical' models.

3.3.1. Description of the Mark-A Ocean Box Model. The Mark-A ocean dispersion box model was designed to be consistent with the North Atlantic GCM developed through an intercomaprison test problem set up between interested parties in the POTG. The North Atlantic GCM used in this study contained high resolution sub-domains describing regional eddy-resolving processes and the bottom boundary layer at the sites of importance. From a simulation of this GCM, specific transports between boxes were extracted for use in driving the Mark-A. Thus, the North Atlantic basins in Mark-A were systematically tied to the more complex state-of-the-art 'dynamical' models.

In the first version of the Mark-A model 14 boxes were used to cover the North Atlantic (12 boxes) and the Arctic (2 boxes), (Figure 4). To compute, in each box, the radionuclide concentration variations with time the following physical processes were taken into account: advection of water representing large scale ocean currents, dispersion, representing turbulent mixing, and radioactive decay. At a later stage this model was considerably improved; the final version included boxes to represent the South Atlantic, continental shelves and the 'rest of the world's oceans'. To increase resolution at disposal sites and to attempt to represent local dispersion and mixing processes, which are characterised by smaller time constants with respect to ocean basin processes, nested boxes were included at the sites.

The most significant improvement, however, was to superimpose upon the physical dispersion boxes a bio-geochemical model to simulate the radionuclide interactions with particulate matter, both in the water column and in bed sediments. A complex series of processes are involved here starting from the radionuclides being scavenged from the water column, transported to deeper ocean layers and a proportion settling out and becoming incorporated into bed sediments through bioturbation and burial. Eventually some radionuclides diffuse into the deeper sediments where, distant from the disposal sites, they become finally removed from the biosphere. In the model, only these scavenging processes and radioactive decay are able to remove radionuclides from the biosphere completely. Incorporating these mechanisms into Mark-A increased the number of boxes to 55.

The final structure of the model is described in Figure 5 where the geographical layout, sub-divisions of the principal compartments and the nested box configurations are illustrated. The four central compartments describing the North Atlantic are divided into five depth layers. For the Arctic, South Atlantic and the remaining ocean water

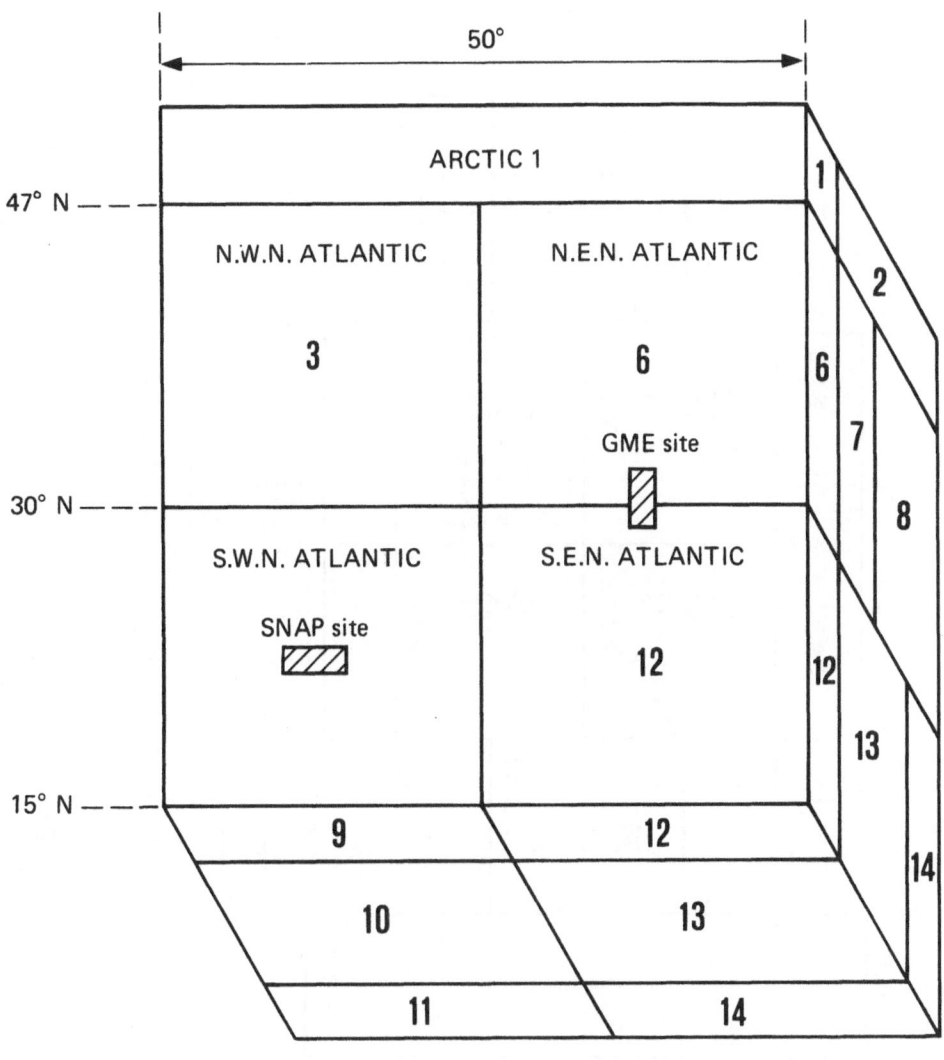

Figure 4: Compartments of Mark-A (first version: 14 boxes) and locations of the SNAP and GME sites

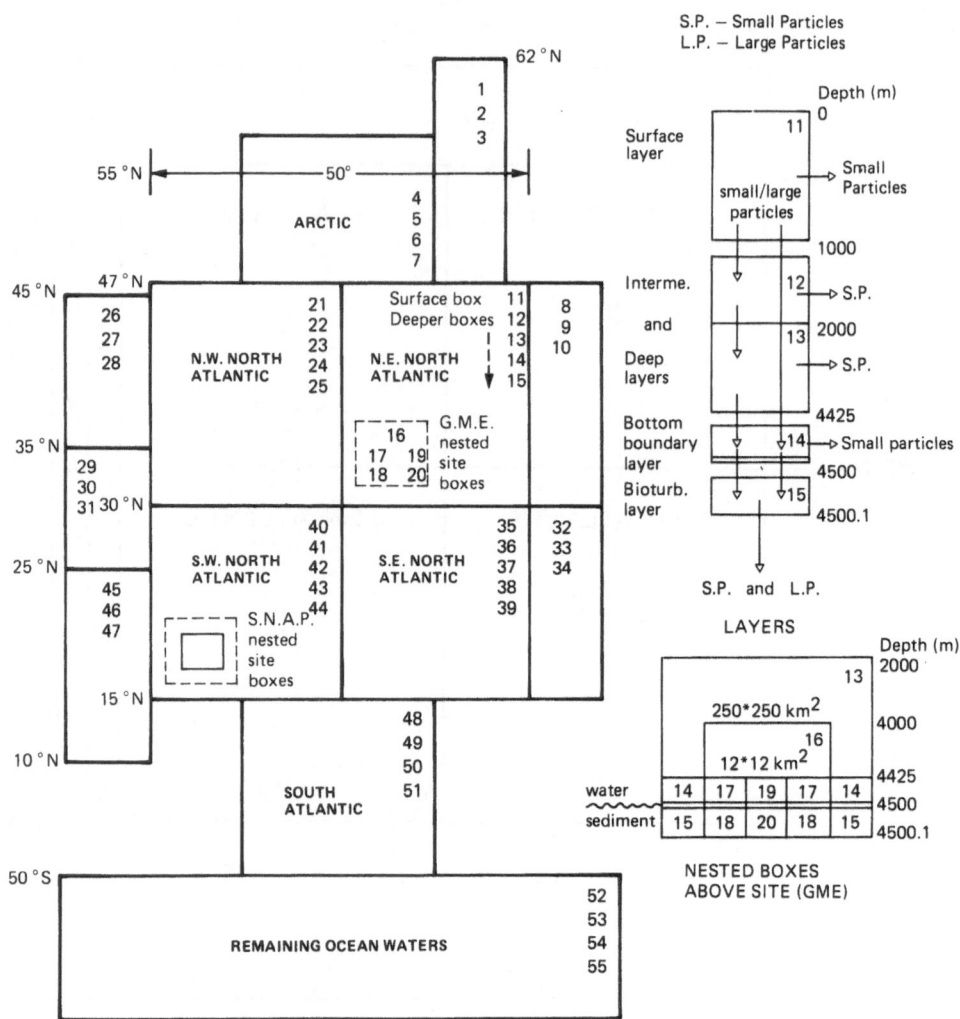

Figure 5: Schematic layout of initial Mark-A ocean dispersion box model used for both sites including positions of coastal boxes, alternative positions for nested site boxes and definition of vertical box numbering.

compartments, four vertical layers are used. Three layers are considered for the coastal compartments. In each compartment, the deepest layer represents the bioturbated sediment (10 cm thick). The box volumes, box interface specifications and water flux values were initially all determined with reference to the standard GCM being studied by the POTG. However, due to difficulties encountered by this group in obtaining a unique, self consistent set of values from this GCM, the advective fluxes between boxes used in the final version of Mark-A were eventually estimated from ocean current measurements.

The final model is able to simulate the slow dispersion of radionuclides in the world's oceans. For an impulse release of radionuclides the model predicts a mixing time in the Atlantic Ocean of 500 years and a complete mixing time with the rest of the world's oceans of around 2000 years. This latter value is in the order of the real mixing time of the oceans. The coarse discretization used in the model means, of course, that the precise pattern of the mixing in the early stages is not described. Thus, the model can be used for scenarios where the release of radionuclides is small and continuous over a long time period and where possible doses to man result from the slow build-up of the concentration in the ocean waters. For rapid releases, as in some abnormal scenarios, other models have to be used.

3.4. Dosimetry

The aim of a radiological assessment is to calculate radiological doses to man (individuals and populations) as a measure of the consequence of the disposal practice. For the seabed option such calculations are based on the concentration levels found in relevant ocean compartments at different times. Doses are computed for ingestion, inhalation and external radiation pathways which requires the following type of data: concentration factors for the radionuclide and ingested material considered, consumption rates of each ingested material, factors of dose per unit intake of consumption and inhalation, etc.

The following sub-pathways were considered in the present study: the consumption of surface and mid-depth fish, crustacea, mollusc, seaweed, sea salt, desalinated sea water, the inhalation of marine aerosols and external radiation from beach occupancy. From these sub-pathways two main pathways are constituted based upon water level, and two hypothetical (future?) pathways are added as follows: Pathway 1. surface water; Pathway 2. mid-depth water; Pathway 3. consumption of zoo- and phyto-plankton, and; Pathway 4. deep sea fish consumption. Pathways 1 and 2 are regarded as the two principal exposure routes while 3 and 4 are used for abnormal scenarios assessing changing contamination pathways.

4. CHECKING THE PERFORMANCE OF THE OVERALL SYSTEM MODEL

In theory, once all the models and data have been gathered together the assessment should be able to be performed straight away. In practice things are not so simple! For the stochastic assessment a number of preparatory stages were needed, mainly for checking purposes, and a series of preliminary simulations were performed in order to optimise the code with a view to making simplifications. Not until then was a definitive version of the encoded models and data arrived at.

Before conducting the assessment it is important to understand the performance of the full systems model and how the constituent sub-models contribute to the overall behaviour. This can be done by following a release pulse through each of the major compartments and noting how it evolves. The results from a preliminary 500 run simulation using the the GME basic data file will be used to illustrate this, concentrating specifically of the results of a single run (number 89) for ease of presentation.

Figure 6 shows, for some important radionuclides, the near-field model output for one particular run, as a function of time. For this run the lifetime of the glass matrix was 470 years and the mean lifetime of the canister was 12200 years. In this case all the radionuclides are released into the sediments by approximately 25000 years. Increasing the canister lifetime to 16500 years (Figure 7), the time of peak release is seen to shift from about 18000 to 22000 years. Since the mechanism of the release model does not depend up on the radionuclide considered, the differences between the radionuclide release rates are due to their initial radionuclide concentrations in the glass and their decay constants. In fact, due to the logarithmic scales and relatively short time period represented by the graphs, the decay effect is not evident.

The far-field output fluxes, resulting from the near-field release rates for some of the radionuclides given in Figure 7, are shown in Figure 8. These are the fluxes of radionuclides that will enter the benthic boundary layer at the sediment/ocean interface before being released into deep ocean waters. The results show how, in the far-field, diffusion spreads the input pulse to form in a very broard output function over the simulation time. Tc-99 appears as the most important radionuclide which can be explained by a number of factors including, technetium's high molar activity value, high molecular diffusion coefficient, small sediment retention factor, and its long half-life.

Following the pulse on through the ocean dispersion and dosimetry models of the biosphere, results in the dose rate versus time profiles shown in Figure 9 (pathway 1, surface water activities); the results are the same for the mid-depth water pathway 2. Tc-99 remains the most important radionuclide over all the simulation time. The profile shapes

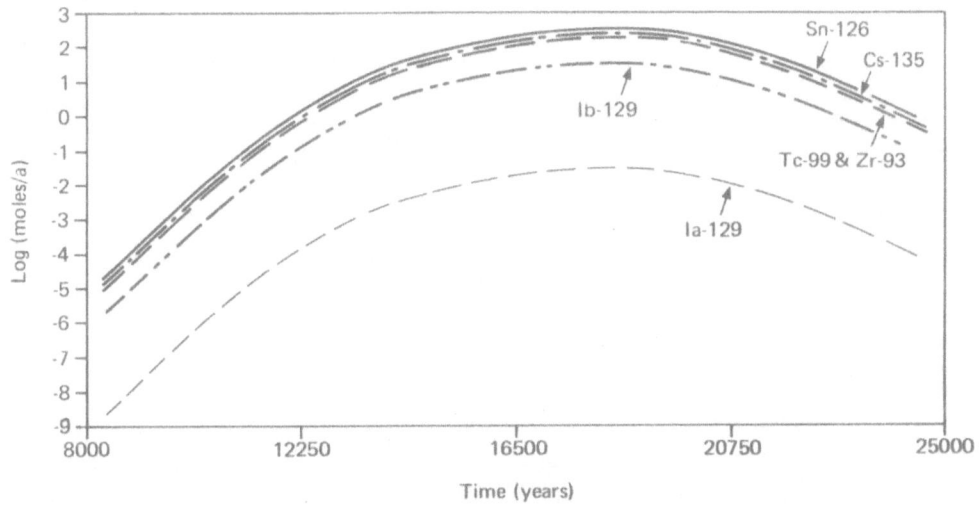

Figure 6: Near-field model output for one particular run obtained from a LISA simulation with 500 runs. (GME site/109 variables)

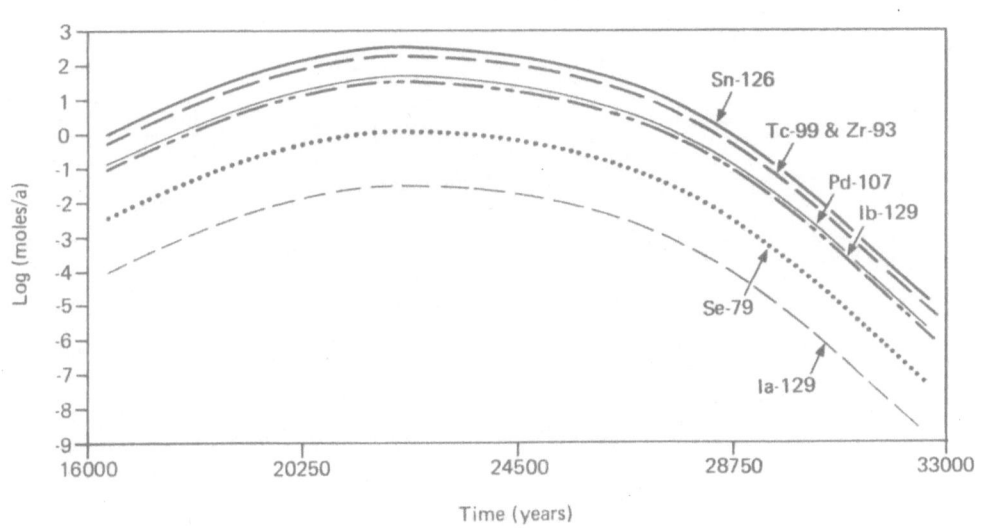

Figure 7: Near-field model output for one particular run with a high canister mean life time (GME site/500 runs/109 variables)

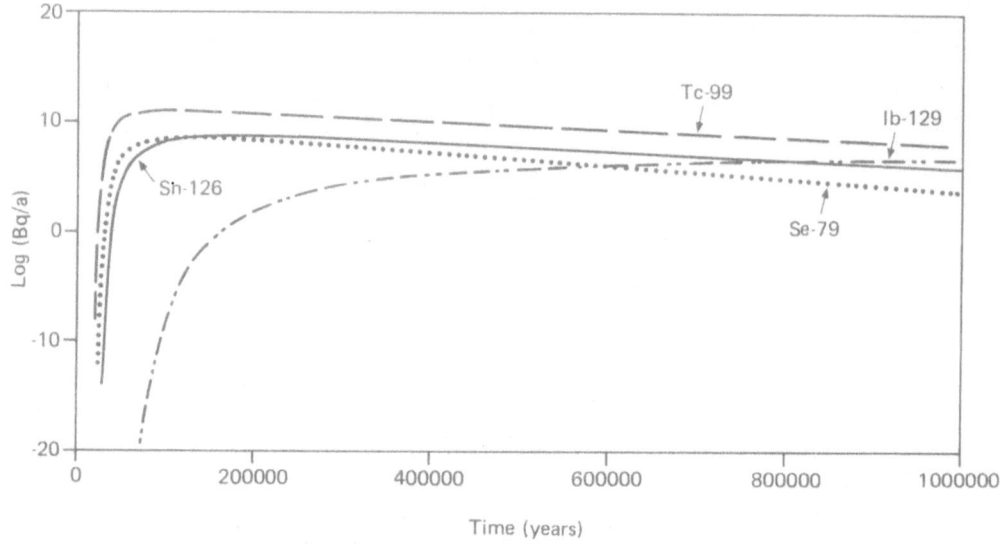

Figure 8: Far-field model output of selected radionuclides obtained using as input the radionuclide release rates given in Figure 7. (GME site/109 variables/500 runs)

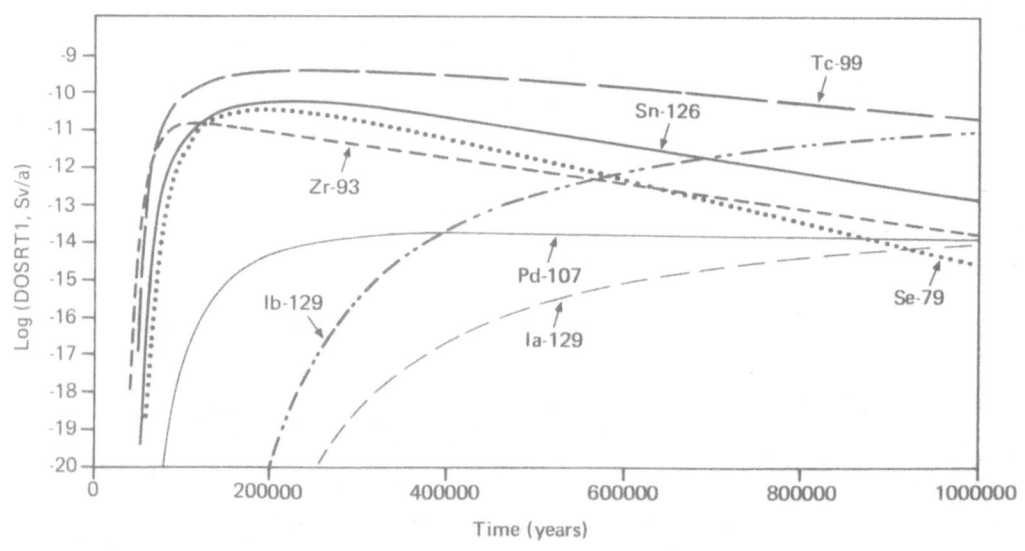

Figure 9: Biosphere model output (individual dose rates versus time for pathway 1) obtained for run number 89. (GME site/500 runs)

are similar to those of the input into the ocean (Figure 8), but a certain separation between nuclides is noticeable. This is mainly due to differences in concentration factors.

To help analyse the results of the whole 500 run test case simulation a contigency table was constructed for nuclides and dose rates (Table 3). Each entry in this table gives, for the GME site and pathway 1, the number of times a given nuclide produced a dose rate of the order of magnitude given on the horizontal axis. For example, the number 504 circled in the table means that from all the maximum values of the Tc-99 dose curves recorded over the 500 run simulation for the various pathways, 504 values are between 3.16×10^{-10} Sv/a (-9.5 on a logarithmic scale) and 3.16×10^{-9} Sv/a (-8.5 on a logarithmic scale). Analysing the results in this table immediately leads to the conclusion that decay chains 2 to 5 in the radionuclide inventory (which includes all the actinides) contribute only negligible amounts to the total dose rates, and are therefore not worth considering! This is also true for the SNAP site simulation. Based upon these considerations, these nuclides were omitted from the data. Clearly, such an analysis is very useful to help decide what simplifications should be made to complex codes.

TABLE 3. Contingency table for GME site, 500 runs and 4 pathways.

| | | LOG (DOSE RATE) | | | | | | | |
		-13	-12	-11	-10	-9	-8	-7	-6
R									
A	Tc-99	109	365	141	220	504	521	52	0
D	Se-79	87	108	272	307	326	58	0	0
I	Pd-107	221	362	27	0	0	0	0	0
O	Cs-135	324	404	518	78	0	0	0	0
N	Ia-129	788	733	0	0	0	0	0	0
U	Ib-129	18	98	358	784	738	0	0	0
C	Sn-126	39	43	98	224	192	354	77	10
L	Zr-93	53	134	330	354	73	0	0	0
I	Th-229	2	7	3	0	0	0	0	0
D	U-233	2	0	0	0	0	0	0	0
E									

The first complete stochastic simulations using LISA (of which the results presented above are a part) required, for 500 runs, a maximum CPU time of 260 minutes using the JRC Amdhal 5868 S and a memory size of about 6000 kbytes. After omitting decay chains 2 to 5 the new input data files contained only 7 decay chains (down from 11) and the number of variable input parameters reduced to 60 for the GME site and 68 for SNAP. This resulted in considerable savings in terms of CPU and memory size requirements for running LISA; for a 500 run simulation the maximum CPU time fell from 260 to 160 minutes, and memory size from 6000 to 2000 kbytes. Such savings can be extremely important since they enable a more efficient use of computer processing time. This in turn can allow an increase in the number of runs performed, with a consequent improvement in the statistics of the results. Also by reducing the number of variable parameters used as input to the code, a more accurate sensitivity analysis can be performed with fewer runs.

At this stage of the analysis additional simplifications were made. These concerned and the radionuclide I-129 and the external radiation pathway (beach occupancy) shown to be insignificant and consequently omitted. For lack of information as to the mode of disposal of the iodine nuclide and the type of its waste form (only 0.1% of the total waste iodine ends up in the glass matrix), it was decided to consider an extreme situation for the initial simulations. In this, the nuclide was effectively considered be in a liquid phase in the canisters – a very conservatively assumption! In the code this fraction was referred to as 'Ib-129'. The results of the preliminary simulations show that were the iodine to be disposed of in this form it would contribute significantly to the overall dose (see eg. Figure 9). It is clear therefore that a more adequate disposal technique is required for this element. This 'liquid' iodine fraction was therefore not considered further in the analysis because the level of sophistication of the model was incoherent with the rest of the system and the results produced would have tended to confuse and possibly mask the effects of other radionuclides. This is a good example of conservatism without realism and the need to have compatibility between models to produce credible results.

5. THE RESULTS

5.1. The 'base case' Stochastic Analysis

The deterministic assessments gives 'best estimates' of the doses. The stochastic assessment, however, provides, in a rigorous statistical manner, an estimate of the range of uncertainty of the doses due to the uncertainty in the input parameters, and the distribution of the doses within the estimated range. Some idea of the extent of this range can also be given by deterministic 'bracketing exercises', although not in a rigorous fashion.

In LISA the dosimetry model provides for each run a curve of the individual dose rates versus time, summed over radionuclides. From these the mean dose rate versus time curve in Figure 10 is computed. These mean values have been obtained from one sample of 500 runs; another sample would probably produce different empirical mean values.

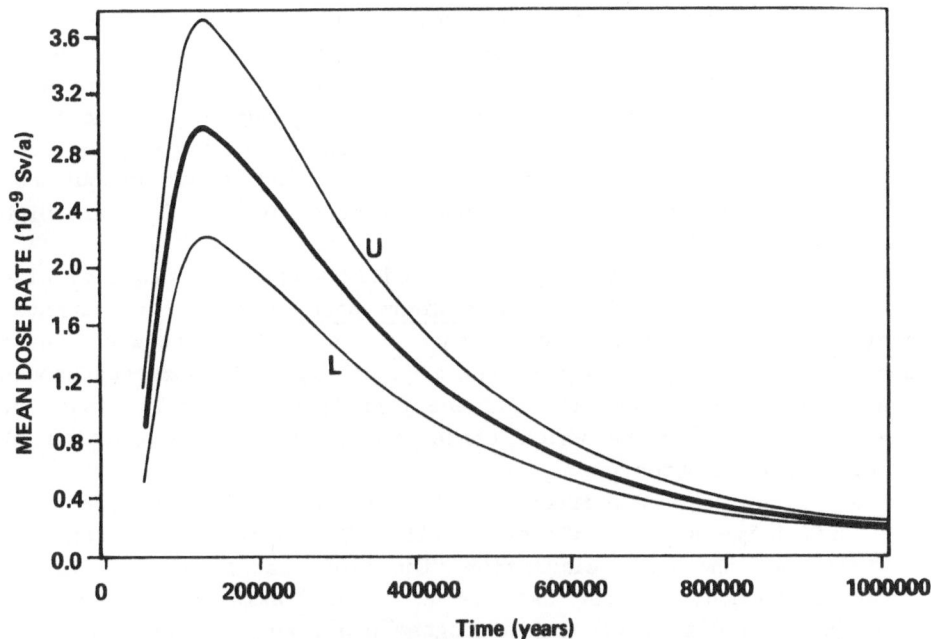

Figure 10: Mean dose rates (Sv/a) versus time for the GME site, base case scenario and pathway 1.
The upper (U) and lower (L) curves define a 95% confidence interval.

To obtain the true mean values a simulation with an infinite number of runs is required but impossible to realise in practice. Fortunately, from the empirical mean values it is possible to define bounds and make the statement that 'the true mean values lie within those bounds' with a certain probability (1-'alpha') that the statement is correct. The level of significance 'alpha' is the maximum probability of rejecting the above statement. In practice, this is done by estimating the confidence bounds on each of the mean values using Tchebycheff's Theorem (see Saltelli (1987) and Saltelli, this volume) for a significance level 'alpha' = 0.05. Such curves are reproduced in Figure 10 as the upper (U) and lower (L) bounds on the confidence interval for the mean dose rates. From this analysis there is a probability of 0.95 that the maximum value of the true mean dose rate ranges between 2.2×10^{-9} and 3.7×10^{-9} Sv/a.

Dose rate frequency histograms are some of the most useful output diagrams from a stochastic code. They can be drawn for specific points in time of the analysis, as often required by regulatory authorities, where they can be imagined as cross-sections of the mean dose curve (Figure 10) at the chosen times. When the output dose rate curves contain significant gradients such sections may prove unreliable as indicators of dose committment; moreover for long time spans, eg. $> 10^4$ years, the actual time of occurrence looses significance. In these situations it is more important to identify the maximum doses calculated up to the specified cut-off time. Frequency histograms can then be constructed from the maximum dose rates of the individual output run curves, independent of when the maximum occurred. In this form all high doses are identified.

An example is shown in Figures 11 and 12 where the frequency histograms for 10^6 years and for the maximum dose rates are illustrated respectively. A significant difference is noticeable between these figures. The peaks represented by the mode of Figure 12 occurred around 10^5 years (see also Figure 10) and are therefore completely missed by the section at 10^6 years which consequently gives an impression of lower overall dose rates.

The maximum dose rate frequency histogram has therefore been taken as the definitive summary of the results. From this we can conclude that throughout the whole simulation time (10^6 years) and for all the 500 runs performed, no dose rate value above 3.2×10^{-8} Sv/a was calculated (upper limit of highest histogram bin). Furthermore, referring to the extra information found in Tables 4 and 5, the highest actual dose rate calculated in the whole set of simulations was 2.53×10^{-8} Sv/a at about 1.09×10^5 years during run number 211 of the simulation, where Tc-99 was the most important contributing nuclide.

These dose rate values have now to be compared with the maximum permitted dose rate for individuals of 10^{-3} Sv/a (ICRP (1985)). This limit is nearly five orders of magnitude above the highest dose rate calculated in this assessment and we can therefore conclude that the 'base case' scenario is radiologicaly acceptable.

If the data in Figure 12 are replotted as a downward cumulative distribution function, adding confidence bands obtained by using Kolmogorov's statistic (see Saltelli (1987), and Saltelli this volume) with a 95% confidence level, Figure 13 is obtained. The value of the upper bound on the highest dose rates computed, now enables an estimate to be made of the probability of exceeding the highest dose rate calculated in the 500 run simulations. For the data in Figure 13 this probability value is 0.06, which can be interpreted that, with a confidence level of 95%, the frequency of doses above 10^{-8} Sv/a will

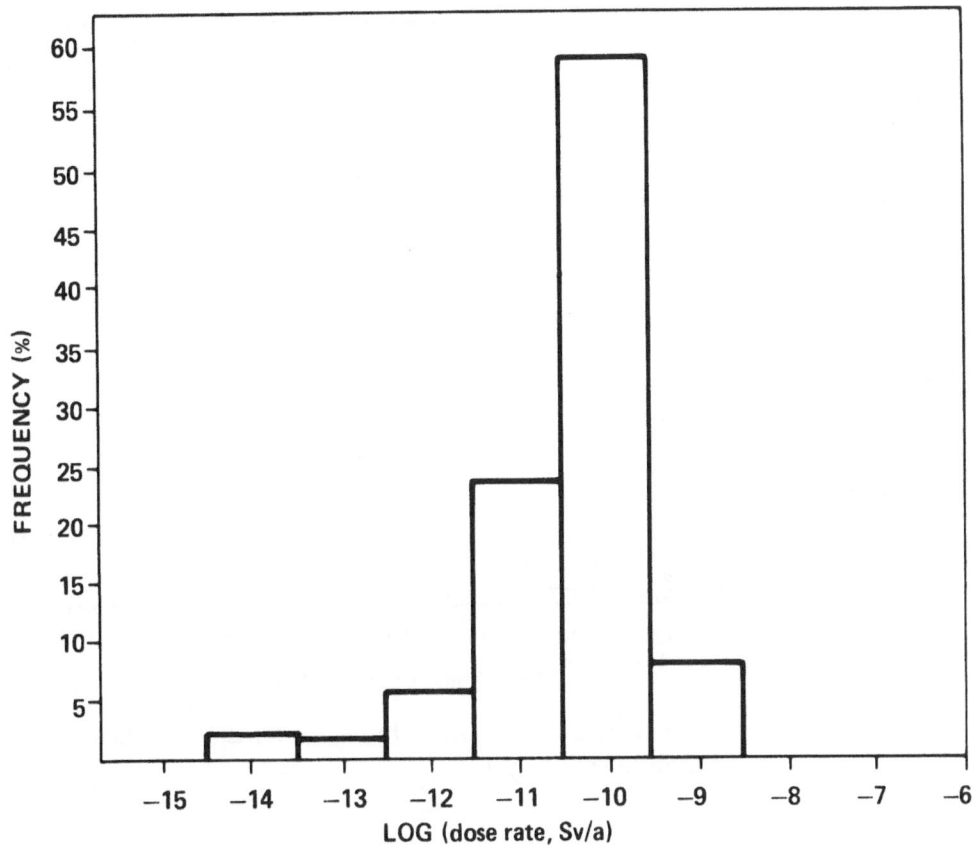

Figure 11: Dose rate frequency histogram for GME site, base case scenario, pathway 1 and at $t = 10^6$ years after vitrification

<u>not be higher than 6%.</u> Such frequency data might now also be used in some form to meet a dose limit criterion. In fact, it has been suggested (AECBC (1985)) that results of this type could be compared with a regulatory prescription by assigning a threshold value to the probability of exceeding a pre-established dose level. However, a properly formulated international criterion has not yet been established.

5.2. Comparison of Deterministic and Stochastic Results

The very nature of the stochastic approach does not allow for an easy comparison to be made with the results of a deterministic calculation. The fact that a stochastic approach allows the selection of parameter value combinations not allowed by the deterministic method means that the results cannot be expected to correspond exactly. Two comparisons were, nevertheless attempted in the seabed assessment. Firstly, the

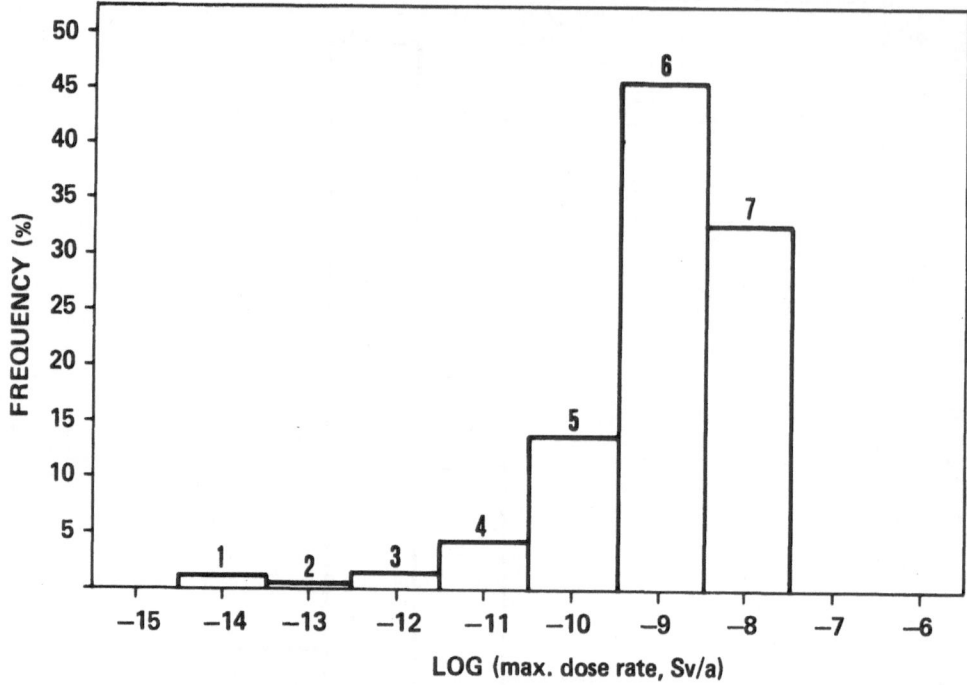

Figure 12: Maximum dose rate frequency histogram for GME site, base case scenario and pathway 1

Table 4: Percentage of runs in each histogram bin of Figure 12 where the given radionuclide is the main contributor.

Histogram unit	Percentage						
	Tc99	Se79	Zr93	Pd107	Sn126	Cs135	I129
1	50	50	—	—	—	—	—
2	75	—	—	—	25	—	—
3	76.9	—	7.7	—	15.4	—	—
4	72	4	—	—	16	8	—
5	58.6	20	7.1	—	14.3	—	—
6	75.8	0.4	1.8	—	22	—	—
7	86.9	—	—	—	13.1	—	—

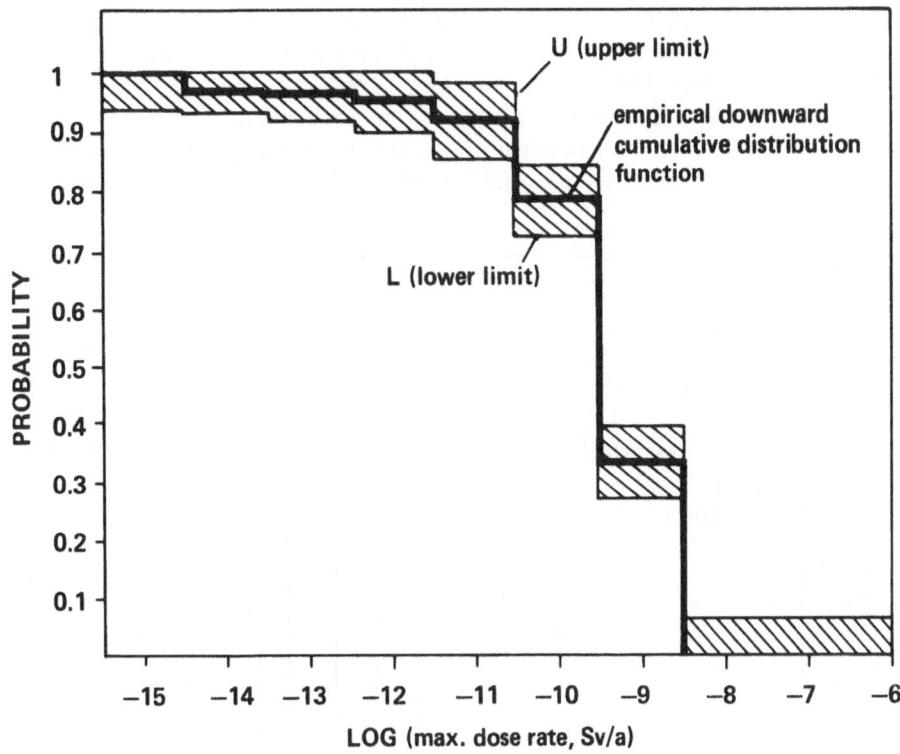

Figure 13: Downward cumulative distribution function of the frequency histogram represented in **Figure 12**
A 95% confidence band is defined by the dashed zone.

Table 5: The ten highest maximum dose rates for the data presented in **Figure 12**

Run number	Maximum individual dose rate (10^{-8} Sv/a)	Radionuclide	Time (years)
221	2.53	Tc-99	108701.2
147	2.39	Tc-99	96656.6
451	1.84	Tc-99	124760.6
470	1.84	Tc-99	152864.6
316	1.83	Tc-99	104686.3
436	1.81	Tc-99	112716.1
254	1.79	Tc-99	88626.9
330	1.65	Tc-99	112716.1
148	1.51	Tc-99	197028.1
332	1.49	Tc-99	124760.6

magnitude of the peak dose rate obtained by the deterministic analysis was compared with the mode bin (most frequently occurring value range) of the maximum dose rate frequency histogram produced by the stochastic analysis (Figure 14). Secondly, the mean dose rate time curve from the stochastic analysis was compared with the dose rate time curve from the deterministic analysis using 'best estimate' values (Figure 15).

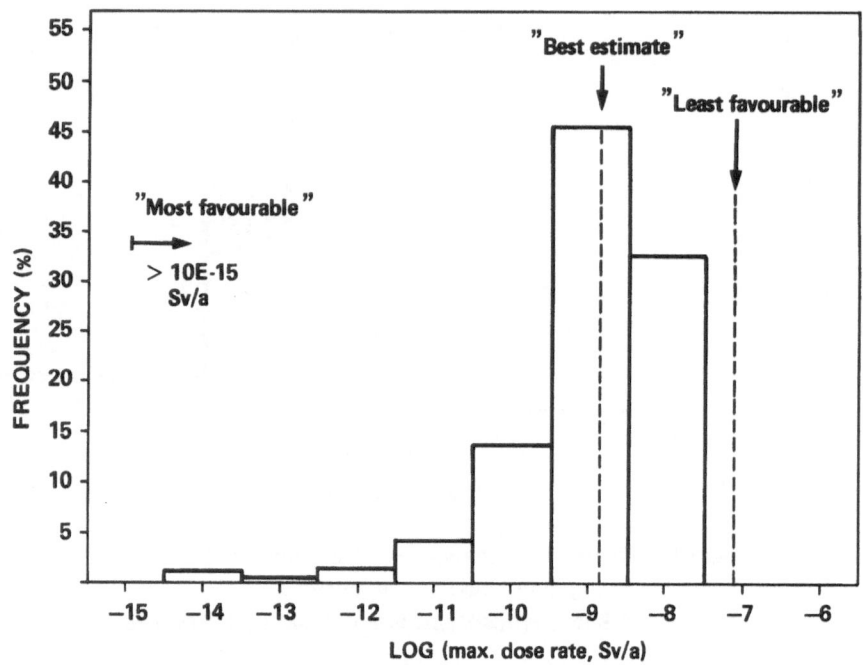

Figure 14: Maximum dose rate frequency histogram for GME site, base case scenario and pathway 1

An excellent agreement is found between the deterministic and stochastic results for both comparisons; the deterministic 'best estimate' value falls on the mode bin of the frequency histogram, and the form and peak position of the deterministic dose time curve corresponds with the stochastic results. For the latter however, beginning at about 60000 years, the mean of the stochastic runs is consistently higher by approximately a factor of two. These differences are not considered very significant; on the contrary, since these calculations were carried out by different workers using different codes, the comparisons are regarded as excellent. Intercomparisons of these type are of tremendous value since they give confidence in the results. This is particularly so when they are performed completely independently by different research groups.

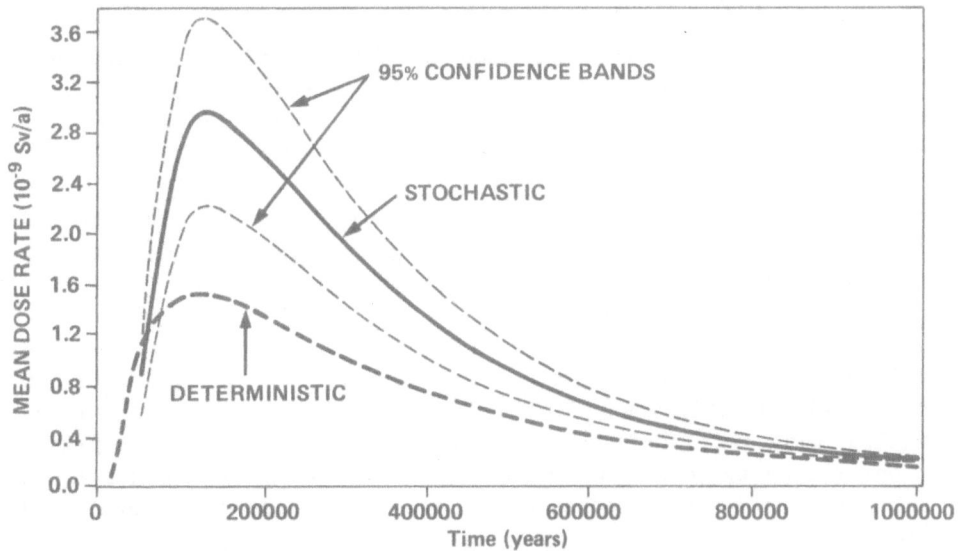

Figure 15: Mean dose rate versus time.
Comparison between stochastic and deterministic assessments
(GME site, base case scenario)

5.3. Sensitivity Analysis

Once a stochastic analysis has been performed it is unnecessary to re-run the assessment to perform a sensitivity analysis since this information is already embedded in the normal output. The problem, therefore, is how to extract the appropriate information from the results. For LISA this is done using the SPOP code (Saltelli (1987)). For brevity the results issuing from the application of this code will be focussed upon, selecting those which highlight significant points and concentrating on interpretation. Because of the simplifications that were found could be made to the overall systems model, reducing the CPU time and the number of variable parameters used (see above), the full simulation was in fact repeated for the sensitivity analysis doubling the number of runs to 1000. This consequently improved the quality of the sensitivity analysis.

Stochastic sensitivity combines the effects of the <u>value</u> of each parameter and the <u>distribution</u> used to describe it. In other words, a variable parameter, identified as sensitive significantly influences the output, either because small parameter variations result in a large variation of the output (high local sensitivity), or because its distribution is very wide (high global sensitivity). The two effects are often combined.

The variables identified as important by the sensitivity analysis are listed below (in order of importance) for both sites separately:
- for the GME site, K_d(Tc), Tc concentration factor for seaweed and crustacea, K_d(Sn), porosity, depth of emplacement XPATH and leach rate;
- for the SNAP site, K_d(Tc), Tc concentration factor for seaweed and crustacea, thickness of the oxidized zone, depth of emplacement, porosity, K_d in the oxidized zone of Zr and Sn.

The appearance in this list of parameters relating to the oxidized surface layer, taken into consideration for the SNAP site but not for the GME simulation (see Section 3.2 above), is an important finding. Excluding this surface layer was a non-conservative assumption, but was deemed expedient due to resolution problems in the far-field model. The reason why these parameters were identified can be explained as follows.

The sorption capacity, or barrier effect, of this oxidized layer is generally weaker than that of the lower mildly reducing zone, particularly for Tc-99. The thickness of the oxidized zone is therefore positively correlated with the output doses since, for a given total sediment column length, an increase of this layer thickness reduces the thickness of the first layer, reducing the total barrier effect of the sediment. When the mean individual dose rates over time are compared for the two sites (Figure 16) no significant differences are found despite the above findings. In fact, the dose rates for the GME simulation are marginally higher than for SNAP. It can thus be concluded that the presence of the surface oxidized layer in the SNAP site does not in itself appreciably affect the results. Leaving this out of the GME simulation was not therefore significant.

The full list of influential parameters identified by the sensitivity analysis can be roughly divided into three groups:
i)   those important over the whole simulation time;
ii) those important only over the initial transient phase; and
iii) those parameters whose influence increases with time.

Taking one example of each and adopting the parameter ranking plots versus time as suggested by Saltelli (1987) to demonstrate the behaviour of the variables, examples of the three cases above can be seen in Figure 17. These show the results for the K_d value for Tc-99, the path length (XPATH) and the sedimentation rate for large particles in the ocean (SL) respectively.

The result for K_d(Tc-99) is not surprising since this has already been identified as the main dose contributor in the base case results.

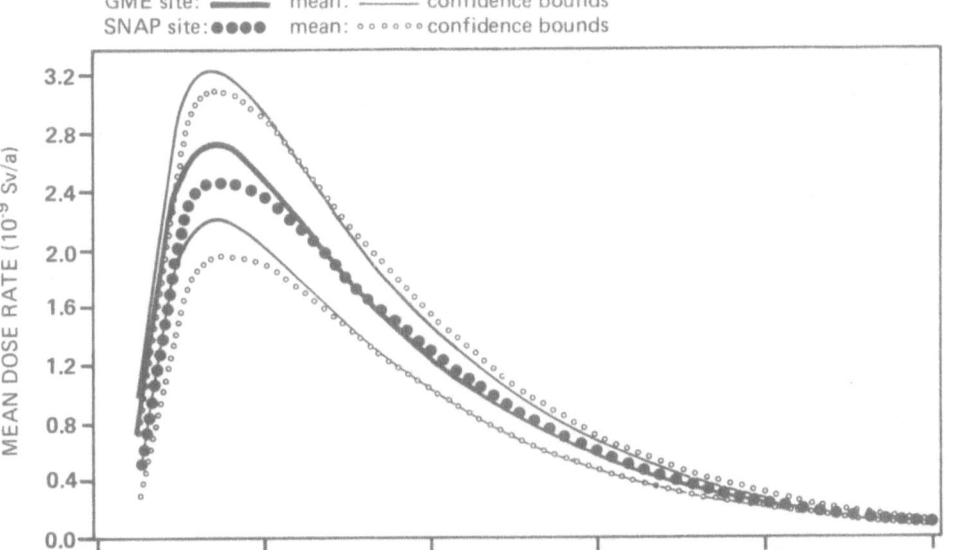

Figure 16: MEAN DOSE RATES VERSUS TIME
with the 95% lower and higher confidence bounds
(Base case scenario · pathway 1 — 1000 runs)

In fact, the sensitivity of the results to Tc-99 is so strong that all but one Tc-99 related variable is identified as important.

The variable XPATH rapidly decreases in influence with time. This can be explained through its influence on the shape of the migration pulse leaving the column. The shorter the sediment column, the higher and narrower will be the pulse at the sediment ocean interface, and thus the higher will be the output doses. Therefore, once the bulk of the dominant radionuclides is released into the ocean, XPATH will become unimportant. This occurs after about $2x10^5$ years which corresponds with the sensitivity results shown in Figure 17b.

Finally, the variable SL increases in influence with time. This relationship is again to be expected from the known behaviour of the models; scavenging in the oceans is a slow process and it is also the last compartment in the overall systems model. Its influence would therefore only become evident after long time periods. There is a problem, however, in understanding why the other settling parameters (concentrations and velocities for large and small particles) do not also display the same behaviour as SL. The reason for this relates to the size and ranges used for their parameter distributions. The sedimentation flux of small and large suspended particles in the water column

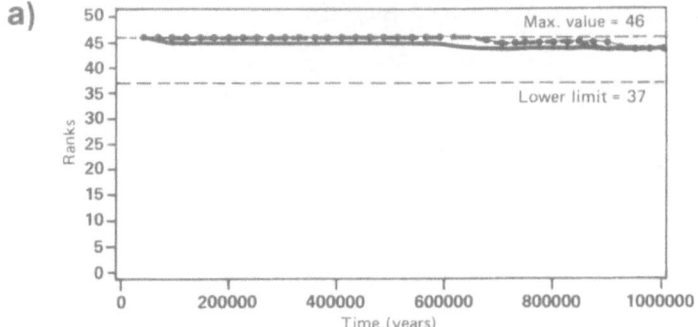

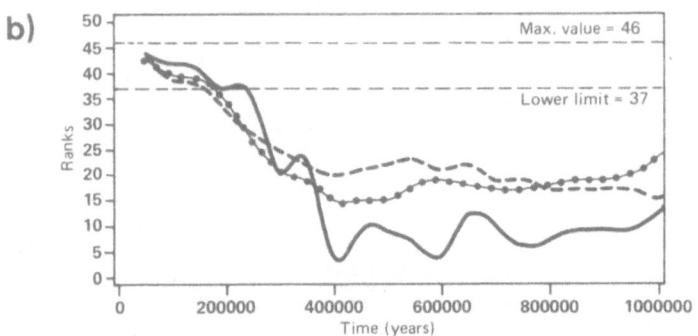

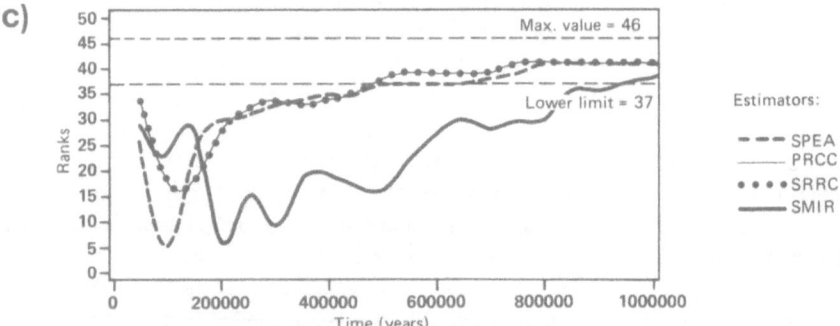

Figure 17: Nonparametric sensitivity analyses (1000 runs/46 variables)
RANK VALUES VERSUS TIME FOR THE VARIABLES: (a) KD (Tc);
(b) XPATH and (c) SL.
GME site — base case scenario — pathway 1

depends on the product of the concentrations and velocities of both sizes of particles. Since the range of the SL values is the most important (most uncertain) this dominates the results.

5.4. Abnormal Scenarios

The generic term 'abnormal scenarios' is used here to cover the following types of events: altered evolution scenarios, the effects of human actions, accidents or human errors. All such events either occur at or post-emplacement. Events relating to the transportation of waste to the site for disposal were dealt with separately. The methodology followed to identify and classify low probability abnormal scenarios was to start by developing a list of events, processes or human actions that could have a negative effect on the confinement of the radionuclides. On analysing the effects of these events on the waste and its environment it becomes apparent that several of these events give rise to the same consequences. A limited number of <u>scenarios</u> were therefore able to be identified from them.

After numerous discussions with the appropriate experts twelve events were identified which were deemed to have an important potential in adversely affecting the confinement of the radionuclides. These included unexpected changes in chemical and physical properties of the sediments, the effects of human intervention or errors, and changing geological and climatic conditions. The full set of scenarios considered which were constructed from such events were as follows:

Case 1: Undamaged penetrator lying on the seabed in deep water immediately after emplacement.

Case 2: Same as Case 1, but damaged canister.

Case 3: Partly buried, undamaged penetrator, lying only 10 m deep in the sediments, immediately after emplacement.

Case 4: Same as Case 3, but with damaged canisters.

Case 5: Enhanced pore water velocity in sediments.

Case 6: Same as Case 5, but with damaged canisters.

Case 7: Enhanced leach rate of glass.

Case 8: Enhanced corrosion rate.

Case 9: Change in contamination pathway – From short cuts or new routes.

Case 10: Change in ocean circulation – This was not studied in detail but has been simulated by increasing regional box concentrations by about 20% in response to reduced advection, diffusion and sedimentation fluxes.

Case 11: Zero K_ds and P_ds – This is a test of the importance of sorption both in the sediments (K_ds) and for water column particulates (P_ds), with all other conditions as for the base case.

Case 12: Changes in K_ds and P_ds - An extension of Case 11, the sensitivity of the total dose, summed over radionuclides, to changes in the K_d values of some specific nuclides was studied.

Case 13 Canister emplaced by the drill option - A sensitivity study on the depth of emplacement.

Case 14: Mining of manganese nodules.

Thus, from a series of specific events and processes a series of cases (or scenarios) were defined for assessment. This simplified the calculations and minimised repetition.

Since all the nuclide transport and dose models are linear with respect to waste quantity (non-linear processes such as the effect of solubility limits were not included), the scenarios were in most cases calculated for a single penetrator or per metric ton of heavy metal (MTHM), allowing scaling up as necessary. The calculations are made in this way because many of the scenarios do not concern all the waste to be disposed of and it is then possible to determine how many penetrators or MTHM could be involved in an abnormal scenario before a prescribed dose limit were exceeded.

On analysing these scenarios it was found that in all but one case, the peak individual doses are so low that the entire inventory could be affected and the dose rates would still be below 10^{-3} Sv/a for the total repository (10^5 MTHM). The exception was a scenario where an extremely high pore water velocity (1 m.a^{-1}) was assumed together with a damaged penetrator (Case 6). Since the probabilities of the scenarios were not assessed, the risk that this poses could not be calculated. In the absence of this information another approach is possible. By back calculating how much of the total inventory could be subjected to this condition before exceeding the dose rate limit it is then possible to judge how likely this is. For the present case the risk from the scenario would be below the 10^{-5} a^{-1} risk limit if the probability of it occurring were 0.55 or less. Since all available measurements fail to show any observable high pore water velocities in the sediments, the probability of this scenario occurring is likely to be much lower than 0.55. Consequently, it is unlikely that the ICRP risk limit would be exceeded.

6. CONCLUSIONS OF THE ASSESSMENT

The results of this seabed assessment lead us to the conclusion that the disposal of high level radioactive waste in sub-seabed sediments is radiologically a very safe option. This statement holds true as far as the assumptions, models and data used in the calculations adequately represent the real system and can be validated. However, even though the models of all major thermal, nuclide transport, biological pathways and radiological effects have been completely verified, they have not been validated. The models produce answers consistent with the input data and equations, but there is no complete assurance that all components are included in the equations and that all the input data are complete and accurate. This is normally true for any assessment and it is therefore important at the conclusion of such work to compile a list of measures which may be taken to minimize radiological impact and reduce the level of uncertainty in the assessment. Apart from the importance attached to validating all major sub-models, the two most important aspects that come out of this study affecting both the safety and overall uncertainty in the assessment are pore water velocity and retention processes. The highest priority should therefore be given to investigating the problems associated with these areas in any future investigations.

7. FINAL OBSERVATIONS

During the performance of this assessment much was learned which could be of use in a wider context. To round off this worked example some of the more important of these points are outlined below:
- Obtaining precise information for assessment purposes from experts was extremely difficult due to the number of unknowns that only the experts really know about. This was tackled in a number of ways:
  . by setting up a 'data base' of that information required to carry out the assessment - this helped to focus attention on the important points;
  . by iteration, showing the experts the results and consequences of their previous decisions on the performance of the models;
  . by asking for considered group opinions.
  The most successful models were those constructed and run in collaboraton with the experts.
- Due to the natural division of the system under study the interfaces tended to be overlooked - these must be properly described.
- The importance of independent verifications and intercomparisons were possibly underestimated; they are of utmost significance leading to increased confidence in the results.
- A careful balance had to be made ensuring that all sub-models and data were compatible and on a similar level of sophistication to maintain coherence. This was particularly difficult when there was disagreement as to how realistic or conservative to be in modelling a particular process.

- A 'devil's advocate' approach, particularly when identifing abnormal scenarios, was useful for 'brain-storming', though annoying at times!
- The performance of the assessment itself took a lot longer than anticipated because many preliminary runs were necessary to ensure matching of results and optimisation of codes. Since this can lead to a considerable improvement in the quality of the assessment it is advisable to plan for this, particularly taking time over the final stages.

ACKNOWLEDGEMENTS

This lecture was based upon an assessment conducted by an international group of researchers made up from many individual contributions (see de Marsily et al. (1988)). The material presented here has been drawn from many sources but mainly from the reports of the international Seabed Working Group under which the overall feasibility study was coordinated. The assistance, directly or indirectly, of all task group members, particularly those of the Radiological Assessment Group, is therefore gratefully acknowledged.

REFERENCES

AECBC (1985) Atomic Energy Control Board of Canada, Regulatory Document R-71, 15 p.

Anderson, D.R. (1987) Tenth international meeting of the NEA coor-dinated program to assess the subseabed disposal option of nuclear waste. Halifax, Nova Scotia, Canada, April 30 - May 3, 1985. Sandia National Laboratories Report, SAND-85-1365, Albuquerque, 146p.

Bishop, W.P. and Hollister, C.D. (1974) Seabed disposal - where to look. Nuclear Technology, 24, 425-443.

Marsily, G. de, Behrendt, V., Ensmineer, D.A., Flebus, C., Hutchinson, B.L., Kane, P., Karpf, A., Klett, R.D., Mobbs, S., Poulin, M., and Stanners, D.A. (1988) Feasibility of Sub-seabed Disposal of High-Level Radioactive Waste into the Seabed: Volume 1, Radiological Assessment, OECD (NEA), Paris, 162 p.

Flebus, C. and Stanners, D.A. (1988) A stochastic assessment using LISA: models, data and results. In: Study of the Feasibility and Safety of the Disposal of Heat Generating Wastes into Deep Oceanic Geological Formations. Part 1: Radiological Safety Assessment. Commission of the European Communities, JRC Report Series. Editor C.N. Murray. EUR 11754 EN.

ICRP (1985) Radiation protection principles for the disposal of solid radioactive waste. ICRP/85/C4-8/12, Publication No. 46, Pergamon Press, Oxford.

Li, Y-H. and Gregory, S. (1974) Diffusion of ions in sea water and in deep-sea sediments. Geochimica Cosmochimica Acta, 38, 703-704.

NEA (1988) Feasibility of Sub-seabed Disposal of High-Level Radioactive Waste into the Seabed: Volumes 1-8, OECD/NEA, Paris.

Robinson, A.R. and Marietta, M. (1985) Research, progress and the Mark a box model for physical, biological and chemical transports. Sandia Report, SAND84-0646, Albuquerque, 202 p.

Saltelli, A. (1987) PREP and SPOP utilities: two Fortran programs for sample preparation, uncertainty analysis and sensitivity analysis in Monte Carlo simulation. Program description and users guide. Commission of the European Communities, EUR 11034 EN.

Shephard, L., Auffret, G., Buckley, D., Schuttenhelm, R.T.E., and Searle, R.C. (1988) Feasibility of Disposal of High-Level Radioactive Waste into the Seabed: Volume 3, Geoscience Characterisation Studies, OECD/NEA, Paris, 318 p.

PERFORMANCE ASSESSMENT OF GEOLOGICAL ISOLATION SYSTEMS (PAGIS)

F. GIRARDI[(*)] and N. CADELLI[(**)]
(*) Commission of the European Communities, J.R.C. Ispra
(**) Commission of the European Communities, D.G. XII-D,
     Brussels

OUTLINE

A. BACKGROUND INFORMATION, PAGIS METHODOLOGY
B. APPLICATION AND RESULTS
   B.1. Clay Formations
   B.2. Granite
   B.3. Salt
   B.4. Sub-Seabed Option
C. CONCLUSIONS

A. BACKGROUND INFORMATION, PAGIS METHODOLOGY

1. In the framework of the Community Plan of Action in the field of radioactive waste (1980-1992) a joint study was launched in 1982, under
   the name of PAGIS (Performance Assessment of Geological Isolation
   Systems), with the participation of national institutions and the
   Commission of the European Communities.
   The main objective of PAGIS was the evaluation of the capability of
   selected geological formations and associated engineered structures
   to confine vitrified High Level Waste (HLW) so as to ensure adequate
   protection to future generations.
   PAGIS was the first attempt at a multinational safety assessment of
   HLW disposal and its particular feature was that approaches and assessment methods were largely common for the various options selected
   for the study. It is a part of the iterative process of consolidating
   the safety assessment along with the advance of scientific and technical knowledge.
2. The basic concept adopted in the study was to choose for each disposal option a "reference case" and one or more "variants", which re-

A. Saltelli et al. (eds.), Risk Analysis in Nuclear Waste Management, 331–355.
© *1989 ECSC, EEC, EAEC, Brussels and Luxembourg.*

flected either variability of environmental parameters in the European Community (E.C.) or variability in design parameters.

It should be stressed that all the above choices were made only for the purposes of PAGIS and do not imply any commitment by the competent national Authorities concerning the site selection for a repository.

This concept favours the evaluation of the impact of the various design parameters and the assessment of their importance, it offers guidance to the R&D programmes on waste immobilisation and confinement. It also resulted in the improvement of the exchange of information among the E.C. Member States on the various options and variants studied in the different countries.

3. Clay, granite and salt were chosen as continental geological options. As a possible alternative to land disposal, sub-seabed was included in the study.

   Geological information on the first three options was largely derived from the European Community catalogue established in 1980, while information on sub-seabed sediments as a potential host rock has been obtained from the programme coordinated by the Nuclear Energy Agency of the OECD.

4. The reference waste arisings and their characteristics refer to the HLW from LWR spent fuel reprocessed by the Purex process. The waste is vitrified as borosilicate glass and contained in stainless steel canisters according to the AVM process as applied at the La Hague plant.

   For this reference waste, the radioactivity decay up to one million years has been computed together with the corresponding total decay heat. A set of radionuclides assumed to be relevant for the safety assessment has been selected.

   Alternative glass composition were also considered in some cases to reflect present national situations (e.g. Magnox and AGR fuels).

5. Repository design was based on requirements of feasibility, safety and economy according to the current status of development in the E.C.

   For continental repositories, the reference designs are based on the shaft-gallery-borehole concept, with dimensions and engineered structures which depend on the local environment, account being taken of the characteristics of the host rock and those of the waste to be disposed.

   For the sub-seabed, the free-fall penetrator is the reference emplacement technique, the drilled hole being the variant.

6. A common methodology has been agreed, which is aimed at providing the general framework for the evaluation of:
   - radiation doses to individuals and populations;

- probabilities that these will be received;
- uncertainties associated with dose and probability results;
- the sensitivity of the computed doses to selected parameters.

Two approaches have been applied to analyse the overall behaviour of a repository system. In the first, detailed models of the barrier system are used to achieve the best available description of the repository and to produce single values of the quantity being predicted, e.g. doses, dose rates for each scenario. These calculations are referred to as "Deterministic".

In the second approach, multiple simulations are performed in order to estimate the range of variability of the quantity being predicted due to possible variations of the input parameters, which are sampled from specified distributions. These calculations are referred to as "Stochastic".

Both approaches have been applied in risk assessment exercises. Advantages and limitations of the two approaches are still a debated matter.

7. Possible failure scenarios have been proposed and selected by expert judgement for each site, and classified as "normal evolution scenarios" representing the gradual changes which may be expected on the basis of the evident geological trends and the information available on repository effects, or "altered evolution scenarios", representing perturbations by events, largely of probabilistic nature, which can modify the parameters determining the normal evolution or generate new evolution scenarios.

The general guidelines indicated above have been adapted to the specific reference cases and variants on a case-by-case basis, depending on information available.

As a result, the scenarios shown in Tab. 1 have been retained for a detailed analysis.

Determinate evaluations of consequences for normal and altered evolution scenarios have been carried out by comprehensive models, while simplified models have been used for uncertainty and sensitivity analysis. The modular approach adopted allowed the choice of the most appropriate model for each particular objective.

B. APPLICATION AND RESULTS

B.1. Clay Formations

Some clay formations can provide an effective barrier for the confinement of HLW due to their particular characteristics. The low permeability of clay results in negligible pore water movements so that radionuclides may only move by diffusive processes. The high retention capacity

TABLE 1. Scenarios selected for the four options in PAGIS.

Option	Normal Evolution	Altered Evolution		
		Short term	Medium Term	Long term
CLAY	. Waste degradation and diffusion Thermal effects		. Human intrusion water wells)	. Tectonic displacement . Climatic changes
GRANITE	. Waste degradation, diffusion in near-field and transport in fissured rock Thermal effects	. Convection in near-field	. Change of fracture pattern (far-field) . Human intrusion	
SALT	. Thermal and convergence effects. Residual uplift. . Subrosion	. Water intrusion. Thermal and convergence effects	. Human intrusion (solution mining, drilling) . Shaft sealing failure	. Climatic changes
SUB-SEABED	. Waste degradation and diffusion Thermal effects	. Incomplete hole closure. . Thermal transients	. Human intrusion	. Tectonic displacement . Human intrusion . Climatic changes

of clay for many radionuclides further reduces radionuclide migration velocities. Provided the integrity of the clay layer is maintained, it is millions of years before radionuclides leave the clay layer and can reach an aquifer. The high plasticity of most clay rocks will reduce the probability of open faults, which might provide a pathway for the accelerated release of contaminated groundwater. Potentially suitable clay formations occur widely within the EC countries. For the purpose of the present study the Mol site in Belgium was chosen as the reference site while the Harwell site in the United Kingdom was taken as a variant.

The capacity for the reference design of a repository in clay derives from a postulated electronuclear installed capacity of 10 GW(e). The corresponding amount of vitrified HLW for a period of 30 years is about 900 m^3, which corresponds to approximately 8200 tonnes of heavy metal reprocessed.

A 200 m depth and a compact clay buffer of 50 m above the repository are expected to provide a safe containment. These figures are satisfied at the Mol site where the subhorizontal clay formation, 110 m thick, is covered by at least 160 m of sediments. However, the geometric characteristics of the formation mean that the repository needs to be limited to a single layer emplacement of waste.

A modular calculation scheme has been employed, in which the transsport of radionuclides through different parts of the migration route (e.g. host clay layer, aquifer, biosphere) has been addressed using appropriate models. The probabilistic approach adopted incorporates submodels for each component of the migration route within a single computer model, and allows input data parameters to be sampled from specified distributions.

A model of "near-field" transport has not been used, since scoping studies of radionuclides migration through clay showed that the predicted dose-rate to man would be virtually independent of repository release rate. The geosphere models for transport in clay and aquifer are conservatively simple, so that many computer simulations can be made, without undue computer resource constraints, maintaining a balance between complexity of the model and knowledge of required data.

Different release scenarios were identified and analysed; they are the Normal Evolution Scenario (NES) and a few altered evolution scenarios.

For the NES, it is assumed that after the sealing of the repository its various components will eventually be contacted by the interstitial water in clay. Natural degradation of the repository will occur and corrosion of canisters and glass blocks will release radionuclides, forming the source term for migration and diffusion through the interstices of the clay formation. Once the aquifers are reached, the radionuclides will be conveyed by the groundwater flow system.

For the Mol site, the normal evolution would be, strictly, for radionuclides to migrate along aquifers and eventually contaminate the
River Nete by artesian seepage. However, because of the high probability
of a well being sunk into the path of the migration plume, a water supply well is considered as an important pathway to the biosphere; this is
treated as an additional pathway, considered to be occurring simultaneously with the river release pathway as part of the normal scenario for
the Mol site.

The transport of radionuclides through the geosphere may be altered
indirectly by medium term climatic changes which influence the hydrological system of the region. The climatic change expected to produce the
worst consequences is a reduction in the effective annual rainfall, for
which there is geological evidence. The main consequence of the change
is that aquifer and river flow rates are altered. Again, the case of water supply wells in the contamination plume was also taken into account.

Another altered scenario is linked to the occurrence of tectonic
displacements (faulting). For the Mol site, displacements in the range
of 5 to 10 meters are assumed to occur in the vicinity of the repository
with a probability of one event per 10 to 100 millions years. They are
not expected to cause the direct release of radionuclides into the aquifers. Nevertheless, these faults can cause local disturbances of the
geological barrier, thus leading to an altered evolution of the system.

For every release scenario, the migration of radionuclides through
the clay and the aquifers is described by the transport equation for
homogeneous porous media, taking into account:
- molecular diffusion and mechanical dispersion (essentially in the clay
  layer);
- hydrodynamic advection, i.e. the transport of radionuclides by moving
  groundwater (essentially in the aquifers);
- radioactive decay and the formation of daughter nuclides;
- sorption processes, characterised by a retardation coefficient, in the
  aquifers and the clay layer.

The pathways to man considered are:
- drinking of contaminated well water;
- consumption of milk, meat, green vegetables and root vegetables;
- inhalation of dust resuspended from contaminated soil.

The annual individual doses calculated for the well pathway of the
normal evolution scenario are illustrated in fig.1 for the most relevant
nuclides. In Fig. 2 the time evolution of the total individual dose rate
for the various scenarios is shown.

It may be seen that the maximum dose rate of 7×10^{-8} Sv/a is due to
Np-237 and occurs at about ten millions years. Cs-135 and Tc-99 make also an important contribution to the total dose.

Using a risk factor of 0.0125 per Sv the corresponding risk for a

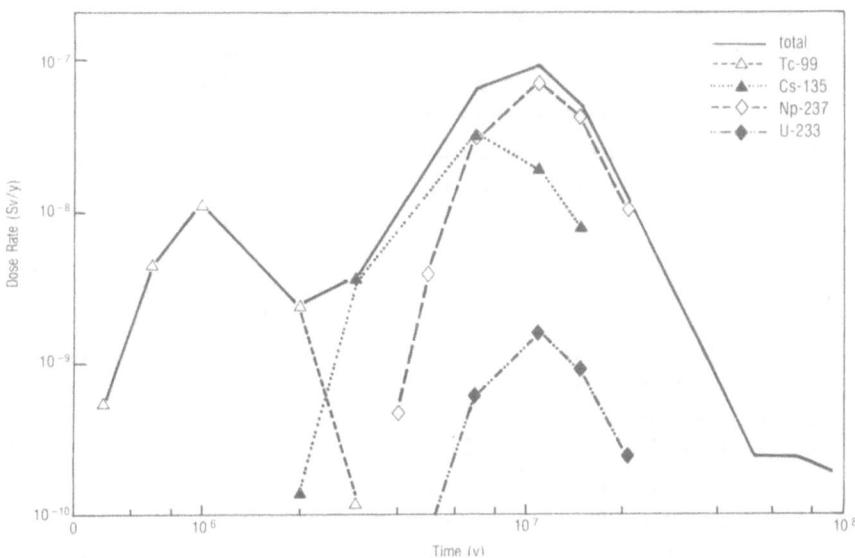

Figure 1. Deterministic evaluation of individual dose rate evolution for the well scenario at Mol with contributions by important radionuclides.

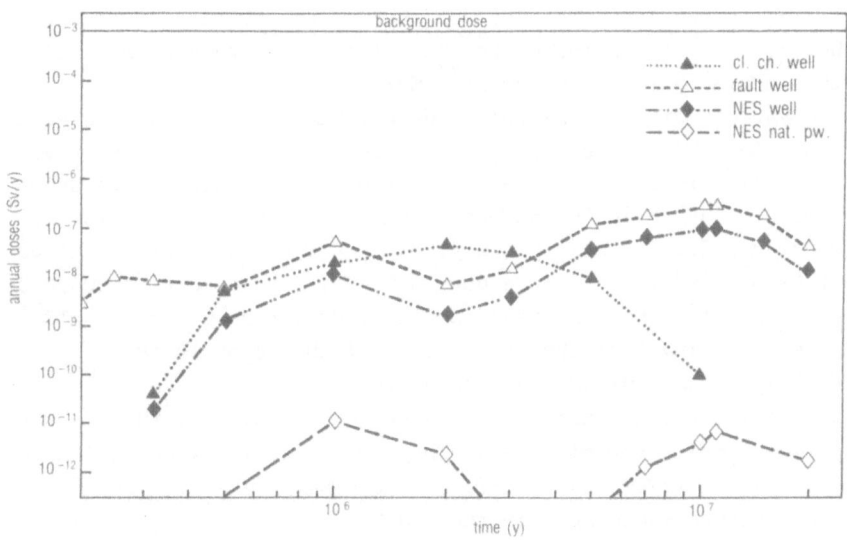

Figure 2. Comparison of total individual dose rates for the climatic change, fault and normal evolution scenarios at Mol.

member of the critical group is about 10^{-9} per year.

The maximum dose rate for the river pathway results to be much lower, of the order of 10^{-11} Sv/a and is equally due to Np-237 and Tc-99. The contribution of the marine compartment is negligible.

For the climatic change considered, the doses calculated are about a factor 4 higher than the corresponding doses in the case of the NES due to the smaller dilution which takes place at the clay-aquifer interface.

Deterministic calculations were carried out for faults occurring at an early time (1000 years) and at a representative time (10^5 years). Only the well water pathway to man has been considered. The maximum dose rate of 4×10^{-8} Sv/a occurs after about 2 million years and is due to Np-237. The maximum dose rates are of the same order of magnitude as those due to the normal evolution scenario. This is because only a relatively small portion of the repository is affected by the fault. The best estimate of probability of occurrence of the scenario is 10^{-4} per year. The associated risk would be about 5×10^{-14} per year.

Local sensitivity studies showed that the calculated dose rates are very sensitive to the parameters which are related to the transport of radionuclides through clay: clay layer thickness, diffusion coefficients, retardation factors.

Stochastic calculations were also executed for the selected evolution scenario, taking account of the intrinsic variability of several uncertain parameters. Fig. 3 shows the histogram of individual dose rate after 5 million years for the case of the normal evolution scenario (well pathway). The figure indicates that the doses vary over a wide range, with a median value of about 10^{-7} Sv/a.

For the variant site of Harwell, deterministic calculations were only performed for the most significant radionuclides. For the NES river pathway the maximum individual dose rate has been calculated to be about 10^{-6} Sv/a, mainly due to Tc-99 and occurring one million years after repository closure. For the well pathway, the maximum dose rate resulted to be about 5×10^{-5} Sv/a, once more due to Tc-99, and occurring 900,000 years after disposal. The corresponding individual risk would be 7×10^{-7} per year, when assuming that the probability of the scenario is 1.

The following conclusions could be drawn :

a) Only extremely low individual doses may be expected, and these occur at very long times after disposal. They are essentially due to three nuclides. Np-237, Tc-99 and Cs-135, as the others decay to insignificant levels before escaping from the clay.

b) The design parameters and the near-field characteristics do not influence significantly the overall performance of the system. Parameters related to the far-field characteristics are, on the contrary, very significant. The clay thickness and the retention parameters are

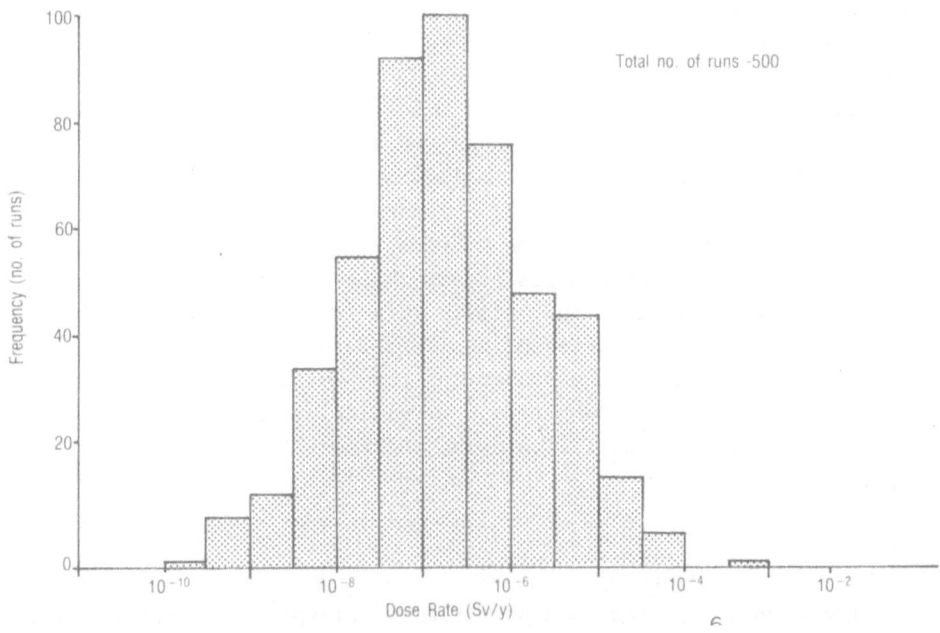

Figure 3. Histogram of individual dose rates at 5x10^6 years for the normal scenario (well pathway) at Mol.

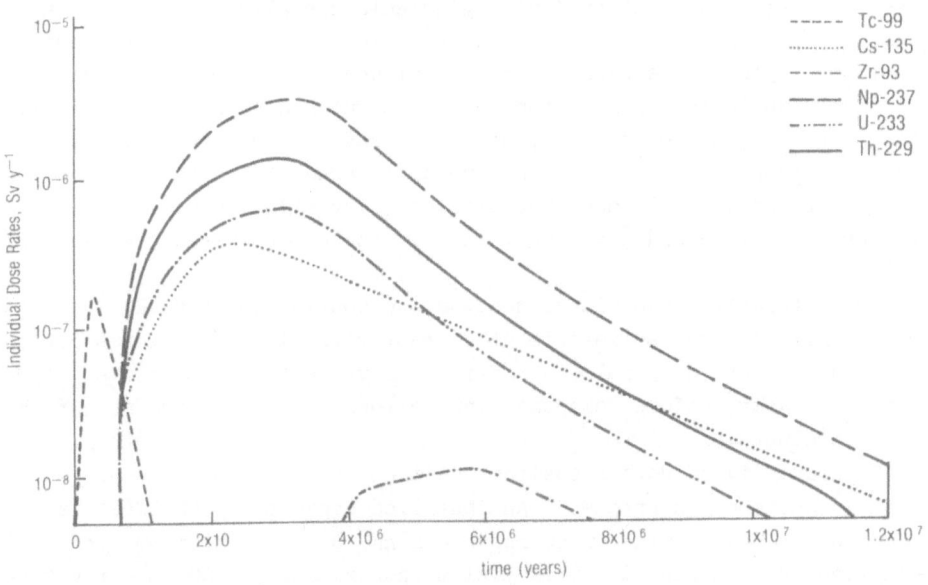

Figure 4. Deterministic evaluation of individual dose rate evolution for the normal evolution scenario at Auriat with contributions by important radionuclides.

the most relevant ones.

c) The uncertainty which affects the results of the analysis has been quantified using statistical methods.

d) The altered evolution scenarios are not of great importance, either because their consequences would not be larger than those calculated for the NES, or because the probability of occurrence is fairly low.

e) For the variant site of Harwell, higher dose rates have been calculated, due to differences in the hydrological properties of the aquifers. Very large uncertainties affect these estimates, due to incomplete characterization of the site. This confirms that a detailed characterization of the adjacent aquifers is of great importance.

f) A clay formation can ensure protection for a high-level waste repository even in the very long term, provided that its characteristics are favourable, and a sufficient clay thickness is present around the repository.

B.2. Granite

Granite has been considered as a potentially suitable host rock for disposal of HLW because it is commonly found in large stable formations and the permeability of the rock, if not fissured, is very low. Thus groundwater flow is expected to be low and the flow path to the biosphere, or to more permeable rocks, is long. Massifs with a small incidence of fissures, and sited so that hydraulic gradients are also small, are preferred. Some degree of fissuring or fracturing is to be expected, but at repository depths these will be mainly closed by lithostatic pressure. Potentially suitable granite formations, which have been geologically stable for long periods of time, occur in many EC countries. The reference site chosen in PAGIS is at Auriat in France, which has been the subject of geological and hydrological investigations. A site at Barfleur in France and a notional site in the United Kingdom were taken as variants considered.

Two alternative repository designs are considered: case A, the reference, which assumes an interim storage of vitrified waste for a period of 30 years before disposal; and case B, a variant, which assumes storage for 100 years before disposal and allows a closer packing of the waste containers.

The capacity of both repository designs is based on the projected nuclear programme in France. An installed capacity of 60 GW(e) is assumed and the HLW arising from reprocessing of fuel from 30 years of production is considered. This yields approximately 5,000 m^3 of vitrified waste and is equivalent to 48,000 t HM.

Different radionuclide release and migration scenarios have been identified and analysed.

In the normal evolution scenario, radionuclides are released following resaturation of the repository by groundwater. Corrosion of the containers and leaching of the glass waste-matrix can then occur. Radionuclides will migrate through the compacted bentonite of the backfill mainly by diffusion. Migration in the granite is expected to proceed by groundwater-mediated transport via fractures and porosity.

Although the granite massifs selected have little known mineralogical interest, it is possible that resource exploration and mining activities could proceed at times in the future and might result in loss of containment or perturbation of the hydrogeological regime. As a consequence potential exposure of workers and to surface dwellers may occur.

Exposures due to human intrusion are evaluated.

Other altered evolution scenarios, like those linked to repository effects and fracturing processes have also been considered.

A number of preliminary studies were carried out in order to obtain a better understanding of the importance of the involved phenomena and to provide a basis for adoption of simplifying assumptions in the calculations of radiological impacts. The individual aspects studied are the:

a) Radionuclide selection;
b) Radionuclide migration in the near-field;
c) Comparison of 2-D and 3-D groundwater flow calculations;
d) The influence of fractures on flow and migration;
e) Biosphere pathways.

Preliminary studies allowed the overall model to be calibrated, in order to estimate dose impact for the various scenarios.

For the Normal Evolution Scenario, individual doses have been calculated for each of the three discharges to the biosphere (Les Pallands, Les Trois Dieux, Le Nouhard). "Les Pallands" is identified as the dominant route. Estimated fluxes of Cs-135, for example, at the other discharge points are four orders of magnitude lower.

Estimates of individual dose due to Cs-135, Tc-99, Zr-93 and the Np-237 chain at the major discharge, are illustrated in Fig. 4. A comparison of total dose from all radionuclides for the two repository concepts is given in Fig. 5. The shift in the time of maximum dose, and its attenuation, in the case of the more compact variant repository concept, is due mainly to the longer trajectory in the geosphere, and to a lesser extent, to additional retardation due to a greater thickness of the bentonite backfill.

Drinking water is the predominant exposure route for an average individual in a region which derives all its water from the contaminated source.

For the reference repository concept the peak dose rate, of 5×10^{-6} Sv/a at three million years is due to Np-237 and its daughters. Np-237 also dominates in the case of the variant concept, for which the

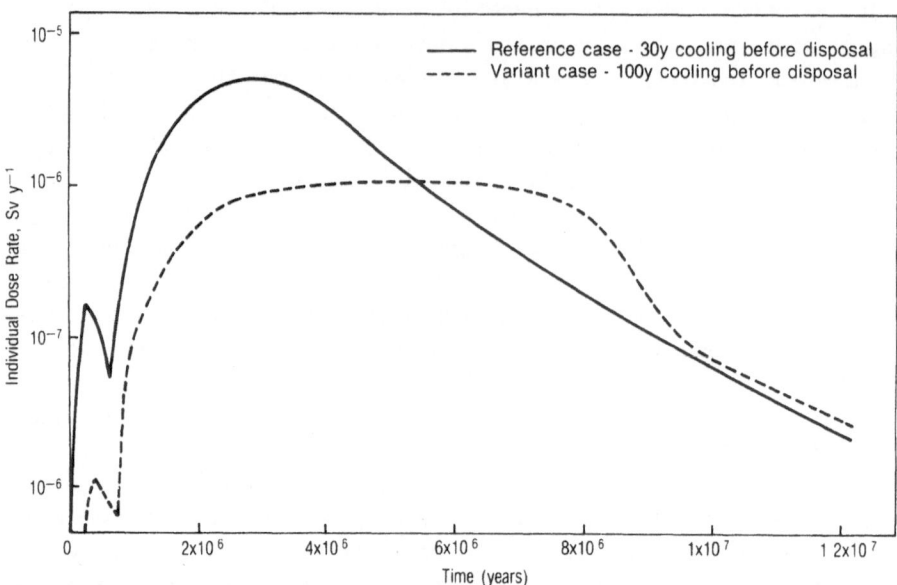

Figure 5. Comparison of total individual dose rate evolutions for reference and variant repository concepts at Auriat.

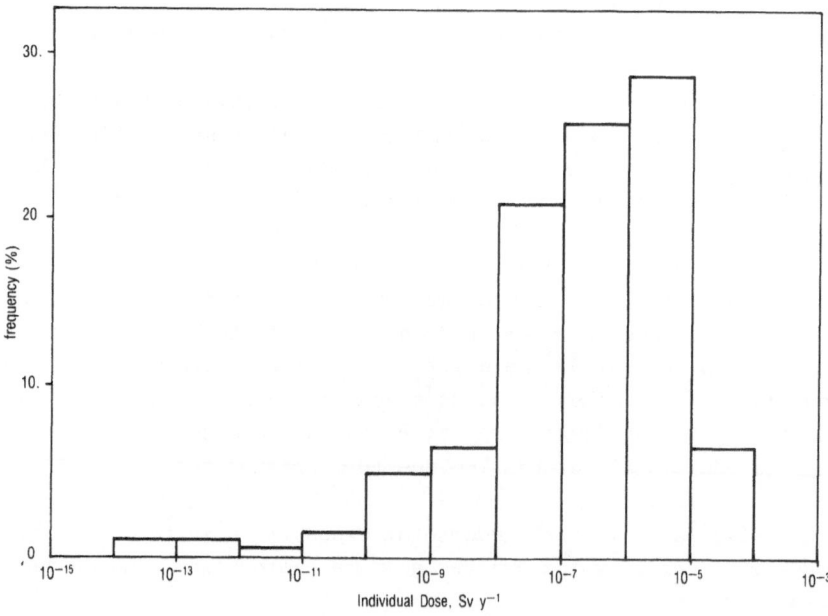

Figure 6. Histogram of individual dose rate due to Np-237 at 5×10^6 years for the Auriat site.

peak individual dose rate is $1x10^{-6}$ Sv/a between three and six million years after disposal. The earliest computed doses are due to Tc-99 and only occur after a few hundred thousand years.

In the human intrusion scenario, the dose for mine workers due to the actinides Np-237 and Th-229 predominates and the maximum equivalent individual dose is $2x10^{-3}$ Sv/a at 100,000 years. This dose rate is high, but comparable with doses received by underground miners due to natural radioactivity in rock. For surface dwellers the maximum individual dose is three orders of magnitude less, $1x10^{-6}$ Sv/a due to vegetable consumption, also at a time of 100,000 years and dominated by Np-237. Dose due to milk consumption peaks at 50,000 years, dominated by Tc-99 but is of over an order of magnitude less than dose by vegetable pathways.

An uncertainty analysis was performed for the normal evolution scenario.

Fig. 6 shows the histogram of individual dose rate due to Np-237 at $5x10^6$ years (Np-237 is the dominant radionuclide in determining total dose). This indicates a median dose of about 10^{-6} Sv/a with no recorded doses above 10^{-4} Sv/a.

For the variant site of Barfleur (coastal granite), it was shown that doses similar to those calculated for the reference site may be expected for the terrestrial discharges, while the doses due to marine pathways are four to five orders of magnitude lower, because of the great dilution which takes place in the sea.

For the notional site in U.K. (granite under sedimentary cover), the important role of the sediments in delaying and decreasing the radioactivity peak was put into evidence. Sediments are capable of decreasing the doses by at least an order of magnitude, due to the high sorption capability of the materials.

In conclusion, the calculations performed show that no significant radioactivity is expected to reach man before one million years and the peak dose rate should occur at 3 million years after disposal. The sensitivity of the results to the retardation factors in granite and to granite permeability is great.

The following remarks can be made on the granite option:
- Uncertainties may be larger in comparison to the other continental options, since, even for the reference site of Auriat, data on the formation properties were extrapolated from measurements obtained from a limited number of boreholes.
- The role of altered scenarios deserves further investigation, especially the consequences of a defective repository sealing.
- It appears that the variant repository design, which is suitable for wastes which have been allowed to cool for a longer period of time before disposal, corresponds to an improved performance of the disposal system.

- The results provided by the analysis suggest that HLW may be adequately confined in granite, even allowing for the present uncertainties on the geological and hydrogeological characterisation.

B.3. Salt

The particular characteristics which make salt a suitable host rock for the disposal of HLW are the dry environment, its low porosity and permeability, and, consequently, the negligible water flow. The natural convergence properties of salt result in the sealing of any faults which might provide a path for the release of radionuclides. Salt formations which are potentially suitable for HLW disposal are widespread within the EC and occur as salt domes or bedded salt formations. The salt dome near Gorleben in the FRG, where extensive investigations are taking place, has been taken as the reference site. As variant site a bedded salt formation in France has been selected.

The HLW characteristics chosen for the salt option reflect the situation in the FRG, where the vitrified glass is assumed to be derived from different fuels and has a different composition from that used for the other disposal options. The main difference from the PAGIS reference is that a higher loading of waste oxides is achieved and that some contribution has come from the reprocessing of MOX (mixed oxide) fuel.

The capacity of the repository is designed to accomodate approximately 9,000 m^3 of vitrified waste expected to arise from the reprocessing of fuel from 50 years operation of an installed nuclear capacity of 50 GWe in the FRG. This will be disposed of in about 60,000 canisters each containing the reprocessed fuel equivalent of 1.15 t HM.

In the reference design HLW canisters are disposed in unlined bore-holes drilled vertically downwards from the disposal level. The stacking height of HLW canisters is limited to 300 m. The bore-holes will be closed, without backfilling, with a plug which consists of compacted crushed salt or salt-concrete. The individual disposal drifts will be backfilled with crushed salt and sealed from the rest of the operating mine by a horizontal sealing of salt-concrete.

The following scenarios were identified as the most relevant and analysed.

The normal evolution assumes that within a short period of time the consolidation of the backfill and sealing will be complete and the existing annular gap between the borehole and steel canisters containing the waste will be closed by convergence due to the high lithostatic pressure and thermal expansion of the salt. Under these conditions a release of radioactivity from the salt in the medium term is not possible because the very low permeability and porosity makes the diffusion and brine migration processes too slow. In the very long term, dissolution of the rock salt by groundwater is assumed to occur at the top of the salt dome (subrosion scenario). As a consequence of subrosion and halo-kinetic uplift the HLW field is continuously raised and eventually reaches the top of the dome where canister, glass and salt are dissolved and radioactivity and salt are transported through aquifers to the biosphere.

As an altered evolution scenario, ingress of brine from an inclusion into a recently filled borehole, coupled with ingress of water via an anhydrite vein might potentially allow activity to leave the repository since convergence in the borehole would be retarded by the presence of the brine inclusion. In this scenario contaminated water is assumed to leave via the anhydrite vein, to traverse the overburden and to reach the biosphere. Dispersion of radionuclides in the overburden is predicted from the knowledge of the groundwater flow. The probability of such a scenario is, of course, very low.

Finally, three kinds of human activities which can lead to an unin-intentional contact with radioactive wastes have been identified. These are mining, borehole drilling and cavern leakage after solution mining. Only this last, expected to be the most significant, has been analysed in detail. For the near future it is assumed that the government of the region would limit disturbance of the repository area. However, such controls cannot be guaranteed to remain effective for more than a few hundred years, since records are liable to be lost.

For the normal evolution scenario, a simple model is used to calculate the radiological consequences. Because of the difficulties of predicting the subrosion rate, conservative assumptions have been made in modelling the structure and properties of the future overburden and groundwater movement. Alternatively prediction uncertainties have been quantified by assigning distribution functions to the system input parameters.

During the first phase of the scenario, the salt dome is eroded to the repository depth. During the second phase, radionuclides (as well as salt) pass into the water and can be transported to the biosphere. Removal of the salt is by way of convective leaching, fluid transport and diffusion and dispersion processes. The concentration of the saturated brine would have to be diluted with 12,000 times the volume of pure water for the salt content to be acceptable as drinking water, for irrigation and for watering livestock. Sorption effects and the above dilution are taken into account in the calculation of radionuclide concentrations in the water consumed.

The subrosion rate in this simplified model determines the time and release of radionuclides from the repository. In the model the dose due to each radionuclide is directly proportional to the activity at the time of release. Therefore, the subrosion rate also determines the ac-

tivity which remains at the time of release. For the deterministic calculations a subrosion rate of 0.033 mm/a was adopted, leading to a time
of release 15 million years in the future.

Although the relevance of a dose impact at such a time span is
questionable the sum of dose rates from all radionuclides that would be
released as a result of this scenario is 10^{-6} Sv/a; the main contribution
to this dose is from Np-237, followed by U-234.

The subrosion rate is the most critical parameter for this scenario, followed by the salt concentration in drinking water. Varying these
parameters within realistic ranges does not result in unacceptable
doses. The low value of the doses and the huge time span involved in the
release show that the normal scenario for the salt option is not the
most critical one.

The extremely conservative altered evolution scenario involving an
undetected brine pocket has been the object of extensive investigation.
Even in the assumption that the brine is so near to the affected disposal borehole to cause contact of water with waste after 10 years, the radioactivity would not be released, since the flank drifts would be completely closed by convergence before the contaminated water reach them.
A release of water to the groundwater would only be possible if - additionally - the water from the anhydrite vein fills the porosity of the
repository galleries immediately after the closure of the repository.

This, in turn, involves contemporary failure of the plug systems.

The peak dose of this scenario (which has been given the name of
"early water intrusion" would be 9×10^{-6} Sv/a at 20,000 years after disposal (Fig. 7), to be associated with a probability lower than 10^{-6}.

For the human intrusion scenario an underground storage cavern is
assumed to be created by solution mining techniques. After its fall into
disuse it becomes filled with water. The water containing the radionuclides released from 115 HLW canisters is then expelled through the entrance shaft by the convergence of the salt. The release rates into the
overburden of a large number of nuclides have been calculated and show
maximum release rates at around 10^5 years. The cavern is estimated to
attain its final porosity at 1.4 million years, after which time releases are minimal.

Fig. 8 shows peak doses from the most important radionuclides. The
maximum dose of 3×10^{-5} Sv/a is given by Np-237 at 700,000 years. The
probability of the scenario is difficult to assess since it depends on
human actions at long times in the future.

Results for the whole set of scenarios are summarised in the Tab.2.

The following remarks can be made on the salt option results:
- Due to the salt convergence, it is difficult to formulate a scenario
  that results in radioactivity release. To produce a release, very pessimistic assumptions need to be made. The scenario probabilities,

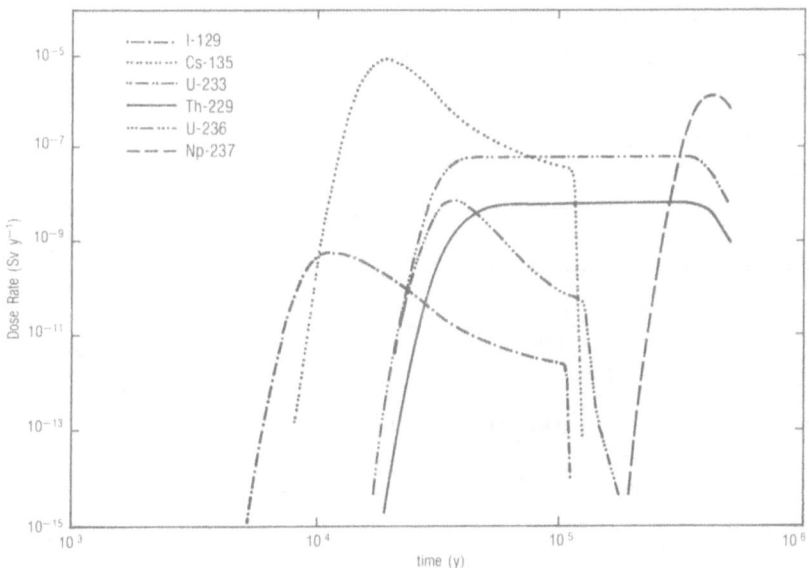

Figure 7. Deterministic evaluation of individual dose rate evolution for the early water intrusion scenario at Gorleben, showing breakdown by radionuclides.

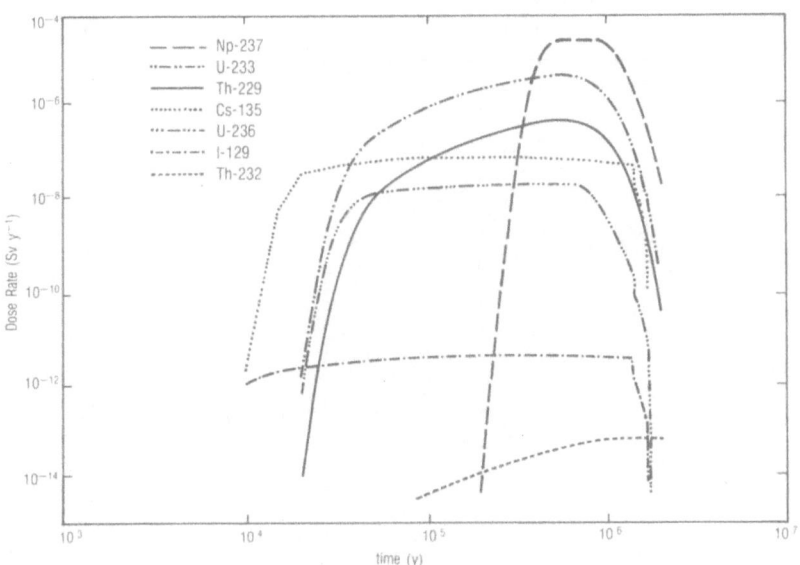

Figure 8. Deterministic evaluation of individual dose rate evolution for the human intrusion scenario at Gorleben, showing breakdown by radionuclides.

TABLE 2. Deterministic evaluation of individual dose rate for normal and altered evolution scenarios for the reference and variant sites.

Site	Scenario	Principal nuclide	Time of maximum (y)	Dose rates (Sv y^{-1})
Gorleben (reference site)	Subrosion (normal evolution)	Np-237	1.5×10^{7}	1×10^{-6}
	Early water intrusion (anhydrite vein)	Cs-135	2×10^{4}	9×10^{-6}
	Human intrusion (storage cavern)	Np-237	7×10^{6}	3×10^{-6}
Bedded salt in France	Standard release 1. Upper bound permeability	Cs-135	3×10^{4}	5×10^{-6}
	2. Lower bound permeability	Cs-135	2×10^{6}	6×10^{-6}

though very low, are difficult to evaluate;

- There is no great difference between doses from normal and altered evolution scenarios; yet the former takes place to such a late time (15 million years) to make its relevance dubious.
- Doses from altered evolution scenarios may take place at earlier times (e.g. 29,000 years for the anhydrite scenario), even in comparison with the other options. These doses appear acceptable for all the scenarios, even neglecting their low probability of occurrence.
- Salt convergence rate is the most relevant parameter for all the altered evolution scenarios. Only one design parameter, namely the dam permeability, was identified as influential (in the anhydrite scenario).
- Because of the overwhelming influence of the convergence rate the near field and the overburden parameters always appear of secondary importance.

B.4. Sub-Seabed Option

The particular characteristics which make the sub-seabed sediments suitable for the disposal of HLW are that they have low permeability, negligible pore water velocity and high absorptive capacity, thus restricting the rate of migration of radionuclides from the waste to the seabed. The long period of time before release from the sediments will result in effectively zero release for most radionuclides. If release occurs, dispersion within the large volume of the oceans will further dilute the radionuclides before appearance in the accessible environment. Layers of sediments of suitable composition and adequate thickness are widely situated in areas of the oceans which are readily accessible. A number of such areas have been identified within the Atlantic Ocean as a result of extensive international research programmes. The reference area chosen, Great Meteor East (GME), is situated on the Madeira plain in the North East Atlantic and has been extensively investigated. The areas chosen as variants are the Cape Verde Rise (CV2) and the Southern Nares Abyssal Plain (SNAP).

For sub-seabed disposal of HLW the repository is not constructed; rather, the portion of the site used and the positioning of the wastes, is determined by the emplacement method employed. Two alternative emplacement methods are considered: use of freefall penetrators and emplacement in drilled boreholes.

For the sub-seabed option the HLW arising from a power program of 3,000 GW(e) years is considered. This is approximately that associated with the next 30 years of nuclear power in Europe assuming an installed capacity of 100 GW(e). The volume of waste to be disposed of is approximately 11,000 m^3, corresponding to 100,000 t of heavy metal.

The normal evolution scenario for subseabed disposal is based on the slow degradation of the vitrified waste, diffusion of radionuclides in sub-seabed sediments, dispersion of any emerging radionuclides in the ocean and exposure to man.

For the reference emplacement (penetrator) option, it is assumed that all the penetrators are successfully buried at a depth of 50 m in the sediment and that hole closure is complete. For the variant emplacement (drilled) option, all the waste is assumed to be emplaced at depths between 350 and 750 m.

As altered evolution scenarios a total of 9 processes that could have a deleterious effect on the containment of the waste in the seabed sediments were identified, and estimates were made of the likely effects, the number of penetrators involved and the probability of the occurrence. The events and their consequences were used to identify a total of 7 possible altered evolution scenarios and their probability of occurrence (Tab. 3).

Ocean models, of different complexity were used for deterministic calculations and for uncertainty analysis.

Best estimate evaluation for the normal evolution scenario gave peak doses of approximately 6×10^{-10} Sv/a, corresponding to peak risks of 10^{-11} a^{-1}.

Fig. 9 shows the breakdown by radionuclide for the dominant fish ingestion pathway. Se-79 and Tc-99 are the dominant contributors to the peak individual dose. Consumption of (mid-depth) fish is the dominant pathway until 2×10^{5} years after disposal, after which consumption of molluscs, and later, external exposure to beach sediments dominate.

The peak doses from the most important radionuclides arise between 10^{5} and 2×10^{6} years after disposal. By the time the maximum individual dose is delivered the radionuclide concentrations are nearly uniform throughout the oceans.

Collective dose rate peaks at 1×10^{5} years, as does individual dose rate and the collective dose commitment is essentially complete at 1×10^{6} years (Fig. 10).

For the altered evolution scenarios, the doses were combined with the probability of the scenario and the probability of a health effect per unit dose in order to give an estimate of risk.

The predicted risks to individuals from these scenarios range from 9.8×10^{-19} y^{-1} to 3.8×10^{-10} y^{-1}. Scenarios 1, 2 and 4 (Tab. 3) give greater risks than the normal evolution scenario. The risks from the first four scenarios are dominated by mining of minerals from seabed sediments and since the probability and consequence of this event are very uncertain, the corresponding risk estimates are very uncertain.

Sensitivity studies were performed using the best estimate models, and varying one parameter at a time. Near field, far field and biosphere

TABLE 3. Altered evolution scenarios of sub-seabed disposal by penetration emplacement.

Scenario number	Description	Time years	Probability per penetrator
1	Undamaged penetrator on seafloor	$<10^4$	$1.2 \cdot 10^{-2}$
2	Damaged penetrator on seafloor	$\leq 10^3$ 10^3-10^4 10^4-10^5 $<10^5$	$2 \cdot 10^{-10}-5 \cdot 10^{-5}$ $1.2 \cdot 10^{-2}$ $6.8 \cdot 10^{-3}$ 0
3	Undamaged penetrator buried 10 m	$<10^4$	$8.6 \cdot 10^{-2}-8.9 \cdot 10^{-2}$
4	Damaged penetrator buried 10 m	$\leq 10^3$ 10^3-10^4 10^4-10^5 $>10^5$	$1 \cdot 10^{-5}-3 \cdot 10^{-5}$ $2.5 \cdot 10^{-1}$ $2.0 \cdot 10^{-1}$ $7.2 \cdot 10^{-3}-4 \cdot 10^{-4}$
5	Open fault near penetrator	$\leq 10^3$ 10^3-10^4 10^4-10^6 $10^{-6}-10^6$ $>10^6$	$8 \cdot 10^{-14}-3 \cdot 10^{-12}$ $5 \cdot 10^{-13}-4 \cdot 10^{-12}$ $8 \cdot 10^{-13}-3 \cdot 10^{-10}$ $7 \cdot 10^{-11}-3 \cdot 10^{-9}$ $7 \cdot 10^{-9}-3 \cdot 10^{-7}$
6	Human intrusion by by drilling (annual probability)		$7.7 \cdot 10^{-8}-3.8 \cdot 10^{-4}$ y^{-1}
7	Human intrusion by mining (annual probability)		1.0 y^{-1}

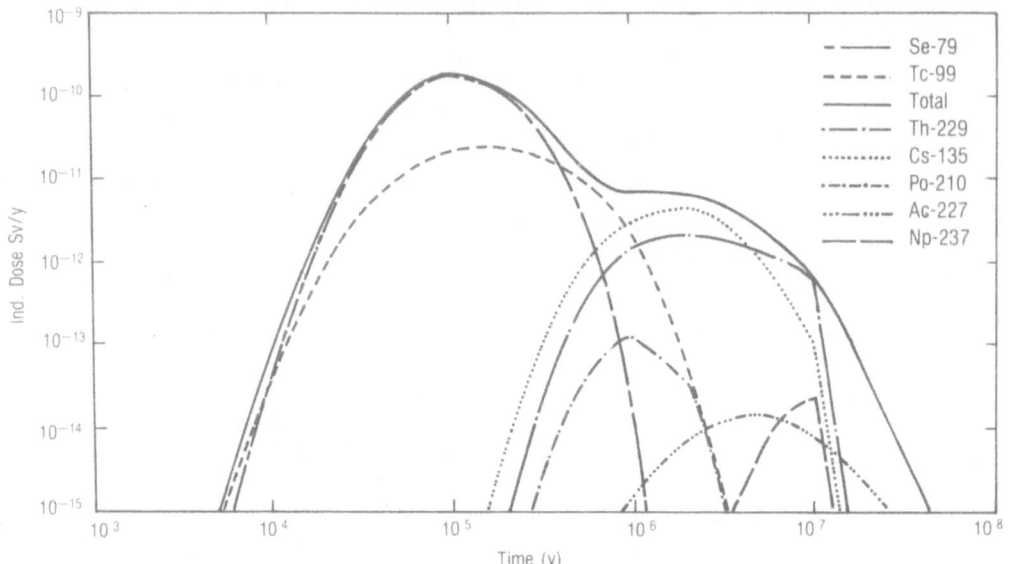

Figure 9. Deterministic evaluation of individual dose rate evolution at GME for fish ingestion pathway (breakdown by radionuclide).

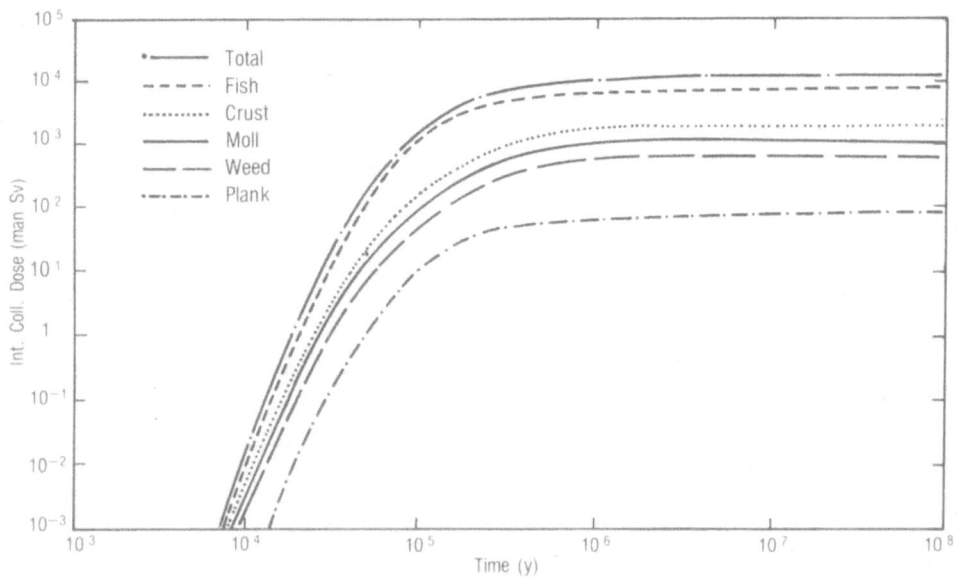

Figure 10. Evolution of integrated collective dose evolution at GME (breakdown by pathway).

were analysed in turn.

Individual and collective doses depended strongly on the various far field parameters. In particular, the retention factors of radionuclides in sediments are of great importance. This is particularly true for short-lived radionuclides, especially some actinide isotopes, which would contribute significantly to the dose if their sorption in the sediments were to be significantly reduced.

The peak individual dose rate for the variant site SNAP is 2.1×10^{-10} Sv/a via ingestion of mid-depth fish which occurs 1.0×10^{5} years after disposal and the dominant contributor to this dose is Se-79. The predicted peak individual doses from SNAP are identical to those for emplacement at GME.

For the alternative drilled emplacement method, the predicted peak individual dose rates are about four orders of magnitude lower than those of the reference method, falling around 10^{-14} Sv/a. This is due to the greater depth of emplacement within the sub-seabed sediments achieved by drilling.

From the above results it may be concluded that the post emplacement doses and risks from the disposal of high level waste in the areas considered, by either free-fall penetrators or drilled emplacement, are negligible.

C. CONCLUSIONS

The exercise is presently being terminated, with the publication of four detailed reports on each option, and a summary report.

An exhaustive analysis of the reports will be required in order to draw conclusions both on methodologies and on relevance of results obtained for possible safety criteria.

A few preliminary and often personal remarks can be proposed for reflection and discussion:

1. From a methodological point of view, the relative merits of the deterministic and the stochastic approach remain a debatable matter. Although in PAGIS a comparison of the results given by the two approaches is made difficult in some cases by differences in the data base, the two approaches are in principle compatible and could be used in a complementary way in risk assessment of real cases, as it has been done in PAGIS and in the Seabed Working Group.

2. The classification of scenarios as "normal evolution" and "altered evolution" scenario has proved effective in PAGIS. A more systematic approach to scenarios selection, justification and analysis is nevertheless a useful objective for future studies.

   Along the same lines, the effort put in analysing a scenario should ideally be proportional to the potential impact of the scenario.

Again a more structured approach to "scenarios analysis" might be a useful objective for future research on methodology. Another theme for reflection is the "cut-off" time, which should be chosen following a common approach.

3. Comprehensiveness and transparency seem competing parameters in risk assessment, even if none of them should be sacrificed in favour of the other. A more systematic and structured approach to presentation of results could be a useful subject for methodological research.

4. Radiation doses delivered to populations appear to be essentially zero for all options, sites and scenarios considered during several ten thousands of years after repository closure.
In the successive period, starting from a threshold year which is obviously dependent on options, sites and scenarios, there might be a slow but continuous increase of environmental contamination, which will eventually lead after hundreds of thousands or millions of years to maxima of radiation which will still be fractions of the present radioactivity background and of the limits accepted for radiological practices.

5. The most important protection appears to be provided by the natural geological barrier, which imposes radionuclide travel times long enough to allow a substantial decay and dilution before the biosphere is reached. Therefore, the factors that determine the travel time, such as distribution coefficients of the few dominant radionuclides in geological media, permeability of the geological media and hydraulic gradients, appear as the most important. This strongly supports the soundness of the geological disposal concept, which assumes that the geosphere will act as a barrier over geological periods of time. It also supports the present orientation of the national and EC R&D programmes on radioactive waste.

6. Repository design and engineering does not play a major role in the long-term safety of the repository, but there is a clear incentive to design the repository in such a way as to minimize the disturbance of the geological barrier. As examples, repositories in clay formations should have a flat design to maintain the largest possible thickness of the clay barrier, and excavation techniques for repositories in granite should not induce major fracturing of the near-field zone.

7. The engineered barriers play a major role in accidental situations, in the initial period of the life of the repository. However, their contribution is masked in the normal evolution of the repository by the predominance of the geological barrier.

8. Possible natural alterations of the normal evolution of the disposal system, due to events of a probabilistic nature like glaciation, tectonic displacement, etc., do not cause a significant increase of

the potential dose. In fact, these events rarely affect the entire repository system, and the long term stability of the waste matrix prevents an abrupt release of radioactivity. If the impact on man is evaluated in terms of risk, as recommended by the ICRP, the increase of dose is partially or even totally compensated by the low probability of the event, so that these scenarios do not seem to have a dominant role in the overall risk of waste disposal.

9. Human intrusions into or near the underground repository are unlikely to provide a dominant contribution to the overall annual risk because of their low probability of occurrence. However, they are important to consider because the resulting doses, although at acceptable levels, are likely to be higher than in the normal evolution scenario, and also because they present the possibility of releases of radionuclides at much earlier times than the normal evolution scenario. Therefore, additional work may be desirable on a site specific basis.

10. The choice of a repository site is very important, whatever the chosen geological medium. As an example the variants considered for granite bring additional safety elements to the isolation afforded by the host rock, namely the radionuclide retardation induced by a sedimentary cover or the special dilution and biological pathways due to groundwater emergence into the ocean.

This observation and the fact that the various geological disposal options are at different stages of scientific and technological development, underline the difficulty, if not the impossibility, of establishing some reliable ranking between these options, as far as radiological safety is concerned.

Section 6 - A smart approach: using natural and archaeological analogues

The Swiss "Project Gewahr" safety analysis is characterised by an extensive use of the so-called natural and archaeological (anthropogenic) analogues. The author succeeds in linking observations arising from widespread sources , from the corrosion rate of human artifacts to the morphology of natural reactors, into a convincing defense of disposal safety.

APPLYING NATURAL ANALOGUES IN PREDICTIVE PERFORMANCE ASSESSMENT
Part I: Principles and Requirements

I. G. McKINLEY
NAGRA
Baden,
Switzerland

ABSTRACT. This paper outlines the rôle of natural analogues in testing the models used in predicitive performance assessment of repositories for nuclear waste. Examples are given which indicate both the use of analogues and the limitations in their applicability.

Introduction

Quantitative performance assessment of a particular nuclear waste disposal concept (or any other industrial process, for that matter) involves utilisation of some kind of mathematical model. Nowadays, such mathematical modelling usually involves computer codes, and the multitude of individual processes which need to be considered are incorporated in sub-components of the safety assessment model chain (cf. Fig. 1). Despite the apparent complexity of such chains, the models represent a great simplification of reality and the question must continually be asked :

Q1. How complete is the model chain - are any important processes missing?

Depending on the particular safety assessment approach adopted, several different model chains may be used for the different scenarios considered. Alternatively, a range of scenarios may be treated by parameter variations within the basic model chain - possibly handled in a probabilistic manner. In either case, it is valid to ask :

Q2. How complete is the list of scenarios?

This question is particularly relevant when considering extrapolations over the timescales of concern which, depending on the waste type, may range from hundreds to thousands or millions of years. The component sub-models are usually derived from experiments or field observations over timescales which are constrained by practicality. These rarely extend over more than a year or so and may be very much shorter. The extrapolation in timescale will thus be from 2 to 5 or more orders of magnitude! A further important question is thus:

Q3. Are the processes considered modelled appropriately for the timescales involved?

Related to the consideration of temporal scales are the extrapolations in physical scale between the size of samples which can be practically characterised and the actual size of the feature to be modelled. Although especially true of the geology of the far-field, scale effects and spacial

A. Saltelli et al. (eds.), Risk Analysis in Nuclear Waste Management, 359–375.

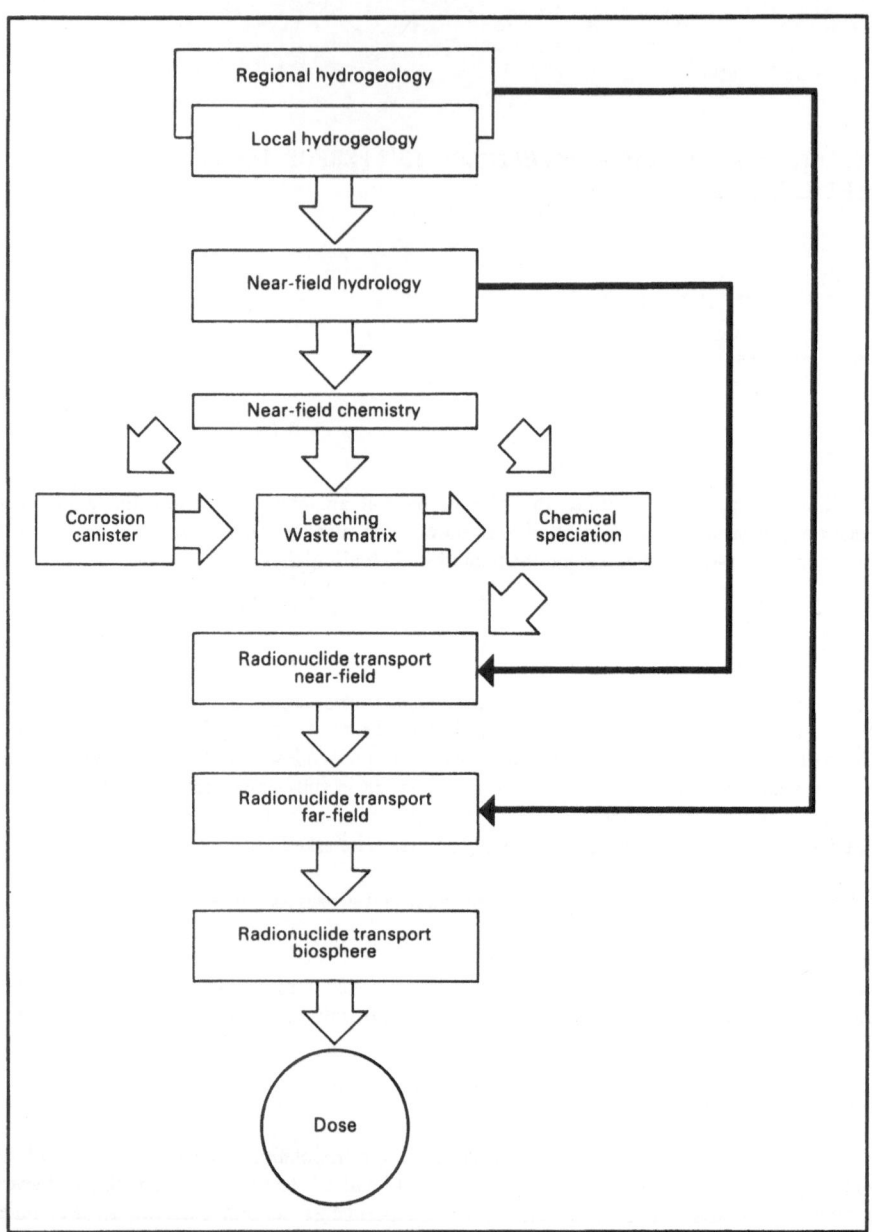

Figure 1. Model chain for safety analysis

heterogeneity are also to be expected in the engineered barriers of the near-field. One may then ask:

Q4. Are spacial heterogeneities taken into account in an adequate fashion?

In practice, most time is usually spent developing the mathematical models and associated computer codes and testing that the latter correctly solve the equations involved - i.e. verification. The problem of testing whether the models adequately simulate reality - i.e. validation - is much more difficult and includes detailed consideration of the questions specified above. No real examples of aged repositories exist, but it is possible to identify natural systems which involve processes which are similar to those considered in the repository assessment and which can be utilised to provide such tests. In the nuclear waste jargon, such systems are usually referred to as "Natural Analogues".

Classification of natural analogues

The definition of natural analogues above corresponds to the common usage but is somewhat vague and can be misleading. The essence of the natural analogue is the aspect of testing of models - whether conceptual or mathematical - and not a particular attribute of the system itself. Many papers can be found in the literature which are fairly conventional geochemical or isotopic studies but which claim to be "natural analogues" due to the relevance of the data produced to some aspect of nuclear waste management. Such usage is not useful and can be often attributed to attempts to obtain funding for a particular piece of research from "nucwaste" sources.

"Real" natural analogue studies can be classified in a number of ways depending on the type of model tested or the type of system examined (Fig. 2). On the model side, the study may be aimed at examining the completeness of the model chain or scenario list (questions 1 and 2 above) in a fairly qualitative manner or may attempt more quantitative analysis of specific processes (questions 3 and 4).

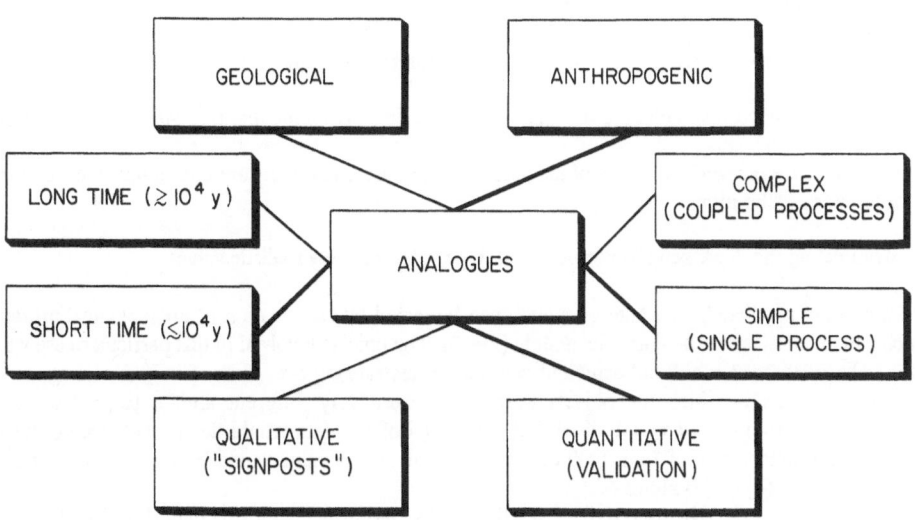

Figure 2. Classification of natural analogues

Again it must be re-emphasised that the approach adopted may be more important than the physical characteristics of a particular system in determining the extent to which it can be considered quantitatively. For example, quite superficial observation of an old, rich uranium ore body can be used to provide support to a model chain which suggests that the rate of leaching of spent UO_2 fuel would be very low in a chemically reducing environment with low water flow rates. Much more detailed characterisation of the same system might allow more quantitative testing of, for example, the temporal extrapolations involved in a fuel-leaching model or the spacial extrapolations in a radionuclide transport model. The quantitative approach may require an extensive amount of site specific support data and, in any case, may be inherently limited by either the time- (or distance-) scales or the complexity of the system studied. These two factors are generally linked but, although long timescales often allow more processes to become significant and hence increase complexity, short timescale processes are not necessarily simple.

As will be seen later, a vast range of "natural" systems involve processes analogous to those modelled for safety assessment. A further subdivision which may be useful at a pragmatic level arises from the nature of the process involved - usually either geological or anthropogenic. Given the emphasis on timescales, most of the latter could also be classified as archeological.

Some examples of analogues

The discussion above is somewhat sterile and the essence of analogue studies is best illustrated by a couple of examples.

One of the simplest examples involves an archeological study of iron corrosion rates. Iron or iron-based alloys are popular materials for the encapsulation of solidified waste of various types. Although the corrosion of iron has been extensively studied in the laboratory, the timescales of such studies are much less than the containment times desired - which may be up to a thousand years or so. The laboratory studies for almost all corrosion studies show an initially high corrosion rate which decreases with time. Although there are good theoretical reasons for such a trend, it is also known that, in a few cases, a further increase with time is observed. The safety assessment modeller could then choose a value for the "most reasonable" corrosion rate by:

a) taking the highest value measured, with the assumption that it will be at least conservative

b) taking the value corresponding to the longest reaction time, assuming that this will be more realistic

c) adding a safety margin (e.g. order of magnitude) onto option (a) to ensure conservatism even in the event of later acceleration

d) extrapolating the decreasing trends observed to get a "completely realistic value"

The corrosion predicted, by whatever model, can be tested by examination of archeological artifacts which can, in many cases, be dated accurately (Fig. 3). The objects involved in this particular case were from 1000 to 3000 years old and exhibited corrosion rates ranging over about 5 orders of magnitude. The environments in which the objects were found were very different and, if those from arid conditions and from sea water are discarded, almost all of the data would lie (to within an order of magnitude) around a rate of 1mm/1000 years. In general, such a value would be around that resulting from approaches (b) or (d) above.

This example can also be used to illustrate the caution with which analogue data should be handled. The numerical values presented (with appropriate error bars!) represent material compositions which may differ somewhat from those considered in disposal applications and, more importantly, the geochemical environment will not be exactly that expected in the near-field of a repository. More fundamentally, the corrosion rates observed may be artificially biased by sampling - objects of appropriate age but with higher corrosion rates might have totally decayed away and no longer be

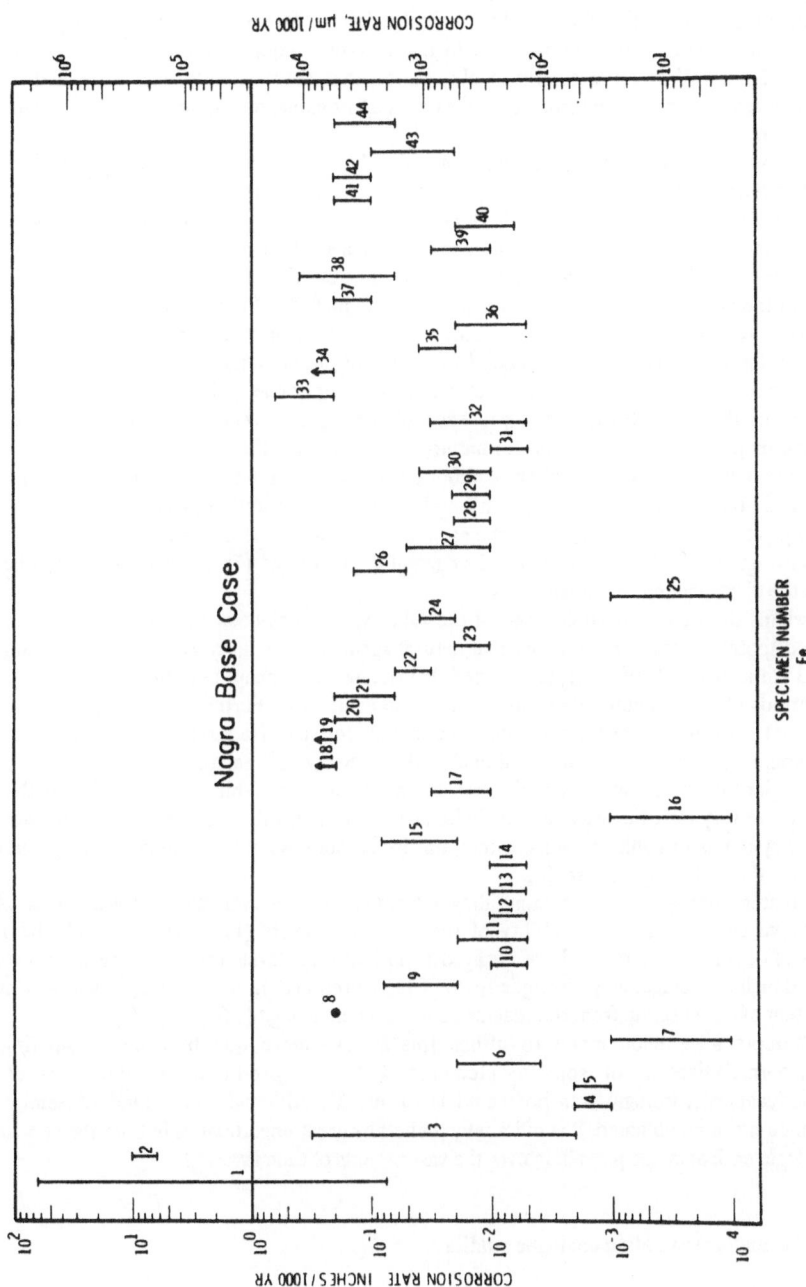

Figure 3. Corrosion rate data for iron artifacts (from PNL 3198, 1980)

recognisable.

These points are further illustrated for a similar data set for copper alloy artifacts (Fig. 4). Very few of these would have compositions close to the very pure copper considered as a possible canister material for HLW and the simplistic observation that there is little difference in the long term corrosion rates of copper and iron could well be unreasonable, due to inappropriate environments and sampling biases.

The examples above are relatively simple - involving a single process and relatively short, easily specified timescales. Nevertheless, it is clear that even here extraction of quantitative data is difficult.

At the other end of the spectrum is the well known Oklo "Precambrian nuclear reactor" which can be considered as a very long timescale analogue of a repository for spent fuel. Oklo is situated in Gabon, West Africa (Fig. 5), and contains a very old, very rich uranium ore body. To understand the origin of the reactor, it is important to realise that the half-life of the fissile isotope of natural uranium (^{235}U) is less than that of the major isotope (^{238}U) and hence its relative concentration was much higher in the past (Fig. 6). At the time of formation of the uranium ore body, about $2 \cdot 10^9$ years ago, the concentration of the fissile isotope was over 3% - comparable to that of enriched reactor fuels! The richness of the ore (50-70%) and presence of appropriate water contents for effective neutron moderation provided the conditions for natural initiation of a self-sustaining chain reaction (criticality). Intermittent operation in a number of different zones over a period of about half a million years consumed about 10 tons of ^{235}U, producing both actinide activation products (e.g. Pu isotopes) and fission products which are effectively the same as those produced in normal power reactor operation. The burn-up of the ^{235}U fuel results in its current depletion (cf. Fig. 6) which was the indicator which led to the identification of this site.

Perhaps the most widespread use of the Oklo system has been as a qualitative analogue of the entire disposal system. From the fact that natural spent fuel has fairly well retained its integrity over a period in the order of 10^9 years, it is argued that the overall concept of ensuring safe isolation of HLW for periods of "only" about 10^6 years ought to be achievable. Further extension of the argument, to imply that the existence of Oklo validates the models (or scenario lists) which suggest that such spent fuel should be very stable, is more difficult. Although several reactor zones have been found in the Oklo region, no other such examples are known elsewhere. While it can be noted that such old deposits are very rare, the argument could be inverted to imply that the lack of other reactors indicates that the system is unstable. To some extent, this is another case of the sample bias problem mentioned above for the archeological samples.

A better approach involves more detailed studies of the individual reactor zones which show impressive evidence for the stability of the uraninite matrix (Fig. 7a). Although the radioactive products of reactor operation have decayed to insignificant levels over the long time involved, their stable daughters remain as an isotopic fossil of their presence. In many cases, it is clearly evident that the extent of mobilisation from the reactor zone has been negligible (e.g. Fig. 7b).

Attempts have been made to utilise this system more quantitatively, comparing observed retention/mobilisation of specific elements with that predicted on the basis of chemical thermodynamic/hydrologic transport models (Table 1). Although some kind of semi-quantitative information can be obtained, it is inherently limited by great uncertainties in both the geochemical and hydrologic environments prevailing over the vast periods of time involved.

Selecting and/or evaluating analogue studies

The examples above may give some idea of the range of possible systems which can be utilised as analogues but they only represent the tip of a vast iceberg of potentially relevant studies. When looking for a suitable analogue for a particular process, it is important that the model to be tested is well defined as this will put constraints on the appropriate systems, e.g. in terms of system complexity, timescales, degree of quantitative treatment possible, etc. To further appraise the range of possible

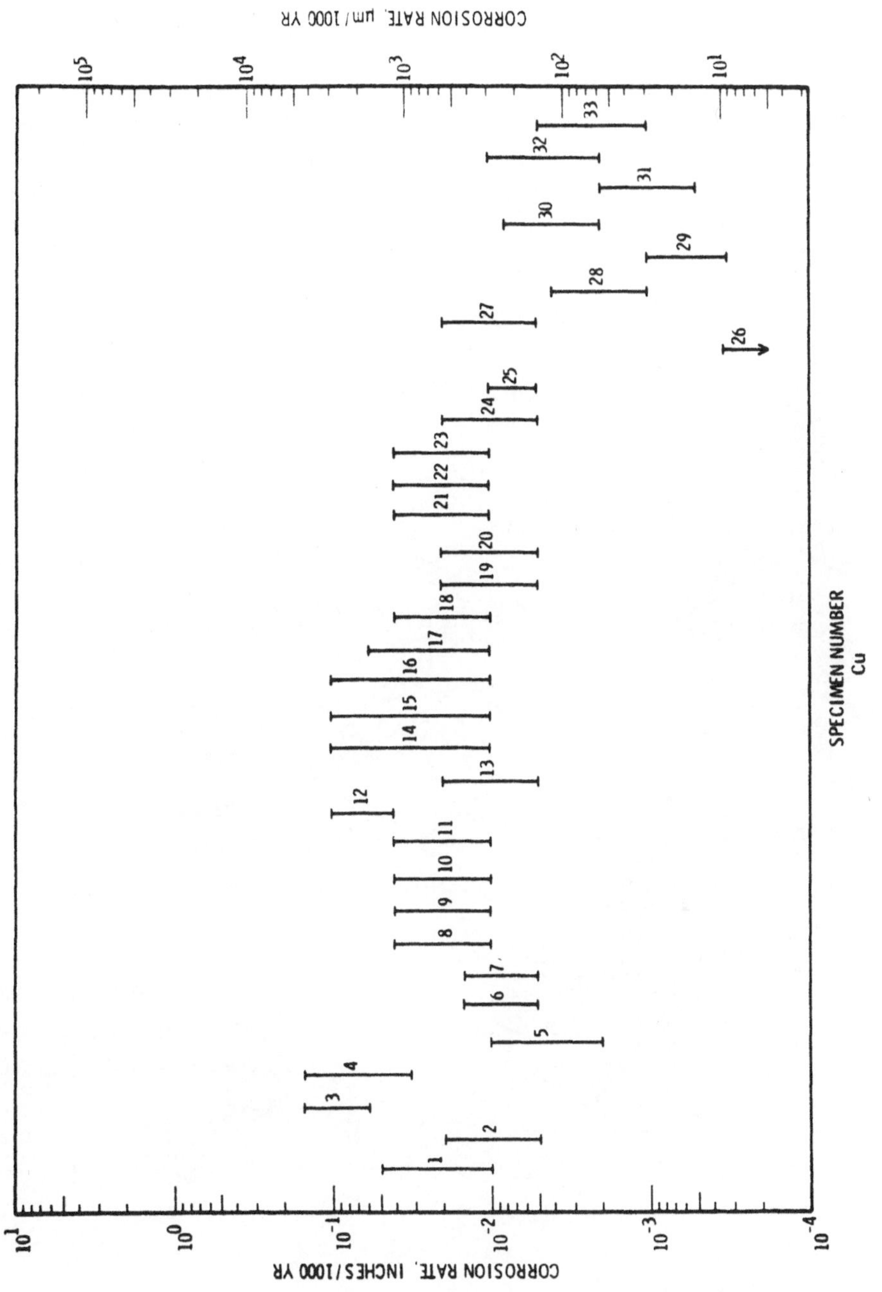

Figure 4. Corrosion rate data for copper alloy artifacts (from PNL 3198, 1980)

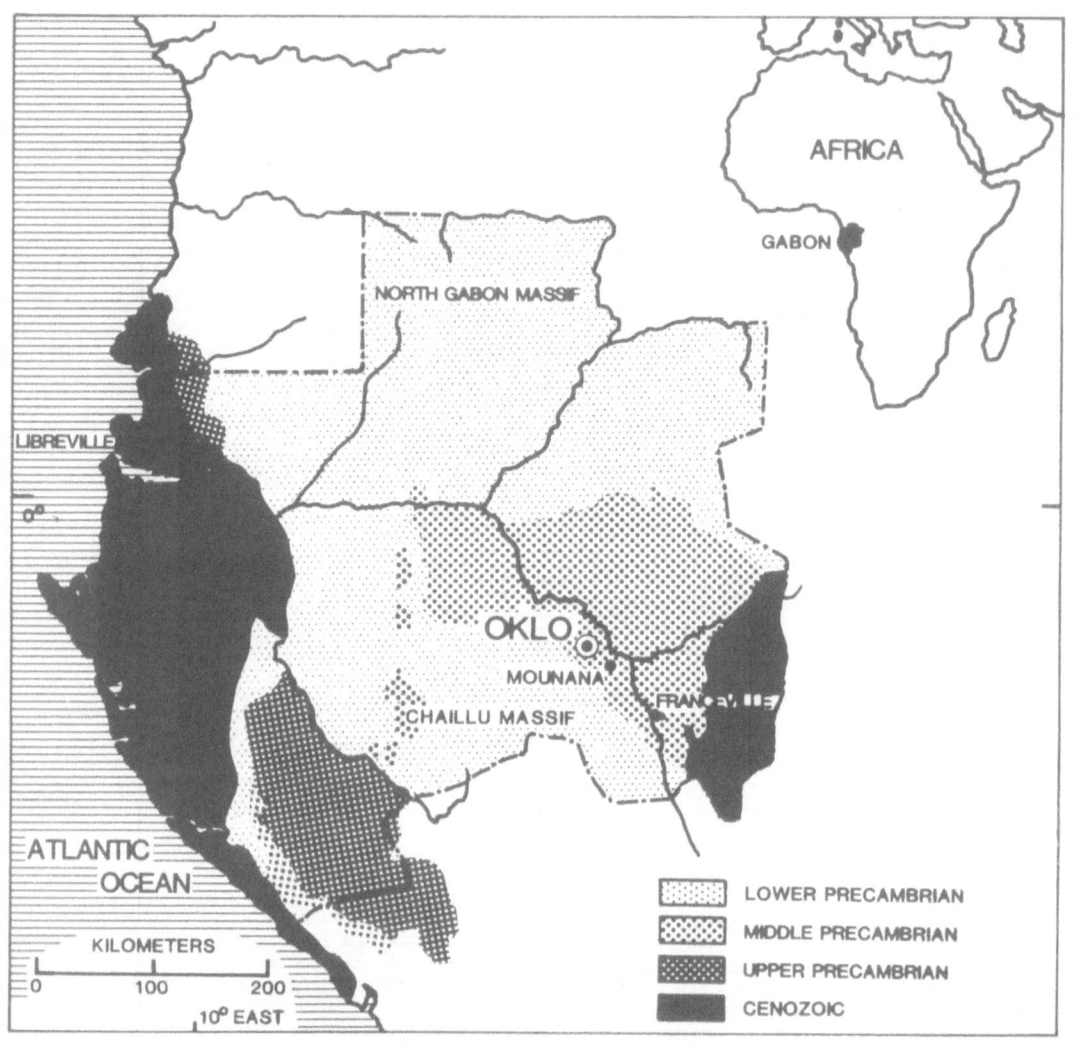

Figure 5. Location of the Oklo uranium mining area in Gabon (Africa)

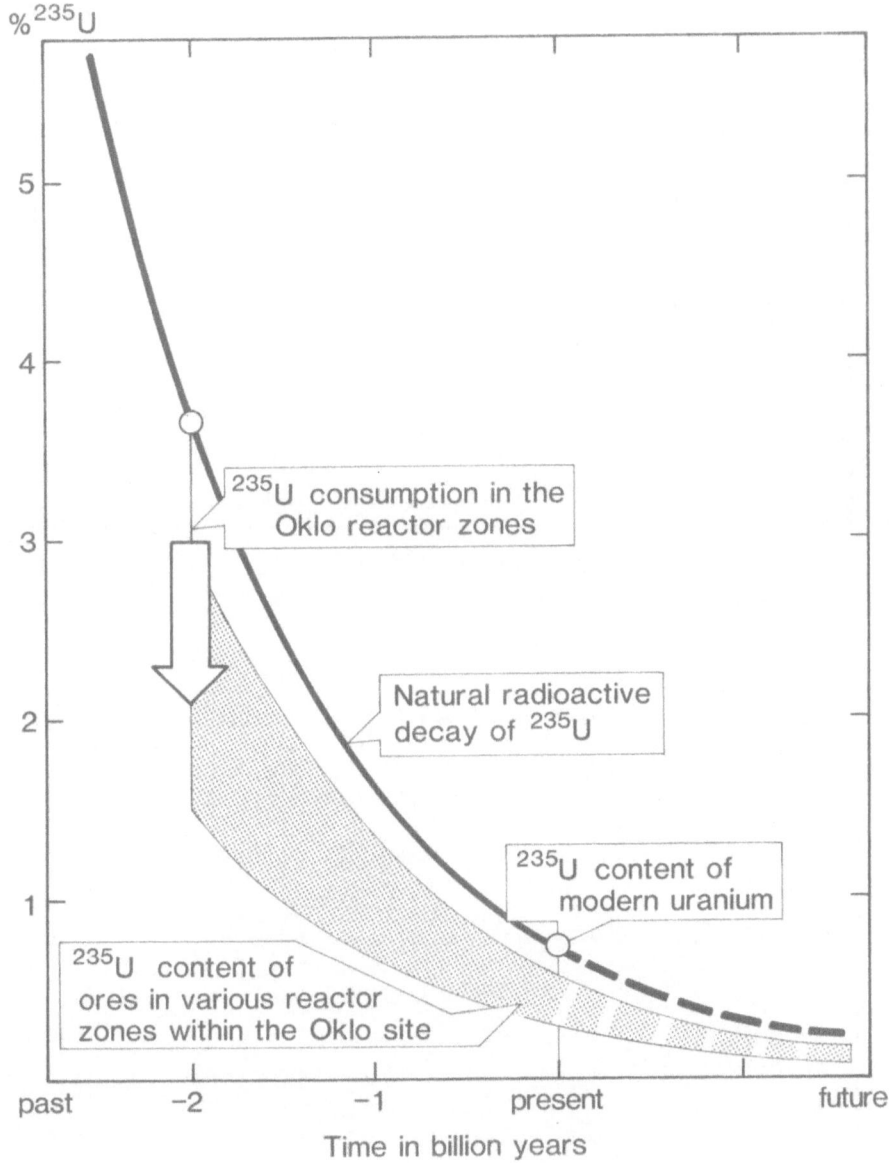

Figure 6. Evolution of the Oklo ore body

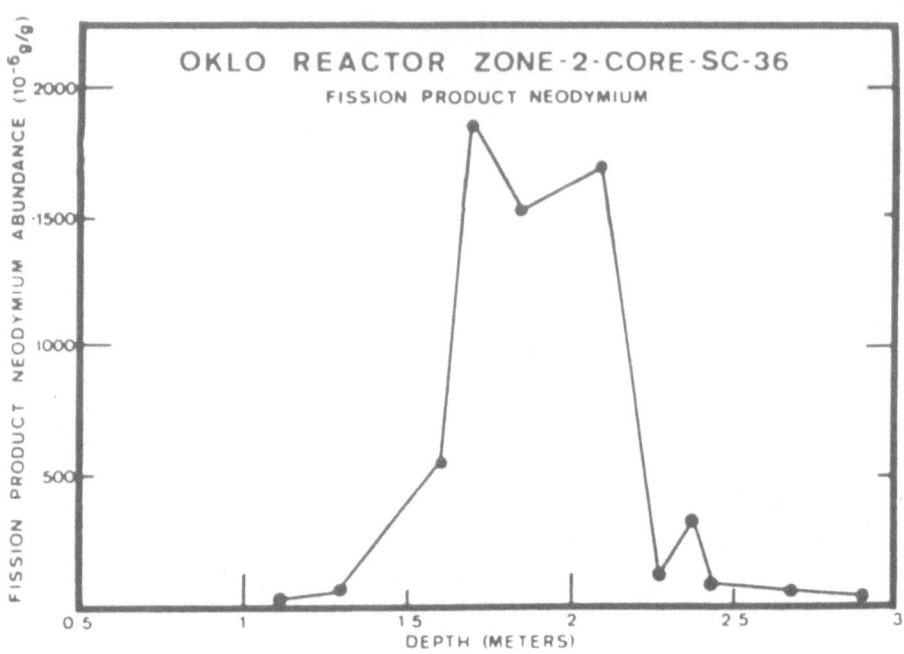

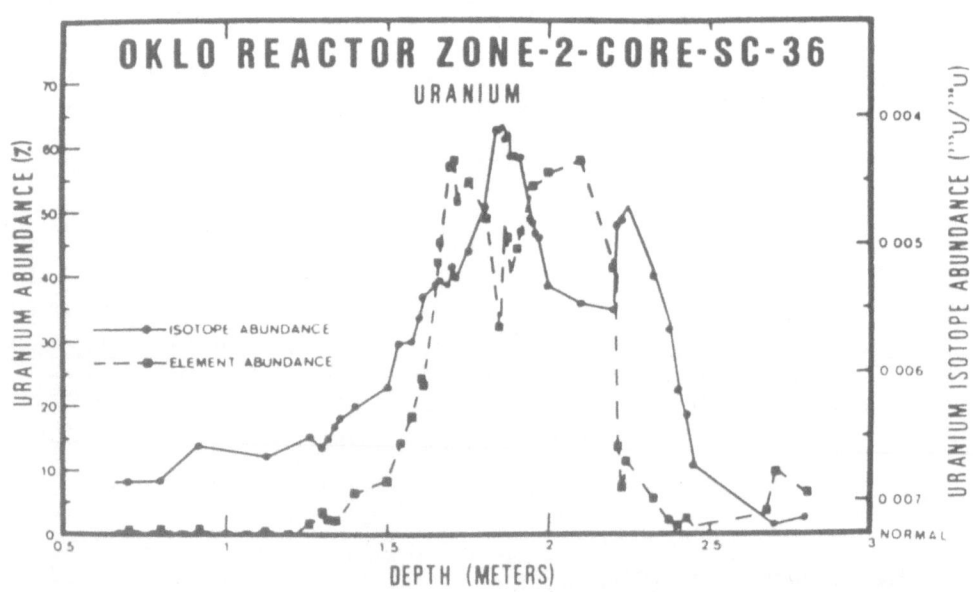

Figures 7a and 7b. Profiles of a) U and b) Nd through an Oklo reactor core (from SKB 85-04, 1985)

TABLE 1: Element retention in the Oklo reaction zones
(after Brookins, 1984)

Element	Comments	Eh-pH Prediction	
		25°C	200°C
Krypton	Most migrated	Not applicable	Not applicable
Rubidium	Probable local redistribution	Not applicable	Not applicable
Strontium	Probable local redistribution	Not applicable	Not applicable
Yttrium	Most retained	Retention	Retention
Zirconium	Most retained: some local redistribution	Retention	Some migration
Niobium	Most retained	Retention	Retention
Molybdenum	Most migrated	Migration	Migration
Technetium	Local redistribution	Retention	Migration (?)
Ruthenium	Local redistribution	Retention	Minor migration
Rhodium	Most retained	Retention	Retention
Palladium	Most retained	Retention	Retention
Silver	Most retained	Retention	Retention
Cadmium	Most migrated	Migration	Migration
Indium	Most retained	Retention	Retention
Tin	Not yet studied	Retention	Retention
Antimony	Not yet studied	Possible migration	Some migration
Tellurium	Not yet studied	Retention	Retention
Iodine	Most migrated	Not applicable	Not applicable
Xenon	Most migrated	Not applicable	Not applicable
Caesium	Most migrated (?: perhaps locally)	Not applicable	Not applicable
Barium	Local redistribution	Not applicable	Not applicable
REE	Most retained	Retention	Retention
Lead	Variable migration	Retention or local redistribution	Some migration
Bismuth	Most retained	Retention or local redistribution	Some migration
Polonium	Most retained	Retention	Retention
Thorium	Most retained	Retention	Retention
Uranium	Some local redistribution	Retention or local redistribution	Some migration
Neptunium	Most retained	Retention	Retention
Plutonium	Most retained	Retention	Retention
Americium	Not measurable	Retention	Retention

systems, the "five rules" for choosing a useful analogue may be helpful (Table 2). Whether an analogue study can be applied quantitatively (as an actual control or benchmark for a mathematical model), or only qualitatively (as a "signpost" to indicate that a conceptual model seems reasonably consistent with reality or checking that trends from mathematical models are going in the right direction) will be strongly dependent on the extent to which these conditions can be satisfied.

TABLE 2: The five rules of natural analogue studies (from Chapman et al., 1984)

1. The process involved should be clear-cut. Other processes which may have been involved in the geochemical system should be identifiable and amenable to quantitative assessment as well, so that their effects can be "subtracted".

2. The chemical analogy should be good. It is not always possible to study the behaviour of a mineral system, chemical element or isotope identical to that whose behaviour requires assessing. The limitations of this should be fully understood.

3. The magnitude of the various physico-chemical parameters involved (P, T, pH, Eh, concentrations, etc.) should be determinable, preferably by independent means and should not differ greatly from those envisaged in the disposal system.

4. The boundaries of the system should be identifiable (whether it is open or closed, and consequently how much material has been involved in the process being studied).

5. The timescale of the process must be measurable, since this factor is of the greatest significance (the raison d'être) for a natural analogue.

It should be noted that the "five rules" tacitly focus on the more geochemical aspects of the model chain (cf. Fig. 1). Hydrogeological modelling (and the required support from geology and geophysics) is completely site specific and, although the mathematical tools involved may be tested at other sites, it is generally not amenable to analogue validation in the sense used above. Similarly, the engineering aspects of repository performance assessment usually involve the analyses of very specific structures which are not found in nature, and the validation of the models involved is part of normal engineering practice. Finally, the biosphere models of transport through the food chain and final dose to man are based extensively on observations of natural systems but, to conform with common usage, these will not be considered as analogues. The main areas of the safety assessment model chain which are amenable to analogue studies can be generally termed material properties, radionuclide chemistry and solute transport. Selection of analogues within these fields are now considered separately.

Materials of interest in a waste disposal context include steel, concrete/cement, glass, copper, bitumen and a range of other metals and polymers dependent on the waste type and disposal concept. An obvious source of analogues in most of these cases would be archeological artifacts as considered previously. The relevance of a particular "find" can be evaluated by considering the "five rules" in turn. For example, applying Rule 1 would focus on samples with an apparently simple history in a simple environment. This would tend to preclude, for example, objects which are known to have been transported between different sites or are found in periodically variable conditions like intertidal sediments. The second rule emphasises the importance of chemical similarity - thus data on pitting corrosion of Roman steels may have no relevance to the specialist steel compositions used for waste canisters or data on the leach rate of volcanic glasses may have little relationship to the leaching of borosilicate glass waste matrices. The third rule notes the need to characterise the physico-chemical environment of the analogue location and the desirability of this being as close as possible to that expected in the repository. Thus, the temporal variation in the chemical environment of samples from a soil zone may be very difficult (or impossible) to establish due to changing climate or land use. Alternatively, samples from the seabed for which the environmental parameters can be well specified may be, nevertheless, totally inappropriate to a land disposal concept. Rule 4 also involves system characterisation, particularly in cases where some kind of transport is involved. Taking another geological example, the presence of native copper in a potential repository host-rock can only be considered an analogue of copper canisters in a repository if it is found in open systems (e.g. water-carrying fissures) and not in geodes which are closed systems, effectively isolated from the prevalent groundwater geochemistry. Even if native copper is found in appropriate environments, application of the analogue remains fairly qualitative unless boundary conditions, such as the original quantity of material at commencement of leaching and the flux of water involved, can be estimated. In all the examples above, application would involve testing the temporal extrapolation of a model and hence, as stated in the fifth rule, the timescale involved needs to be defined for any quantitative use.

It should be emphasised that the "rules" are really, in practice, merely guidelines and may not be appropriate to particular applications. For example, if the leaching of volcanic glasses in nature is compared to their leaching in a particular experimental set-up, then conclusions may be drawn about the general applicability of the data produced by the experimental method even though the actual "analogue leach rates" as such have no direct relevance.

In terms of radionuclide chemistry, a typical analogue study might involve determination of the aqueous concentration of an element in a particular geochemical environment in order to test the predictions of a chemical thermodynamic model. Again, consideration of the 5 rules can help in the planning/evaluation of such a project. For example, taking rules 1,4 and 5 together would suggest that in order to test predicted solubility limits, a geochemical system with a large, well-defined mineral source of the element of interest and known to be closed with respect to water exchange (or have long water residence times) would be sought. This would avoid potential complications of source term constraints or kinetic limitations. Wherever possible, the study should be based directly on the element of interest - even though the isotopes considered may be different (isotope effects will generally be neglegible). In some cases, it may not be possible to study any isotopes of relevant elements in a particular environment and hence an element with similar chemistry might be considered. Here very great caution must be exercised in fully defining the chemistry of the system (rules 2 and 3) and, in general, it is much better to examine trends than individual values. For example, an absolute concentration measured for Th gives little information on the solubility of reduced Pu(IV), but an observed trend in concentration of Th with increasing dissolved organic carbon concentration would be, at least, qualitatively relevant. The consideration of trends is especially important in the examination of aqueous phase speciation - e.g. the observed disequilibrium of sulphur species with regard to other redox couples in the absence of microbiological catalysis could be expected to hold also for the chemically similar selenium (even though the dominant species may not be the same for given geochemical conditions).

A grey area between elemental chemistry and radionuclide transport involves the use of analogues to test assumed or modelled sorption properties. Sorption terminology, as used in the nuclear waste literature, is a veritable minefield of contradictory definitions and inconsistent usage. The source of the problem is probably the fact that many different processes can be involved in the partitioning of a radionuclide between a rock and a groundwater, and that such processes are simplistically grouped together under the global term "sorption" - which is often simply assumed to be a constant (a Kd) for a particular rock/water system. Although this is known to be a gross over-simplification for most cases, the situation gets even worse when such simple constants are used directly to calculate a retardation constant for a dynamic system. This problem of terminology has been emphasised because it is at the source of many misapplications of analogues. Many examples can be found in the literature of the concentration of a particular element (or natural radionuclide) being measured in the solid and aqueous phases, and the resultant ratio called a Kd and compared to those used in repository safety assessment. In most cases such a comparison is totally meaningless. The "Kd" here could include major contributions from either components of the rock which are not exchanging with solution or solid phase concentrations which have resulted from precipitation rather than any type of sorption. In both these cases, the apparent Kd is higher (possibly by orders of magnitude) than that appropriate to radionuclide transport calculations and hence, in safety assessment terms, would be highly non-conservative. Other approaches exist in which ratios of natural series isotopes are used to derive Kd values (and also information about the kinetics of sorption). For specific cases, e.g. gravel aquifers, such methodology may be acceptable but application to more relevant formations is fraught with difficulties which are often unrecognised.

Such discussion leads on to the topic of solute transport. Here the mechanisms of retardation can be roughly broken up into those dependent on the bulk rock (Fig. 8a), those related to properties of the flow path (8b), surface sorption (8c) and retardation due solely to changes in aqueous chemistry (8d). In order to sensibly examine such a complex system, a simplifying model must be used but, as emphasised above, this model must be compatible with that used in safety assessment. Despite the difficulties involved, migration analogues represent a major component of the work in the analogue field. The simplest type of migration analogue involves analysis of the mobilisation of a datable geochemical anomaly - for example as previously considered for Oklo. It is clear, however, that the age and complexity of the Oklo reactors limit the extent to which the rules can be applied and hence the output tends to be fairly qualitative (e.g. Table 1). Many other geochemical anomalies can be used as source terms covering the range of relevant timescales - e.g. ore bodies, igneous intrusions, sedimentary anomalies from vulcanism or glacial rebound, anthropogenic pollutants, etc. In cases where the system is simple and the chemistry and timescale can be well specified, such studies can allow quantitative testing of migration models - but these cases are rare exceptions. In general, system complexity or insufficient characterisation limit the output of these studies to a qualitative level.

An alternative approach to that above involves the study of steady state systems where a stable profile in concentration is maintained by the competing processes of transport and decay. This is particularly appropriate for studies involving the natural decay series where a mobile daughter is produced by the decay of a very long-lived, immobile parent. The supply rate and the decay rate are defined by the well known constants of radioactive decay and hence transport rates can be derived from isotopic profiles - particularly when the migration process is relatively simple - e.g. diffusion only. Such analysis is a fairly standard technique for the study of marine sediments and can also be applied to hard rocks - e.g. to study matrix diffusion in crystalline rocks. In the latter case, although isotopic profiles readily indicate the extent of pore interconnectedness in a fairly qualitative manner (e.g. Fig. 9), derivation of more quantitative output is difficult due to the problems of fully defining the boundary conditions involved.

Conclusions

This first paper has attempted to provide a background of the fundamentals of analogue studies - showing up both their advantages and limitations. It is indicated that this approach can be very powerful when properly applied but that there are many potential pitfalls involved. In the following paper, the foundations provided will be built upon to show how analogues can be applied as part of a repository safety analysis.

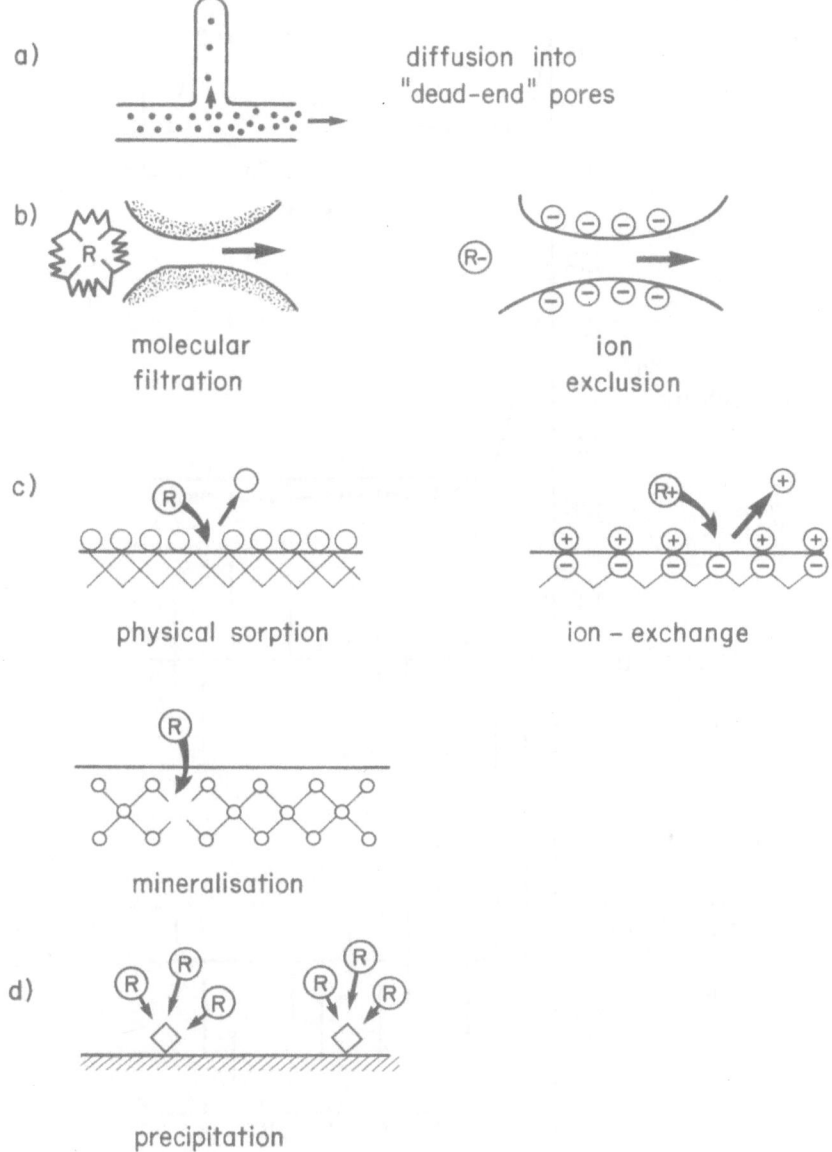

a)   diffusion into
         "dead-end" pores

b)

molecular
filtration

ion
exclusion

c)

physical sorption

ion – exchange

mineralisation

d)

precipitation

Figure 8. Retardation mechanisms

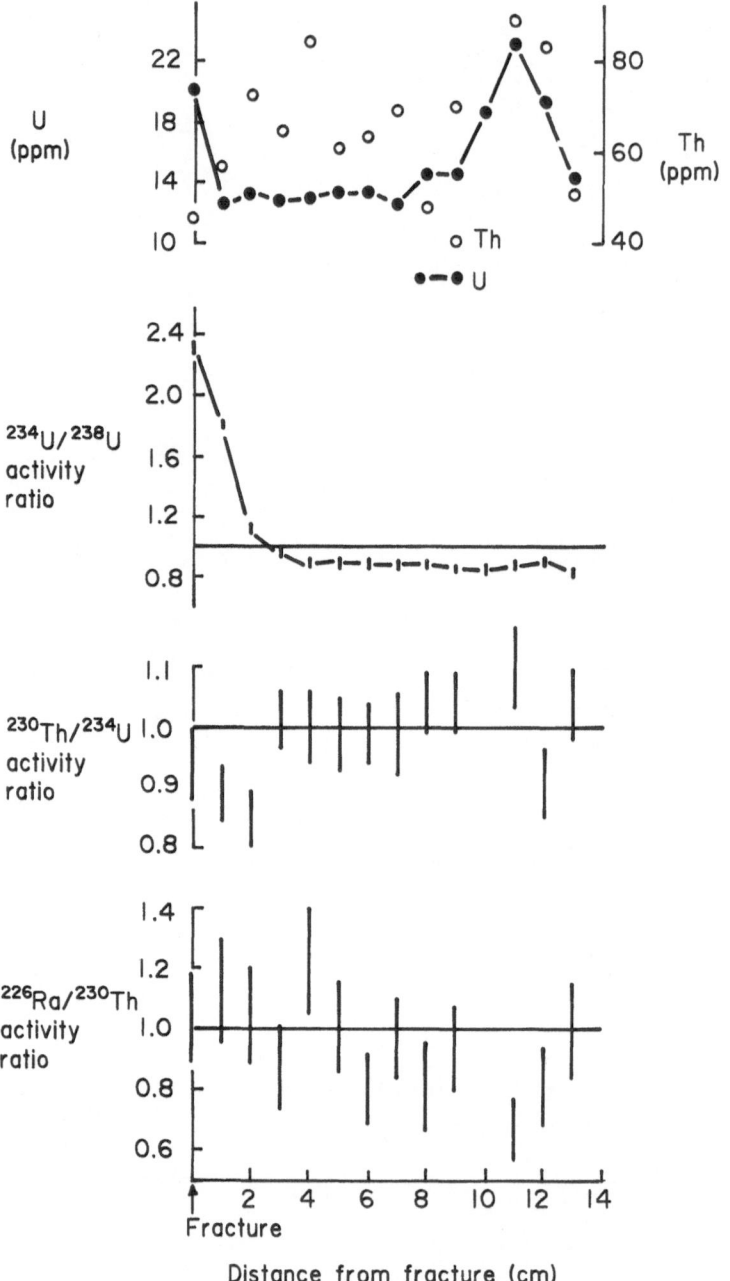

Figure 9. Natural serial radionuclide profile from a water-carrying fissure into bulk granite (from Nagra NTB 87-08, 1988)

Bibliography: General information

REVIEWS

Chapman, N.A., McKinley, I.G., Smellie, J.A.T. (1984) The potential of natural analogues in assessing systems for deep disposal of high-level radioactive waste, Nagra NTB 84-41 / EIR Ber. 545 / KBS TR 84-16.

Brookins, D.G. (1984) Geochemical aspects of radioactive waste disposal, Springer-Verlag, chapter 11.

Chapman, N.A., McKinley, I.G. (1987) The geological disposal of nuclear waste, Wiley, chapter 11.

CONFERENCE PROCEEDINGS

I.A.E.A., Proceedings of a symposium on the Oklo phenomena, STI/PUB/405, I.A.E.A., Vienna, 1975

I.A.E.A., Proceedings of the technical committee meeting on natural fission reactors, STI/PUB/475, I.A.E.A., Vienna, 1978

Smellie, J.A.T. (ed.) (1984) Natural analogues to the conditions around a final repository for high-level radioactive waste. Proceedings of the natural analogue workshop held at Lake Geneva, Wisconsin, USA (October 1-3, 1984). SKB TR 84-18

Come, B. and Chapman, N.A. (eds.) (1985) Natural Analogue Working Group, Final Meeting Report from First Meeting, CEC report No. EUR 30315

Come, B. and Chapman, N.A. (eds.) (1985) Natural Analogue Working Group, Final Meeting Report from Second Meeting, CEC report No. EUR 10671

Come, B. and Chapman, N.A. (eds.) (1987) Natural analogues in radioactive waste disposal. CEC report No. EUR 11037

APPLYING NATURAL ANALOGUES IN PREDICTIVE PERFORMANCE ASSESSMENT
Part II: Examples and Discussions

I. G. McKINLEY
NAGRA
Baden,
Switzerland

ABSTRACT. This paper illustrates the application of natural analogues within various performance assessments carried out in Switzerland. The examples considered cover both high and low/intermediate level waste disposal in both crystalline and sedimentary host rocks.

Introduction

In this paper, the actual application of natural analogues as part of a predictive performance assessment is discussed. The examples considered here are all taken from safety analyses carried out in Switzerland, but it should be emphasised the work is fairly generic and, indeed, the same studies may also have been used in other national programmes. As was emphasised in the previous paper, the first stage in setting up an analogue study is defining the waste disposal system considered and hence the models to be tested.

In the Project Gewähr 1985 (PG'85) [1] safety analysis, two main repository systems were considered:

i) A "type C" repository for vitrified HLW at a depth of approx. 1300m in the crystalline basement of Northern Switzerland. It is envisaged that the waste will be encapsulated in a massive cast steel canister and horizontally emplaced into tunnels which are subsequently backfilled with compacted bentonite (Fig. 1).

ii) A "type B" repository for L/ILW. Waste of various types from reactor operations, decommissioning, fuel reprocessing, medicine, industry and research will be solidified in a range of matrices including cement, bitumen and resins. Horizontal emplacement in the side of a hill is envisaged (Fig. 2). Potential host rocks include marl, granite and anhydrite although the safety analysis was focused on the first of these.

Since 1985, a further extensive study of the feasibility of HLW disposal in agillaceous sediments has been carried out [2]. Here the repository layout would be basically similar to (i) above but a shallower emplacement depth (ca. 800m) and use of tunnel liners would be required for geomechanical reasons.

Near-field: HLW

A layout of the near-field considered in PG'85 is shown in Fig. 3. The sequence of processes occurring

A. Saltelli et al. (eds.), Risk Analysis in Nuclear Waste Management, 377–396.

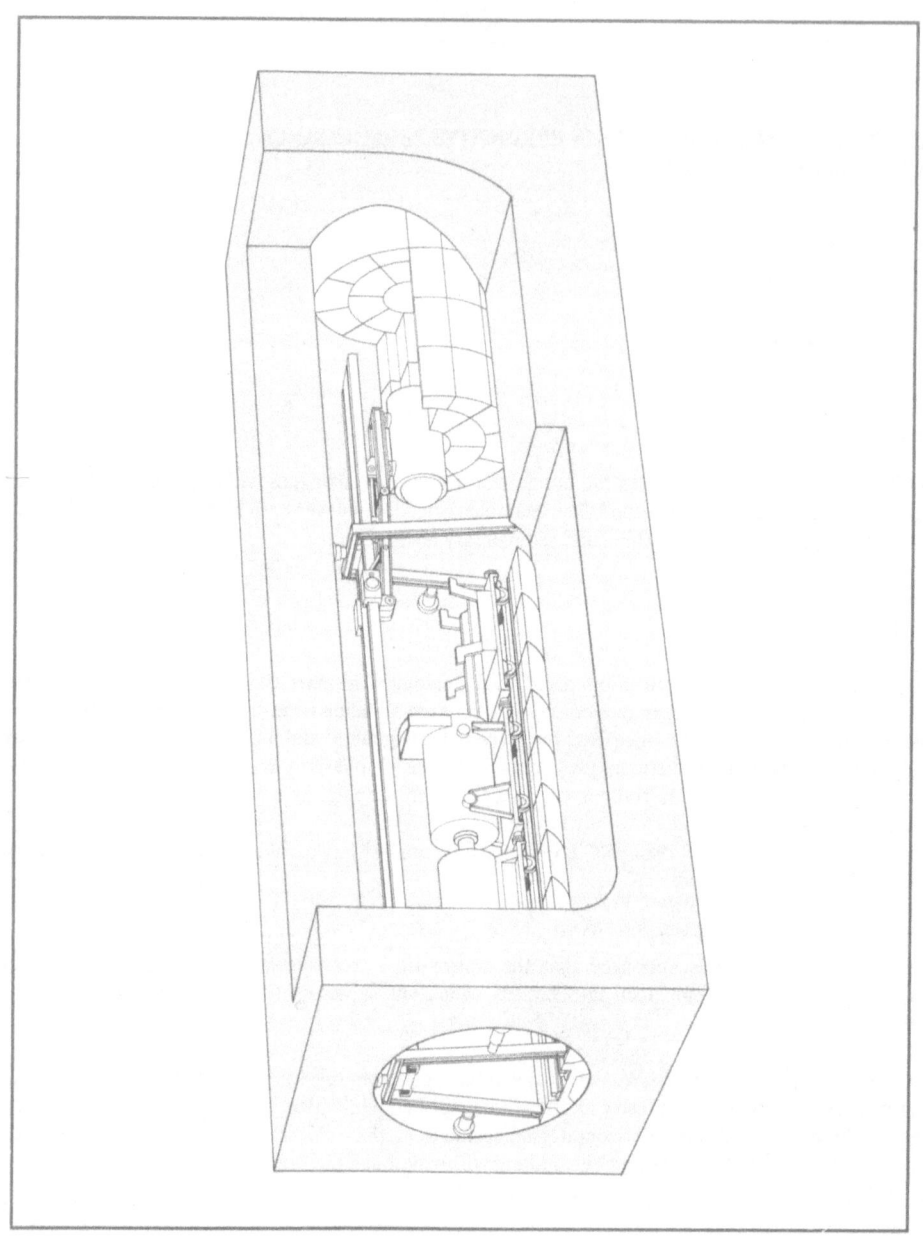

Figure 1. The emplacement procedure for HLW. The space between waste canister and tunnel-wall is backfilled with blocks of compacted bentonite.

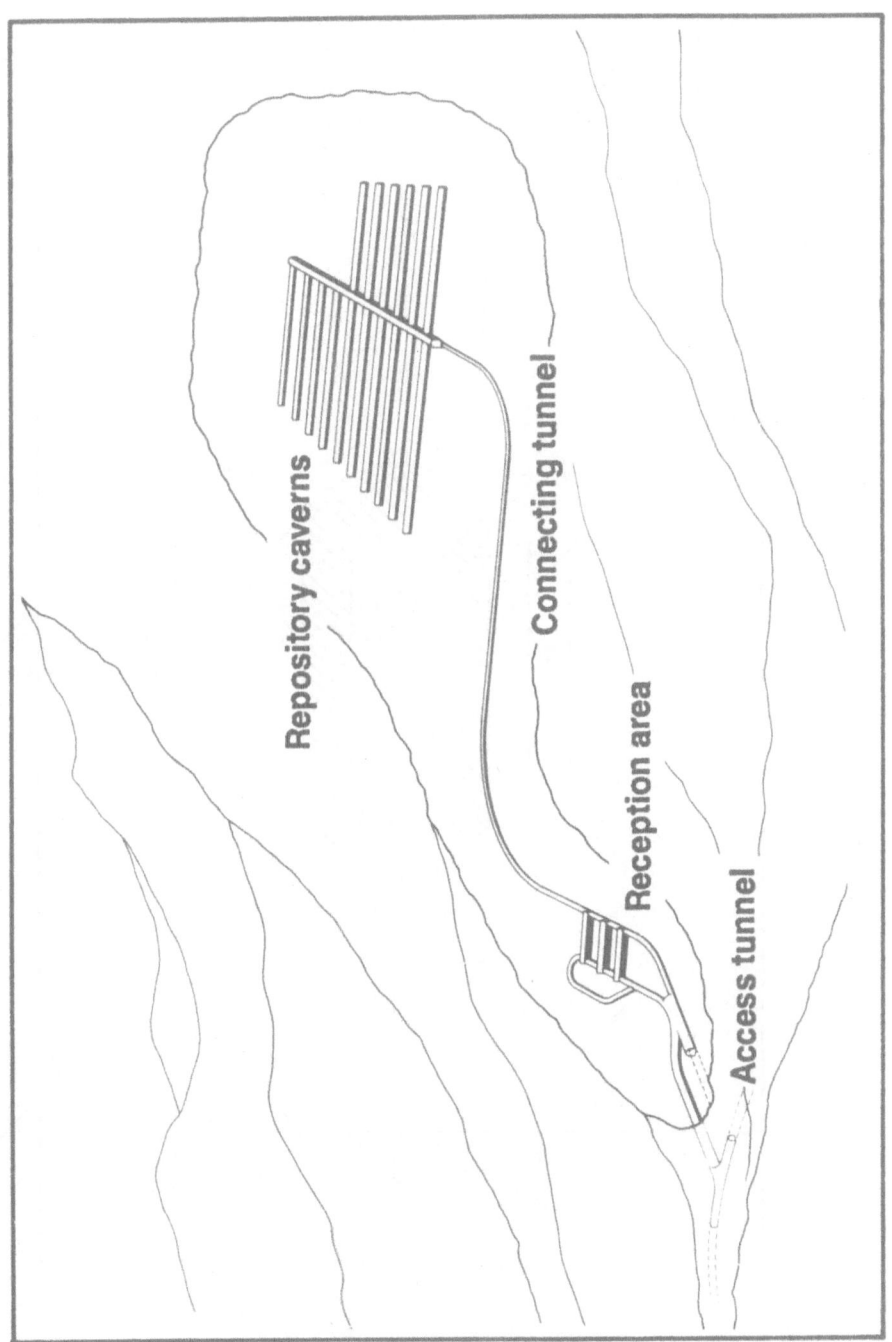

Figure 2. Perspective overview of type B repository installations

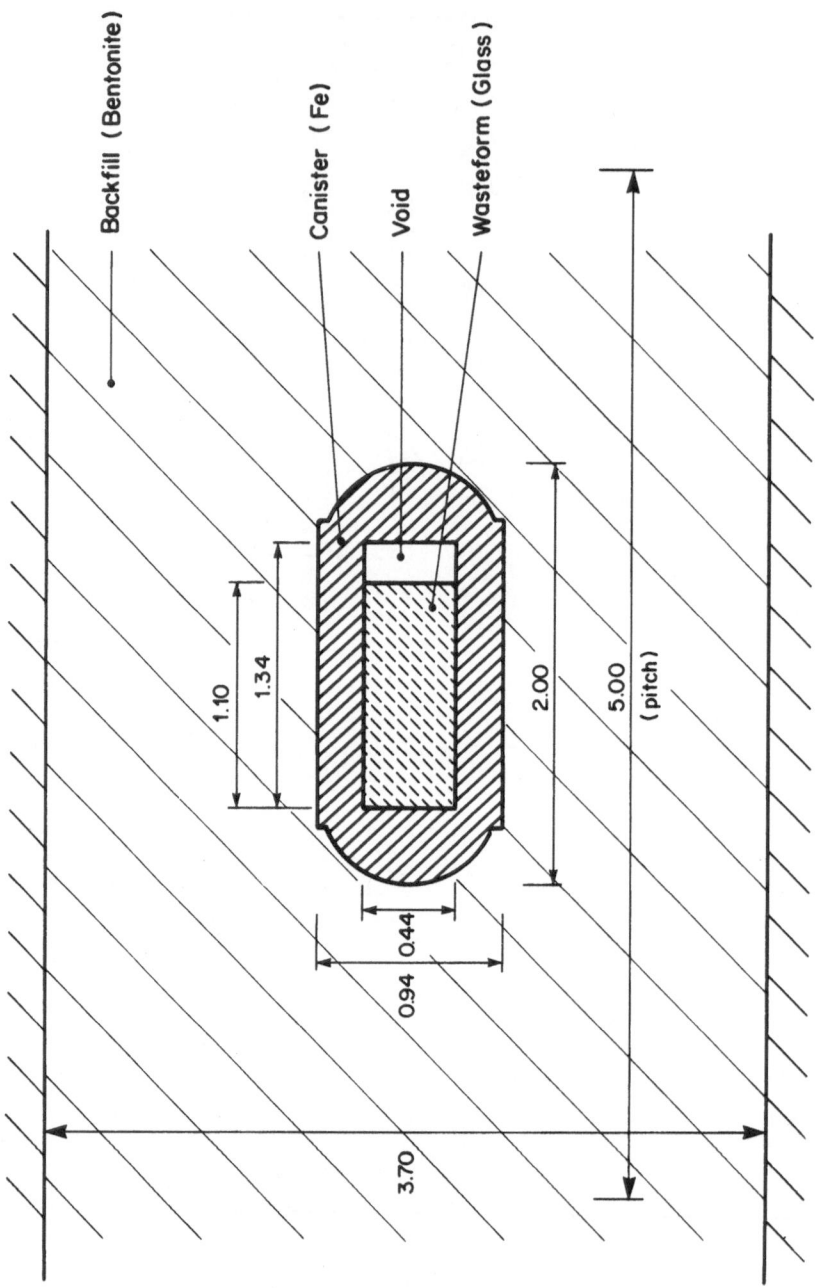

Figure 3. Waste emplacement geometry (dimensions in metres)

in temporal evolution of the near-field which have to be considered in the safety analysis [3] are:

1. Water re-invades zones of the repository host-rock which have been partially drained during the operational phase. The hydrogeology of the fissure system begins to stabilise and the bentonite buffer begins to absorb water and swell.

2. The gradual wetting of the buffer is coupled to spreading of radiogenic heat from the waste through the buffer. After a thermal peak which occurs within the first few years, the temperature profile through the bentonite gradually decays away over a period of several hundred years in line with the decrease in thermal output from the waste. During the same time period the bentonite becomes saturated and exerts a swelling pressure on the surrounding rock and the waste package, sealing any remaining "construction gaps" in the backfill structure.

3. From first wetting, hydrothermal alteration of backfill (e.g. montmorillonite to illite) will commence, the rate of which will be controlled by reaction kinetics (temperature dependent) and the rate of supply of particular solutes (e.g. K^+ for the conversion above). When the wetting front contacts the iron waste canister, corrosion will occur which will initially be oxic but quickly become anoxic as trapped air is used up. The rate of corrosion may be affected by the build-up of layers of solid reaction products (oxy-hydroxides) or by the limiting rate of supply of corrodents or loss of dissolved or gaseous products by diffusion through the backfill.

4. After the canister has been corroded to a certain extent (more than 5 cm) the canister will fail mechanically and fluid will then contact the waste matrix. Even after canister failure, however, continued corrosion of unreacted Fe and buffering by corrosion products will exert an influence on the chemistry of the water in contact with the waste glass.

5. Corrosion of the vitrified waste will occur by a process of congruent dissolution with some associated precipitation of secondary mineral phases. Preferential leaching of particular nuclides would only occur in the very early stages of corrosion. During steady-state leaching the rate of release of specific nuclides would be controlled either by the rate of matrix dissolution or by limits set by the total solubility of the element in question.

6. The leachate thus formed is transported to the far-field by diffusion through the remnant canister, corrosion products and the bentonite backfill. The rate of such transport will be dependent on diffusion rates in the bentonite pores, the magnitude of retardation processes occurring (e.g. sorption, ultra-filtration, ion-exclusion, etc.) and on the water flow field in the heterogeneous host rock. The resultant radionuclide release rates into water conducting zones provide the input functions for the far-field transport calculations. If the supply by diffusion of radionuclides through the backfill is relatively small and significant quantities of water flow in the water conducting zones, an additional dilution may result.

From the principles presented in the previous paper, material properties of the bentonite, steel and glass, speciation and solubility of leached radionuclides and radionuclide transport through the bentonite are identified as areas in which natural analogues may be applicable. The short-term re-saturation and swelling of the bentonite and the thermal transient are more appropriate for laboratory studies while near-field hydrology will be very site-specific.

In the safety analysis [1, 3, 4], illitisation is identified as being the most likely bentonite alteration reaction. This reaction is inherently limited by the rate of supply of K^+ which has been calculated to ensure a barrier life in excess of 10^6 years for the expected hydrology/geochemistry in either a granitic or sedimentary host rock. Illitisation is also known to be kinetically slow and hence a number of analogue studies have attempted to estimate the rate of alteration of natural bentonites. These studies have been reviewed in several reports [3, 4, 5, 6, 7] and show a consistent picture of alteration rates below those assumed for the safety analysis, <u>even in cases with much higher K supply rates</u>. At ambient repository temperatures ($<60°C$), indeed, the studies suggest that the maximum alteration will be around 30% [5] and hence the assumption in the safety analysis that bentonite properties will remain constant for at least a million years seems to be well justified. It might also be noted that, although extensive alteration is considered highly unlikely for the expected repository conditions, a recent study has examined the properties of natural clays which have a range of compositions corresponding to possible alteration assemblages. This indicates that, even then, acceptable performance with respect to physical properties could be expected [8].

The steel canister used in PG'85 is designed to survive approximately 1000 years - the 25 cm wall thickness includes 5 cm corrosion allowance and 20 cm for mechanical strength [1, 3, 4, 9]. The corrosion model used is based on an average corrosion rate of 10 microns/year and a pitting factor of 3. The use of archeological analogues to support steel corrosion models was considered in general terms in the previous paper and Fig. 4 shows the data previously presented (cf. also [10]) compared with the modelled penetration rate. It is clear that if samples from an oxidising marine environment are excluded (samples 1 and 2), the safety assessment assumptions are shown to be reasonably conservative. Rare native iron deposits and iron meteorites offer further possibilities to provide analogue support for the corrosion models [6]. Although some very low corrosion rates could be derived from these sources, they probably cannot be used in a quantitative manner due to the peculiar geochemistry of the native iron deposits and inappropriate composition (Fe-Ni alloys) of the meteorites. More recent work has focused on a more detailed analysis of the extent of pitting and the corrosion products found on archeological artifacts [11] which promises to allow more sophisticated tests of the corrosion models.

In PG'85, a simple congruent dissolution model was used to describe glass leaching which incorporated a leach rate of 10^{-7} g/cm²/d. The use of natural analogues to test glass leaching models has been mentioned in the previous paper and, although the limitations in the quantitative application of the data were stressed [6], data from natural glasses in relevant environments indicated that the rate assumed seems to be reasonably conservative [6, 12, 13]. For the more recent sediment study, the leach rate adopted has been further supported by more mechanistic modelling work within the JSS (Japan/Sweden/Switzerland) study [14]. This has been tested in an analogue study of both the rate of degradation of a range of natural volcanic glasses and the alteration products observed [15]. This body of work suggests that, although the safety analysis model is very simplistic, it provides a conservative estimate of the extent of matrix degradation.

Speciation and solubility of radionuclides in the near-field are evaluated by standard chemical thermodynamic models. Although these model predictions are, in principle, testable by natural analogue approaches (cf. I/LLW near-field below), this was not included in PG'85 base-case because, at the time, the near-field porewater chemistry was thought to be insufficiently well defined for such calculations to be justified. The actual source-term model, therefore, incorporated far-field solubility limits [16]. Although this overlaps with the far field discussion, it can be noted here that the predicted solubilities for U and Th, at least, were compared against field observations in relevant conditions and a database selected which ensured conservative over-prediction of solubility [17]. In the sediment study, near-field solubilities are used in the source-term model. Although no quantitative analogue validation has been attempted, it is noted that the generally very low solubilities predicted in what is effectively a reducing clay porewater is in qualitative agreement with observations of redox halos in Permian sediments (Fig. 5). The latter show very large enrichments of relevant elements such as U, Ni, Se, Mo, Pd and Pb in their reducing cores [18] which are concentrated by redox-driven diffusion and are retained over periods of millions of years.

The final important part of the near-field analysis is the modelling of transport through the backfill. Transport is assumed to occur predominantly by diffusion and, although the backfill is shown to be an important barrier, full credit for its performance is not taken into account in the base-case safety

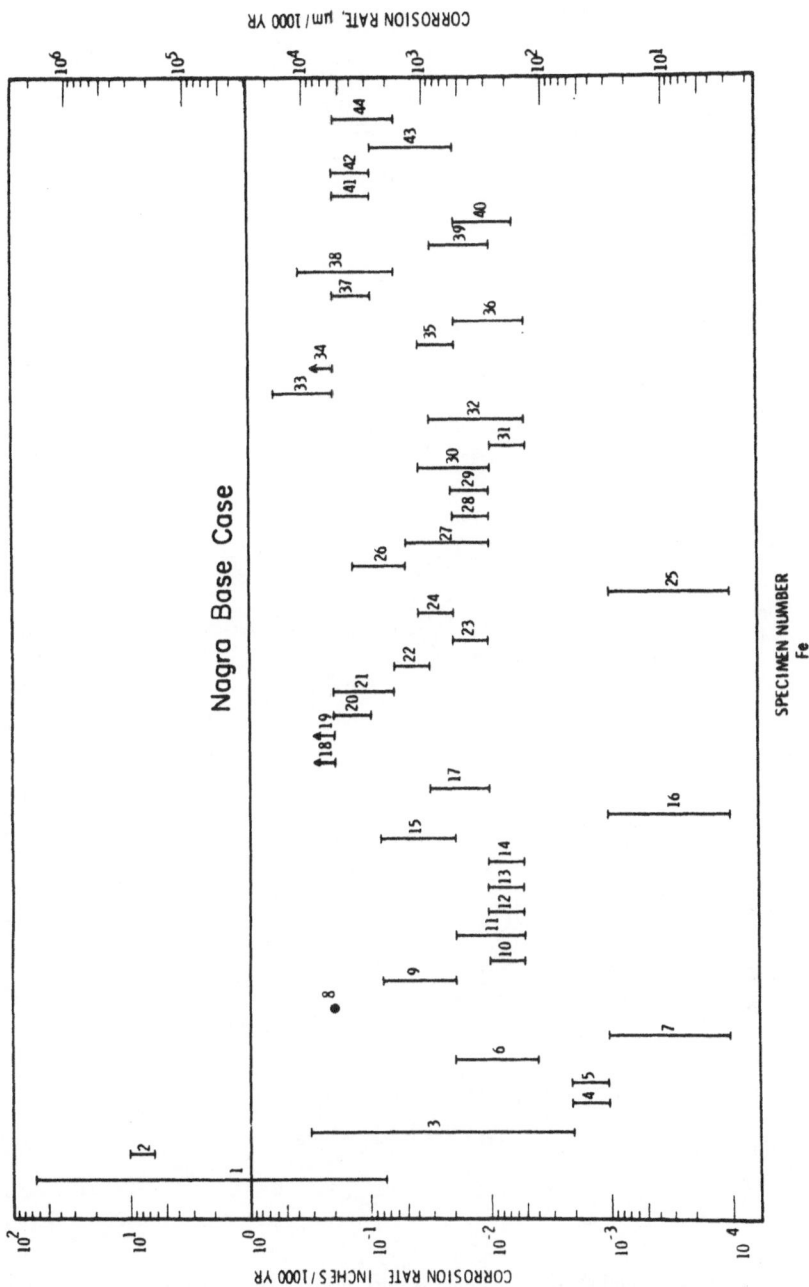

Figure 4. Corrosion rate data for iron artifacts (from [10]) compared with the Nagra base case

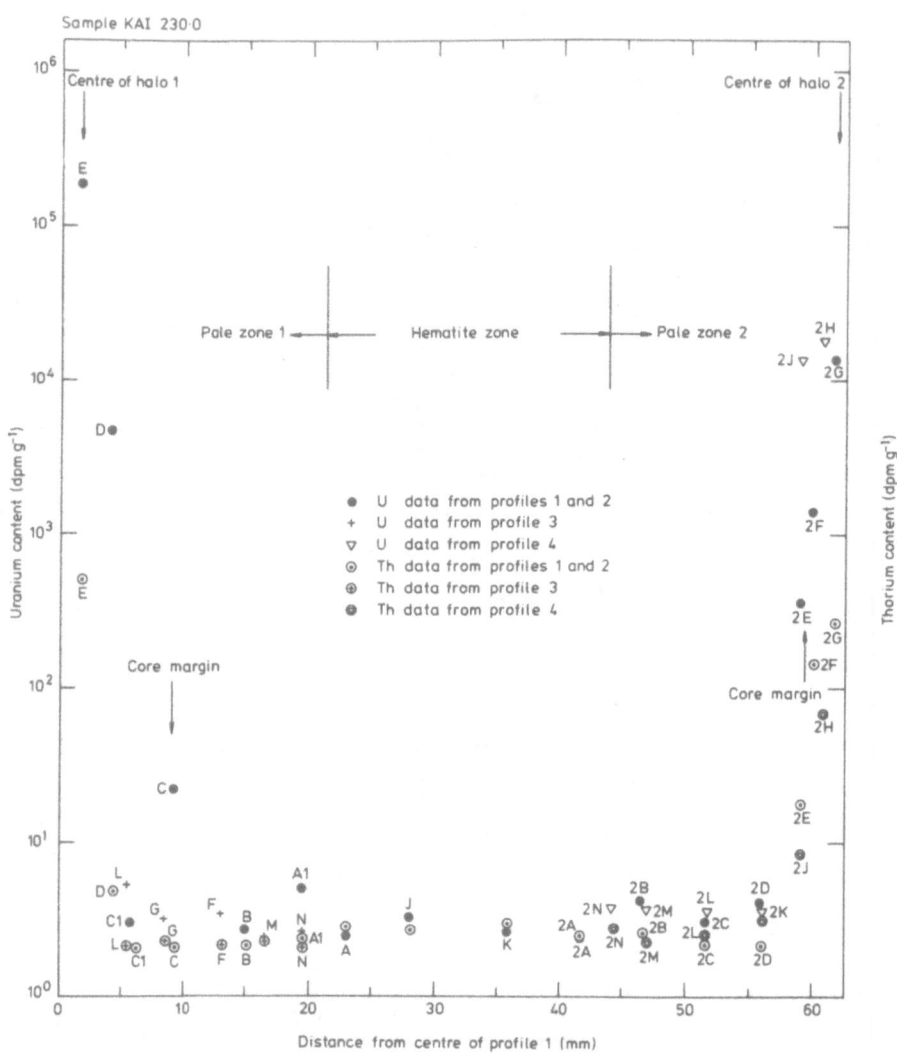

Figure 5. Plots of uranium and thorium contents against distance from centre of profile 1 for all four profiles (from [18])

analysis models. In the latter, the most important role of the backfill is as a colloid filter, although it is also accepted as providing total containment of shorter-lived, highly sorbed species (which simplifies the far-field transport calculations somewhat). Although a fair amount of laboratory data exist to support the diffusion models, analogue support has also been useful here. Although many migration analogues exist [6], the Loch Lomond study has been particularly useful here [19, 20].

The sediment in Loch Lomond in West Scotland shows a very distinct band of material of marine origin in the middle of freshwater sediments. This is attributable to the changing sea level after the last ice-age resulting from "fast" ice melt followed by "slow" isostatic rebound of the land (Fig. 6). The marine phase has been accurately dated to a period of between approx. 6900 and 5400 years ago by a suite of radioisotope (^{14}C, natural series) paleomagnetic and palynological methods.

The principle of the study is that the marine sediment (and associated pore water) has a distinct chemical signature which will alter with time due to diffusion processes (into or out of the surrounding freshwater sediment). This diffusion is analogous to that of radionuclides from a waste package into a clay backfill or an argillaceous host rock and can be modelled by techniques used in repository safety assessment given appropriate input data (timescales, sorption constants, porosity, etc.). The results of the study are described in detail elsewhere but showed that, in general, observed mobility was compatible with laboratory data (although sorption of many relevant elements appeared to be poorly reversible and hence difficult to describe with a simple diffusion model). The halogens I and Br provided a particular surprise, however, as they appeared to be much less mobile than expected. Using a simple model (Fig. 7), diffusion constants could be calculated which, at about 10^{-7} cm^2/sec, were at least an order of magnitude smaller than expected. It is possible that organic complexation may considerably decrease the mobility of these and other relevant elements (e.g. Ra, U).

This work indicates that the calculated diffusion rates through compacted bentonite (which would be expected to have a lower porosity and diffusivity) are reasonable while the assumption of efficient colloid filtration is also qualitatively supported. Further indications of the low mobility of macromolecules in compact clays are provided by observations of loss of only small molecular weight organics during the weathering of clays with apparently complete retention of larger species [21, 22].

Finally, it might be mentioned that analogues have also been reviewed which cast some qualitative light on the models used to consider potential problem areas such as radiolysis or redox front mobility [6] but, for the Swiss case of vitrified HLW and a massive steel canister, these processes are expected to be unimportant [3].

Near-field/far-field interface: HLW

A region will exist outside of the engineered barriers which is physically or chemically perturbed by the presence of the repository and thus can either be considered as a part of the near-field or as an interface between the near- and far-field. In the granite host-rock, the role of this disturbed zone is expected to be primarily associated with hydrologic perturbations. The sediment study has indicated, however, that for this host-rock significant geochemical perturbations may arise from the oxidation of pyrite during the operational phase [4]. First scoping calculations were based on an assumption of complete oxidation of pyrite by oxygen diffusing through the decompressed zone surrounding the emplacement tunnels, but recent information from the Poços de Caldas analogue project [23] indicated that this approach could be overly simplistic and non-conservative, as considered below.

The C-09 uranium mine near Poços is a roll front deposit which shows very marked redox fronts where the main reaction occurring is the oxidation of pyrite [24]. Surprisingly, water samples from the redox front have low sulphate concentrations (and also low sulphide) and no evidence of sulphur oxidising bacteria. Nevertheless, when such waters flow to the surface they are rapidly oxidised to give high sulphate and very low pH values (ca. 2). This has been provisionally interpreted in terms of partial oxidation of pyrite by iron oxidising bacteria to produce ferric iron (as an oxyhydroxide) and polymeric or colloidal sulphur. If such a mechanism occurred in the repository, the extent of pyrite oxidation

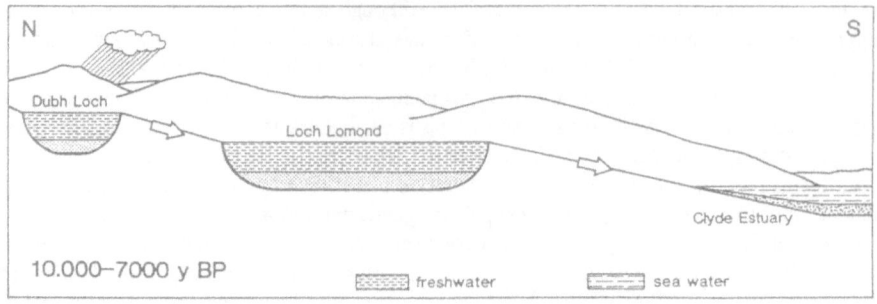

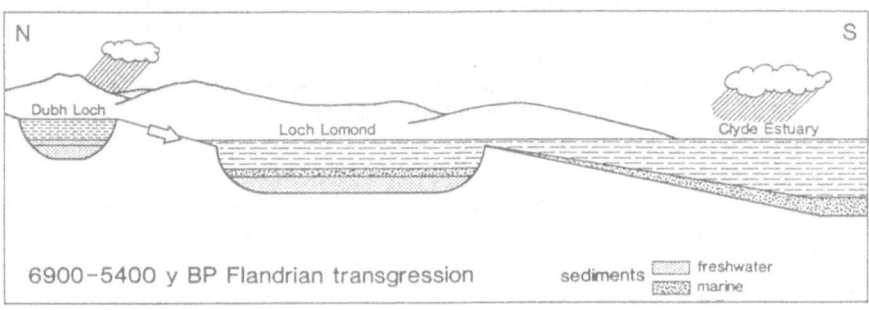

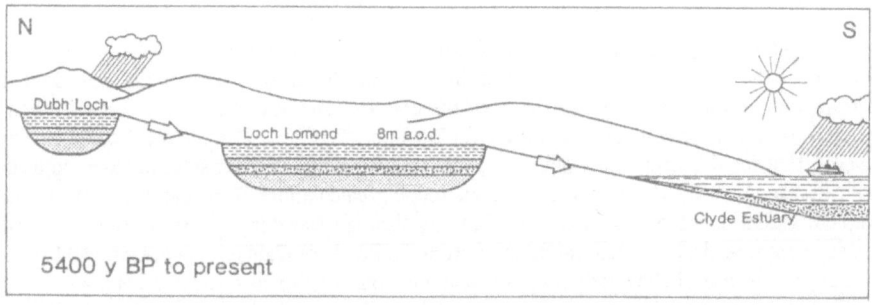

Figure 6. The evolution of sediments in Loch Lomond

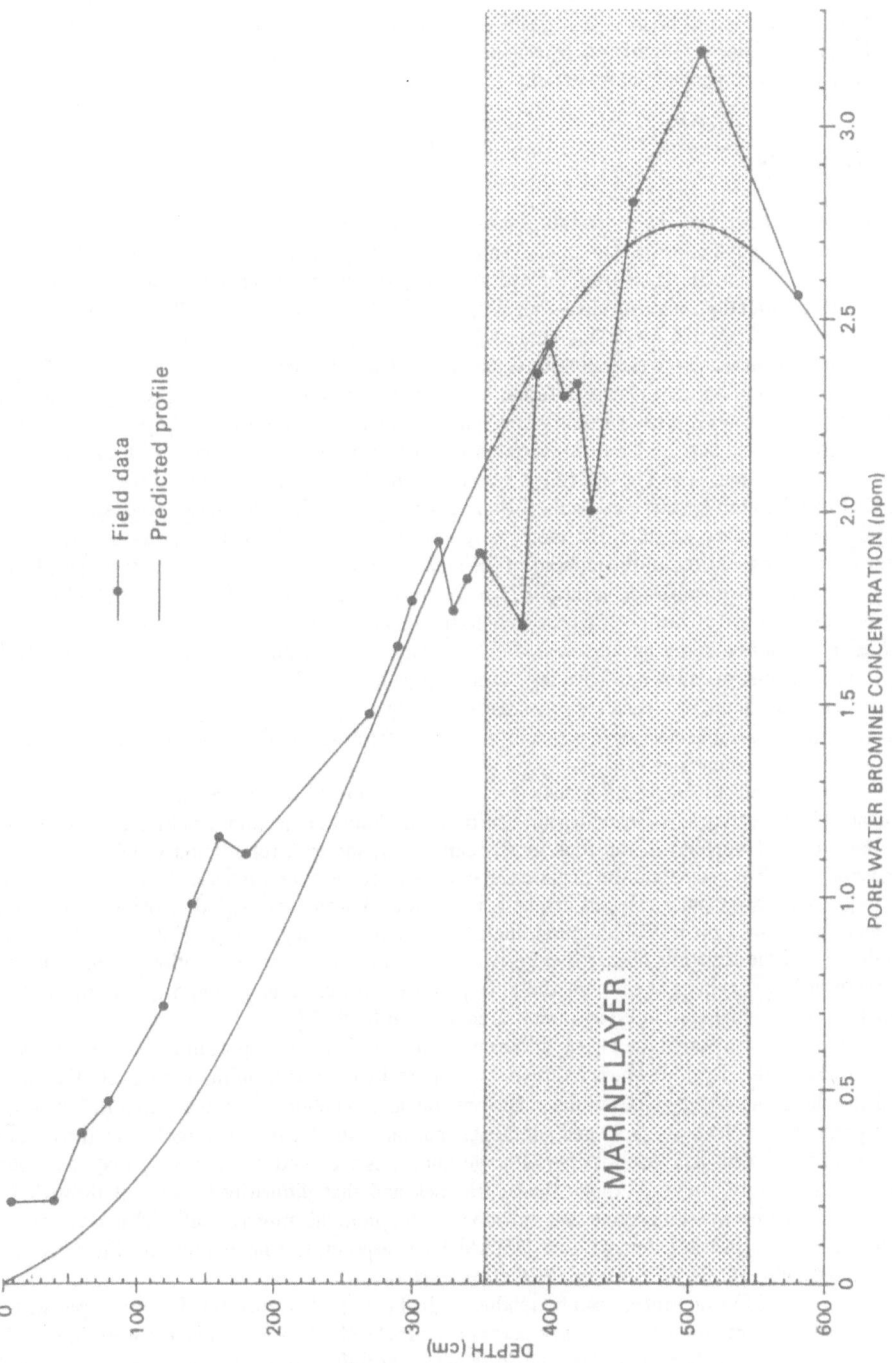

Figure 7. Depth profile of Br in Loch Lomond sediments (from [20])

would be much greater than the original calculations suggested and shows the possibility of acid (and biomass) production at or near the tunnel walls which could have detrimental effects. This problem could, in principle, be minimised by engineering design but further study at Poços is foreseen to examine the phenomenon in more detail.

Far-field: HLW

Both in the granite and the sedimentary host-rocks, transport is assumed to occur predominantly in discontinuities (e.g. fissures, shear zones, sand channels, etc) and not as a flow through the entire body of the rock (e.g. Fig. 8). Apart from the hydraulics of the system, the far-field transport models contain 2 important retention mechanisms - "sorption" within the discontinuity itself and diffusion into the surrounding rock matrix. Modelling studies [25] have shown that sorption alone can provide a very significant barrier when the discontinuity is massive (e.g. a wide shear-zone or "kakirite" or a large clay-rich sand channel) but that matrix diffusion plays a critical role when the water transporting feature is smaller (e.g. narrow fissures or sand layers). The models used incorporate a simple Kd description of sorption (cf. previous paper) but the sorption database was explicitly selected with the limitations of this concept in mind [26]. The Kd values selected were thus chosen to ensure conservative over-prediction of migration by the models used even in cases where the Kd concept was known not to be applicable. Wherever possible, therefore, sorption measurments from laboratory experiments were compared to observations of the mobility of the elements involved in relevant natural systems [26].

The term matrix diffusion is applied to the process by which solute in a hydrologic system dominated by flow in distinct fissures penetrates the surrounding rock [27]. The cases of matrix diffusion of interest in the nuclear waste field represent an extreme case of the general dual porosity hydraulic medium in which advective flow occurs entirely in the fissure (or primary) porosity while all solute transport in the bulk rock (or secondary) porosity occurs by diffusion [28]. The diffusion into the rock occurs in a connected system of pores or microfractures, diffusion through the solid phase is insignificant by comparison.

The importance of matrix diffusion is that it provides a mechanism for enlarging the rock surface in contact with advecting solute from fracture surfaces and infill to a portion (or all) of the bulk rock. The sorption of radionuclides in the rock matrix can greatly increase retardation which, if the resulting transport time is greater than the nuclide half-life, can decrease total releases by orders of magnitude [1, 29]. Additionally, pulse releases can be spread over a longer time period - thus decreasing release concentrations by a process of temporal dilution. Even non-sorbing nuclide releases can be decreased by dilution in the reservoir of matrix porosity. Although radionuclide retardation is generally of most concern, it has also been noted that the bulk rock can also serve as an important chemical buffer - especially for oxidants produced by radiolysis in the near field [29].

In order to incorporate the matrix diffusion mechanism into a transport model two key parameters are required - the depth to which interconnected porosity extends from the fissure into the bulk rock and the diffusivity of the solute in the rock (possibly as a function of distance from the fissure). The first parameter is obviously rock- and site-dependent but, even for the same rock type, the conceptual model adopted can vary considerably. For granite, it is assumed by KBS [29] that a network of connected porosity extends through the entire rock and that diffusivity is constant throughout this region. Alternative models assume that diffusion occurs predominantly (or entirely) in dead-end pores [30] or in a microfractured damage zone [25] which corresponds to a limited matrix diffusion extending only in the order of centimetres from the flowing fracture.

The matrix diffusion process can be readily studied in the laboratory only for rock types with large diffusivities (e.g. fractured sandstones) whereas most rocks of relevance to nuclear waste disposal (e.g. crystalline rocks, argillaceous sediments) have very low diffusivities. The main problems arise from sampling and preparation. Sampling almost inevitably involves allowing the rock to destress which, for a microporous system, could considerably increase the effective diffusivity of the sample. The pore

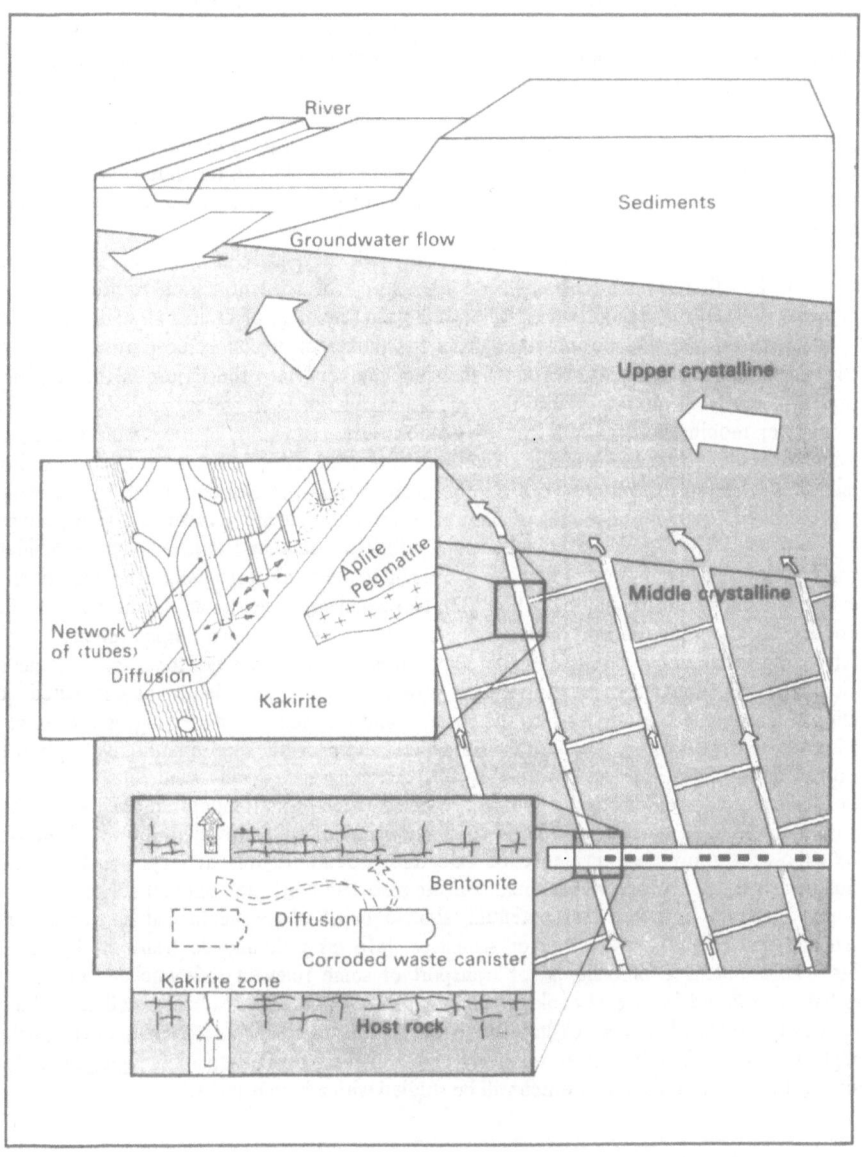

Figure 8. Schematic overview of transport and retention processes in the base scenario

structure may not recover completely, even if the rock is re-stressed prior to measurement. Sample preparation usually involves either cutting or splitting along a natural fissure which can cause extensive surface damaging and, in the latter case, loss of secondary hydrothermal alteration products which may naturally seal some (or all) of the fissure surface porosity. Surface damage is particularly important for crystalline rocks as it will inevitably extend to at least a depth comparable with the grain size (as the rock minerals are generally stronger than the rock matrix). For a coarse grained rock, this depth would be comparable to the diffusion distance of a tracer on a manageable laboratory timescale. It should be noted that these problems also, in principle, apply to field tracer tests to examine diffusion [31].

Further, the potential errors caused by these perturbations all tend to cause overestimation of the diffusivity. As overestimation of matrix diffusion leads to better performance of the geosphere, it is non-conservative in the safety assessment sense.

One approach which does, however, circumvent both the problems of sample perturbation and slowness of the diffusion process involves the study of natural series radionuclide profiles perpendicular to water carrying fissures. The preferential mobility of U and Th daughters relative to their parents allows them to be mobilised from the bulk rock if connected porosity extends to an advective flow. The diffusion of such nuclides from the rock into the fissure is thus analogous to diffusion of solute from the fissure into the rock as considered in the safety analysis models. Natural series daughter mobilisation, as reflected by isotope ratios, can thus, at the simplest level, provide unambiguous evidence of the minimum depth of interconnected porosity. Additionally, however, information on the rate of diffusion can also be derived from the clock provided by the radioactive decay process. In principle, a model of natural series radionuclide release and transport into the fissure could be derived which used laboratory diffusivity data and hence could be validated by observations. In real life, however, it is very difficult to determine the boundary conditions in the system in sufficient detail for full validation [32]. Nevertheless, at least semi-quantitative data on the order of magnitude of the diffusion rate might be derived by curve-fitting a simpler model.

Matrix diffusion has been discussed in detail because it has been a major focus for analogue studies in Switzerland [33 - 36]. The results to date present a consistent picture of connected porosity extending at least some centimetres into the rock around relevant water-carrying fissures (e.g. Fig. 9) which is consistent with detailed mineralogical analysis and provides semi-quantitative support for the transport models [37].

Matrix diffusion in relevant sediments has received much less attention although a natural series study along the lines of those in granite is currently planned. Qualitatively, however, observations of surface exposures show clear evidence of oxidation around fissures in clays which indicate that diffusion of oxygen is occurring which could indicate associated diffusion of other solutes.

Finally, it should be noted that potential "short-circuits" of the geological barrier are of great concern and one of the current "horror scenarios" involves radionuclide transport by non-sorbed colloids. The importance of colloids for transport of some relevant radionuclides in near-surface waters has been noted in several analogue studies [38, 39]. Recently, extensive colloid mobility over distances in the scale of kilometres has also been demonstrated in a reasonably deep granite [40]. Although the granite in this case is relatively highly altered, this observation highlights a potential weakness in the safety model chain which will be studied with a high priority.

L/ILW

Within PG'85, the safety analysis for L/ILW was less detailed than that for HLW and the models used tended to be simpler and more conservative. Since then there has been an effort to treat this waste more realistically, in particular taking more credit for the near-field performance. Studies of archeological artifacts made of cement [41] and bitumen [42] have indicated that more credit for the barrier role of these materials could be accepted. Additionally, the rôle of near-field chemistry could be very important, and here the Oman study has been particularly useful.

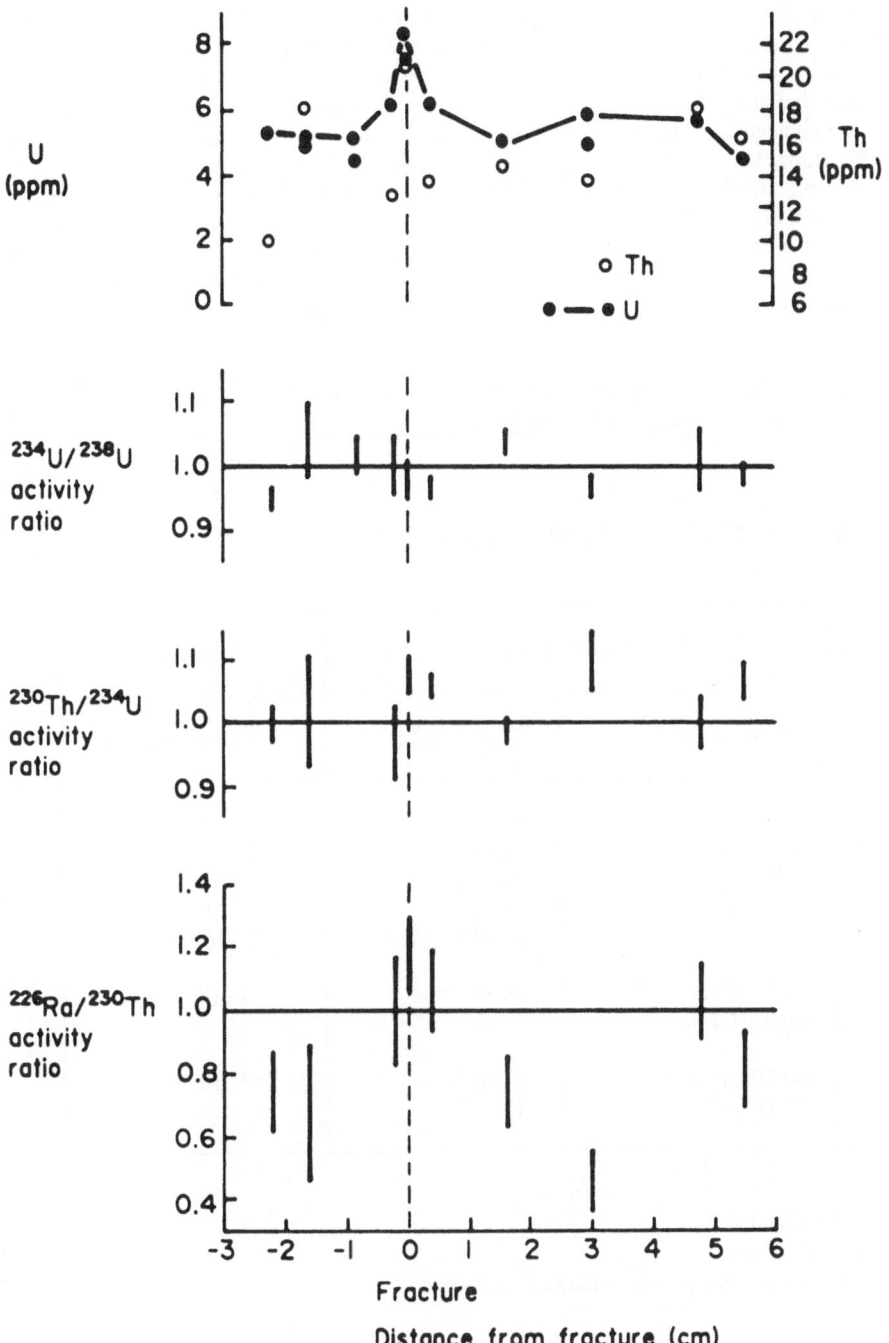

Figure 9. Natural series profiles around a water-carrying fissure in granite (from [37])

In the desert environment of the Semail Ophiolite Nappe in Northern Oman, spring waters are found with remarkable chemistry [43]. They are, in effect, Ca-OH/Na-Cl waters which have very high pH (up to approx. 12) and can be so reducing that they evolve hydrogen gas. Such chemistry is remarkably similar to that expected in aged concrete or cement and hence, on a chemical basis, the springs are analogous to the near-field of a low/intermediate-level nuclear waste repository. In a joint UK/Swiss study, the microbiology and trace-element chemistry of such springs were studied and compared with model predictions. To avoid unconscious biases being introduced, the chemical analysis was performed in the UK independently from the chemical modelling at EIR. The results of the study are described in detail elsewhere [44 - 47], but the key comparison of predicted solubility limits and measured concentrations for a number of relevant elements are shown in Table 1 (for one of the five springs studied).

TABLE 1. Comparison of predicted and measured concentrations
in Oman groundwater (from [47])

Karku well pH = 11.44, Eh = -363 mV, Temp. = 35.7°C

Element	Model predictions		Observations
	Dissolved species	Solubility (M)	Concentration (M)
Se	Se^{2-}/HSe^-	$5 \cdot 10^{-7}$	$< 3 \cdot 10^{-9}$
Pd	$Pd(OH)_2$	10^{-22}	$2.8 \cdot 19^{-9}*$
Sn	$Sn(OH)_3^-$	10^{-18}	$< 2 \cdot 10^{-9}$
Zr	$Zr(OH)_5^-$	$2 \cdot 10^{-3}$	$< 10^{-9}$
Ni	$Ni(OH)_3^-$	$3 \cdot 10^{-7}$	$2 \cdot 10^{-8}$
Th	(a) $Th(OH)_4$	10^{-10}	$< 2 \cdot 10^{-10}$
	(b) $Th(OH)_4$	$5 \cdot 10^{-10}$	
U	(a) $U(OH)_4/UO_2(OH)_3^-$	$3 \cdot 10^{-9}$	$< 4 \cdot 10^{-11}$
	(b) $U(OH)_5^-$	10^{-6}	

(a) = NEA Database
(b) = EIR Database
* = Maximum value, possible analytical problems

In general, it is notable that measured concentrations are below predicted solubilities. In itself this might only indicate that insufficient input of this element exists to give rise to larger concentrations. Given that U and Ni are also present in fine particulate/colloidal phases and that Ni is relatively plentiful in the host rock the models really do seem to be "conservative" for at least these species. More detailed analysis of the results can yield more information - implying that U predictions may be over-conservative for more oxidising conditions and that the sulphur system needs to be better modelled before NiS can be included as a possible controlling solid.

One notable exception to the compatibility of observation and prediction is Pd. Here, however, there are considerable doubts about the measurement itself but, if it were taken at face value, it would suggest that an important aqueous species was missing from the Pd database - probably a hydroxy complex such as $Pd(OH)_3^-$ which could be significant in terms of both solubility and expected migration properties.

The microbiological study was less quantitative but, nevertheless, was designed to test a specific conceptual model of the gradual build-up of microbial tolerance to extreme conditions [48]. The observed presence of relevant microbial groups (e.g. sulphate reducers) in such a hyperalkaline environment indicates that this potential perturbation must be studied further and that the development of models of microbial activity [49, 50] are worthwhile.

As far as the far-field is concerned, the discussion above for the HLW cases are directly applicable to the potential granite and marl L/ILW host-rocks. So far, the anhydrite option has not been studied in detail.

Global analogues

The discussions above have focussed on the more technical applications of natural analogues and it can be seen that they can supply considerable support to the construction of a safety case. In PG'85, analogues were also used in a more "public relations" type of manner - presenting Oklo (cf. previous paper) and rich ore bodies (such as Morro de Ferro [51]) to show that immobilisation of radionuclides for geological periods was possible. Calculated releases of radionuclides were also put in perspective by comparison with concentrations of natural radionuclides in relevant groundwaters.

Natural analogues are thus an important component of repository safety analysis and, as more performance assessments are completed and the demands for quality assurance and model validation increase, their importance is only likely to increase with time!

References

1. Nagra, Project Gewähr 1985 English summary (1985) Nagra NGB 85-09, Nagra, Baden, Switzerland.

2. Nagra (1988), Sedimentstudie Zwischenbericht, Nagra NTB 88-25.

3. McKinley, I.G. (1985) The geochemistry of the near-field, Nagra NTB 84-48 / EIR Ber. 563.

4. McKinley, I.G. (1988) The near-field geochemistry of HLW disposal in an argillaceous host rock, Nagra NTB 88-26.

5. Grauer R. (1986) Bentonit als Verfüllmaterial im Endlager für hochaktiven Abfall: Chemische Aspekte, Nagra NTB 86-12.

6. Chapman, N.A., McKinley, I.G., Smellie, J.A.T. (1984) The potential of natural analogues in assessing systems for deep disposal of high-level radioactive waste, Nagra NTB 84-41 / EIR Ber. 545 / KBS TR 84-16.

7. Müller-Vonmoos, M., Kahr, G. (1985) Langzeitstabilität von Bentonit unter Endlagerbedingungen, Nagra NTB 85-25.

8. Pusch, R., Börgesson, L., Erlström, M., (1987) Alteration of isolating properties of dense smectite clay in repository environment as exemplified be seven prequaternary clays, SKB 87-29.

9. Nagra Working Group on Container Technology, 1984. An assessment of the corrosion resistance of the high-level waste containers proposed by Nagra, Nagra NTB 84-32.

10. Johnson, A.B., Francis, B. (1980) Durability of metals from archaeological objects, metal meteorites and native metals, Battelle Pacific Northwest Laboratory, PNL, 3198, Hanford.

11. van Orden, A.C., McNeil, M.B. (1987) The archeological data base relevant to long-term corrosion behaviour of a-216 mild steel, paper presented at Sci. Basis Nucl. Waste Manag., XI, Boston.

12. Grauer, R. (1983) Glasses used in the solidification of high-level radioactive waste: their behaviour in aqueous solutions, Nagra NTB 83-01.

13. Grauer, R. (1985) Synthesis of recent investigations on corrosion behaviour of radioactive waste glasses, Nagra NTB 85-27.

14. Grambow, B. (1987) Nuclear Waste Glass Dissolution: Mechanism, Model and Application, JSS Technical Report 87-02, SKB Stockholm.

15. Jercinovic, M.J., Ewing, R.C. (1988) Basaltic glasses from Iceland and the deep sea : natural analogues to borosilicate nuclear waste-form glasses, JSS report 88-01.

16. Hartley, R.W. (1985) Release of radionuclides to the geosphere from a repository for high level waste - mathematical model, results, Nagra NTB 85-41.

17. Schweingruber, M.R. (1983) Actinide solubility in deep groundwaters - Estimates for upper limits based on chemical equilibrium calculations, EIR-Bericht Nr. 507 / Nagra NTB 83-24.

18. Hofmann, B., Dearlove, J.P.L., Ivanovich, M., Lever, D.A., Green, D.C., Bärtschi, P., Peters, Tj, (1987) Evidence of fossil and recent diffusive element migration in reduction halos from Permian red-beds of Northern Switzerland, Natural analogues in radioactive waste disposal, in B. Come and N. Chapman (eds.), pp. 217-238.

19. McKinley, I.G., MacKenzie, A.B., West, J.M., Scott, R.D. (1984) A natural analogue study of radionuclide migration in clays, Sci. Basis Nucl. Waste Manag., VII, pp. 851-857.

20. Hooker, P.G., MacKenzie, A.B., Scott, R.D., Ridgeway, I.M., McKinley, I.G., West, J.M. (1985) A study of long term (103 - 104 y) elemental migration in saturated clays and sediments Part III, Rep.Fluid Processes Res. Group Brit.Geol.Surv.,FLPU 85-9.

21. Schaefer,R.G., Leythaeuser, D. (1984) Genese von Kohlenwasserstoffen in den Lias- und Dogger-Sedimenten"; Nagra NTB 84-34, pp. 65-77.

22. Wittwer, C. (1988) Sediment Zwischenbericht Geochemistry, Nagra Technical Report Series, in press.

23. Smellic J.A.T., Barrozo Magno L., Chapman N.A., McKinley I.G., Penna-Franca E. (1987) The Poços de Caldas project feasibility study. Natural analogues in radioactive waste disposal, in B. Come and N. Chapman (eds.), 118-132.

24. Smellic J.A.T., Status report on the Poços de Caldas project, Proc. Snowbird meeting NAUG, CEC EUR 1988

25. Hadermann, J., Rösel, F. (1985) Radionuclide chain transport in inhomogenious crystaline rocks - limited matrix diffusion and effective surface sorption, EIR Ber 551/Nagra NTB 85-40.

26. McKinley, I.G., Hadermann, J. (1984) Radionuclide sorption database for Swiss safety assessments, Nagra NTB 84-40.

27. Neretnieks, I. (1980) J. Geophys. Res., 85, pp. 4379-4397.

28. Grisak, G.E., Pickens, J.F. (1980) Water Resources Res., 16, pp. 719-730.

29. KBS (1983) Final Storage of Spent Nuclear Fuel - KBS 3.

30. Glueckauf, E. (1980) Harwell, Rep. AERE R 9823.

31. Abelin, H., Birgersson, L., Gidland, J., Moreno, L., Neretnieks, I., Tunbrant, S. (1986) Sci. Basis Nucl. Waste Manag., IX, pp. 627-639.

32. Herzog, F. (1987) Modelling isotope distributions in boreholes, Natural analogues in radioactive waste disposal, in B.Come and N.Chapman (eds.), pp. 239-248.

33. Smellie, J.A.T., MacKenzie, A.B., Scott, R.D. (1986) An analogue validation study of natural radionuclide migration in crystalline rocks using uranium-series disequilibrium studies, Sci. Basis Nucl. Waste Manag., IX, pp. 91-98.

34. Alexander W.R., Scott R.D., MacKenzie A.B., McKinley I.G. (in press) A natural analogue study of radionuclide migration in a water carrying fracture, Radiochimica Acta.

35. Alexander W.R., Scott R.D., MacKenzie A.B., McKinley I.G. (1987) Natural analogue studies at Grimsel, Southern Switzerland. Natural analogues in radioactive waste disposal, in B. Come and N. Chapman (eds.), pp. 473-484.

36. McKinley, I.G., Alexander W.R., Bajo, C., Frick, U., Hadermann, J., Herzog, F., Höhn, E. (in press) The radionuclide migration experiment at the Grimsel Rock Laboratory Switzerland, Sci. Basis Nucl. Waste Manag., XI, pp. 179-187.

37. Alexander W.R., MacKenzie A.B., Scott R.D., McKinley I.G. (in press) Natural analogue studies using the natural decay series radionuclides : I The influence of water-bearing fractures on radionuclide immobilisation in a crystalline rock repository. II Uranium remobilisation in the Grimsel Felslabor, Nagra NTB 87-08.

38. Mickelcy, N., Jesus, M.C., da Silveira, C.L.P., Kuechler, I.C. (in press), Colloid investigations in the "Poços de Caldas" natural analogue project. Submitted to Sci. Basis Nucl. Waste Manag., XII.

39. Ivanovich, M., Longworth, G., Wilkins, M., Duerder, P., Payne, T. Nightingale, T., Edgehill, R., Cockayne, D., Davey, B. (1987) Natural analogue study of the distributon of uranium series radionuclides between the colloid and solute phases in the hydrogeological system of the Koongara uranium deposit, N.T. Australia, 300-313, in B. Come and N. Chapman (eds.) Natural analogues in radioactive waste disposal, pp. 167-178.

40. Hofmann, B. (in press), Natural analogue studies in the Krunkelbach mine - Southern Germany: geology and water-rock interaction, submitted to Sci. Basis Nucl. Waste Manag. XII.

41. Shin, W.S. (1982) The long-term stability of cement and concrete for nuclear waste disposal under normal geologic conditions, Nagra NTB 82-03.

42. E.W.I. (1983) Bitumen, ein Verfestigungsmaterial für radioaktive Abfälle und seine historischen Analoga. Nagra NTB 83-11.

43. Neal, C., Stanger, G. (1985) Past and present serpentinisation of ultramafic rocks; an example from the Semail Ophiolite Nape of Northern Oman in J.I. Dever (ed.) The Chemistry of Weathering, pp. 249-275.

44. McKinley, I.G., Berner, U., Wanner, H. (1987) Predictions of radionuclide chemistry in a highly alkaline environment, PTB-SE-14, pp. 77-89.

45. Bath, A., Cave, M., Berner, U., McKinley, I.G., Neal, C. (1987) Testing geochemical models in a hyperalkaline environment in B. Come and N. Chapman (eds.) Natural analogues in radioactive waste disposal, pp. 167-178.

46. McKinley, I.G., Bath, A., Berner, U., Cave, M., Neal, C. (in press) Results of the Oman analogue study, Radiochimica Acta.

47. Bath, A., Christofi, N., Neal, C., Philp, J.C., Cave, M., McKinley, I.G., Berner, U.(1987) Trace element and microbiological studies of alkaline groundwaters in Oman, Arabian Gulf - a natural analogue for cement porewaters, Rep. Fluid Processes Res. Group, Brit. Geol. Surv., FLPU 87-2/Nagra NTB 87-16.

48. McKinley, I.G., Grogan, H.A. (1987) Natural analogue support for geomicrobiological modelling, EUR 10671, pp. 153-159.

49. McKinley, I.G., West, J.M., Grogan, H.A. (1985) An analytical overview of the consequences of microbial activity in a Swiss HLW repository, Nagra NTB 85-45 / EIR Ber. 562.

50. Grogan, H.A. (1987) The significance of microbial activity in a deep repository for L/ILW, Nagra NIB 87-05.

51. Eisenbud, M., Lei, W., Ballad, R., Penna Franca, E., Miekely, N., Cullen, T., Krauskopf, K. (1982) Studies of the mobilisation of thorium from the Morro do Ferro, Scientific basis for nuclear waste management, V, pp. 735-744.

Index

salt domes 340
Saltelli, A. 74, 79, 80, 84, 94, 95, 116, 151, 152, 227, 259, 315, 316, 321, 322
sample size 78, 80
sampling 72, 150, 156
sampling methods 133
Sartori, E. 116, 152
SAS 95
savannah 239
scatterplots 87
scavenging processes 291
scenario likelihood 43
scenario list 361
scenario probability 219
scenario's consequences 215
scenarios 38, 58, 62, 69, 75, 167, 234, 246
Schaefer, R.G. 394
Schweingruber, M.R. 394
SCICON 259
SCK/CEN 154
Scott, R.D. 394, 395
sea-level changes 186, 239
seabed abyssal plains 286
second generation p.r.a. code 231
sediment pore waters 298
sedimentation 179
seismic activity 276
seismicity 177
semail ophiolite nappe 392
sensitivity analysis 32, 60, 61, 63, 69, 70, 83, 105, 129, 130, 149, 155, 159, 226, 314
sensitivity studies 290
sequences of events 43
Shackleton, N.J. 260
Shaeffer, D.L. 95
shallow site 251
Shephard, L. 305
Sherman, G.R. 95, 102, 116, 152, 259
Shin, W.S. 396
Short, J. 14, 18
simulations 102
Sinclair, Jim 88
site selection 332
SKB 154
SKB, SKI 154
SKBF/KBS May 1983 283
SKN 154
Slovic, P. 7, 8, 9, 10
Smellie, J.A.T. 375, 393, 395
Smirnov test 86, 148
So, A. 116

societal risk 161, 171
software quality assurance 157
solubility 70
solute transport 370, 372
sorption 26, 388
sorption coefficients 302
Space Shuttle Commission 30
SPAM 100
Spearman coefficient 84
SPOP 74, 88, 136
Stallen, P.J. 11, 13
Stamberg's law 218
Standardized Regression Coefficient 146
Stanger, G. 396
Stanners, D.A. 94
Starr, C. 5, 6, 12
steady state 231, 242
Stephens, M.E. 72, 95, 152
stochastic 162
stochastic approach 317
stochastic health effects 162, 163
stochastic range 62
stochastic sensitivity analysis 285
storage 57
stratified sampling 73
stress field 197
structure 181
Student's t test 78
sub-seabed option 285, 343
sub-seabed repository 287
subduction 190
subjective immunity 27, 28
subseabed option 161
Sumerling, T.J. 259
Sutherland, D.G. 260
Swatson 11
Swedish KBS 153
systems variability analysis (SVA) 57
systems variability analysis codes 97
SYVAC (SYstems Variability Analysis Code) 60, 61, 99, 153, 231, 250
SYVAC 1 234
SYVAC/AC, TIME2 99
SYVAC/SU 99
SYVAC2 100
SYVACA 242
Tanguy, P. 283
Tchebycheff's theorem 78, 79
technological disasters 28
technostructure 27
tectonics 177, 276